FOUNDATIONS OF STATISTICAL INFERENCE

THE UNIVERSITY OF WESTERN ONTARIO SERIES IN PHILOSOPHY OF SCIENCE

A SERIES OF BOOKS
IN PHILOSOPHY OF SCIENCE, METHODOLOGY,
EPISTEMOLOGY, LOGIC, HISTORY OF SCIENCE
AND RELATED FIELDS

VOLUME 35

ADVANCES IN THE STATISTICAL SCIENCES

Festschrift in Honor of Professor V. M. Joshi's 70th Birthday

VOLUME II

FOUNDATIONS OF STATISTICAL INFERENCE

Edited by

IAN B. MacNEILL and GARY J. UMPHREY

Department of Statistical and Actuarial Sciences,
The University of Western Ontario

Associate editors:

M. SAFIUL HAQ

Department of Statistical and Actuarial Sciences,
The University of Western Ontario

WILLIAM L. HARPER

Department of Philosophy,
The University of Western Ontario

SERGE B. PROVOST

Department of Statistical and Actuarial Sciences,
The University of Western Ontario

SPRINGER-SCIENCE+BUSINESS MEDIA, B.V.

Library of Congress Cataloging in Publication Data

Foundations of statistical inference.

(Advances in the statistical sciences; v. 2)
(The University of Western Ontario series in philosophy of science; v. 35)
1. Mathematical statistics—Congresses. 2. Probabilities—Congresses. I. MacNeill, Ian B., 1931– . II. Umphrey, Gary J., 1953– . III. Series. IV. Series: University of Western Ontario series in philosophy of science; v. 35.
QA276.A1A39 vol.2 519.5 s [519.5] 86–29670

ISBN 978-94-010-8623-3 ISBN 978-94-009-4788-7 (eBook)
DOI 10.1007/978-94-009-4788-7

TABLE OF CONTENTS

CONTENTS OF THE OTHER VOLUMES OF THE JOSHI FESTSCHRIFT

VOLUME I

Applied Probability, Stochastic Processes, and Sampling Theory

VOLUME III

Time Series and Econometric Modelling

VOLUME IV

Stochastic Hydrology

VOLUME V

Biostatistics

VOLUME VI

Foundations of Actuarial Science

PREFACE

On May 27–31, 1985, a series of symposia was held at The University of Western Ontario, London, Canada, to celebrate the 70th birthday of Professor V. M. Joshi. These symposia were chosen to reflect Professor Joshi's research interests as well as areas of expertise in statistical science among faculty in the Departments of Statistical and Actuarial Sciences, Economics, Epidemiology and Biostatistics, and Philosophy.

From these symposia, the six volumes which comprise the "Joshi Festschrift" have arisen. The 117 articles in this work reflect the broad interests and high quality of research of those who attended our conference. We would like to thank all of the contributors for their superb cooperation in helping us to complete this project.

Our deepest gratitude must go to the three people who have spent so much of their time in the past year typing these volumes: Jackie Bell, Lise Constant, and Sandy Tamowski. This work has been printed from "camera ready" copy produced by our Vax 785 computer and QMS Lasergraphix printers, using the text processing software TEX. At the initiation of this project, we were neophytes in the use of this system. Thank you, Jackie, Lise, and Sandy, for having the persistence and dedication needed to complete this undertaking.

We would also like to thank Maria Hlawka-Lavdas, our systems analyst, for her aid in the layout design of the papers and for resolving the many difficult technical problems which were encountered. Nancy Nuzum and Elly Pakalnis have also provided much needed aid in the conference arrangements and in handling the correspondence for the Festschrift.

Professor Robert Butts, the Managing Editor of *The University of Western Ontario Series in Philosophy of Science* has provided us with his advice and encouragement. We are confident that the high calibre of the papers in these volumes justifies his faith in our project.

In a Festschrift of this size, a large number of referees were needed. Rather than trying to list all of the individuals involved, we will simply say "thank you" to the many people who undertook this very necessary task for us. Your contributions are greatly appreciated.

Financial support for the symposia and Festschrift was provided by The University of Western Ontario Foundation, Inc., The University of Western Ontario and its Faculties of Arts, Science, and Social Science, The UWO Statistical Laboratory, and a conference grant from the Natural Sciences and Engineering Research Council of Canada. Their support is gratefully acknowledged.

Finally, we would like to thank Professor Joshi for allowing us to hold the conference and produce this Festschrift in his honor. Professor Joshi is a very modest man who has never sought the limelight. However, his substantial contributions to statistics merit notice (see Volume I for a bibliography of his papers and a very spiffy photo). We hope he will accept this as a tribute to a man of the highest integrity.

INTRODUCTION TO VOLUME II

The scientific investigation of unknown characteristics of a system is the fundamental object of Statistical Inference. A number of different approaches to statistical inference have evolved over the years; these include: estimation and hypothesis testing by Fisher, Neyman, Pearson and others, Bayesian procedures, the decision theoretic approach originated by Wald, Fisher's fiducial inference and Fraser's structural inference.

It is worth noting that for a considerable period of time the foundations of statistical inference has been a topic of substantial interest to the Canadian statistical community, perhaps due to the development of structural inference by Fraser and many of his students. The landmark conference on "Foundations of Statistical Inference" organized by V. P. Godambe and D. A. Sprott held in 1971 also stimulated much local interest. Furthermore, a series of conferences on this and related topics have been held by the U.W.O. Philosophy Department and have been published in *The University of Western Ontario Series in Philosophy of Science;* this volume and its five companion volumes are the latest to appear in this series.

As a contribution to the foundations of statistical inference, the method of structural inference was introduced by Fraser during the mid-sixties. It plays an important role in the context of Canadian contributions to the understanding of the inference process. The structural method of inference exploits the internal sources of variation inherent in the generating system, rather than making assumptions. These variations are described as error-effect variations by Evans, Fraser and Monette in this volume. Godambe illustrated the criterion of the "data based choice of an ancillary statistic" through an example given by Fraser. Haq demonstrates the use of structural relations between observations and of error variables for prediction distributions.

During the past three decades the Bayesian procedure has gained enormous popularity among statisticians. Among Canadian statisticians, Guttman's contribution is most significant. In his paper with Srivastava, the Bayesian method is applied to "detect a change point of a regression model when some of the parameters are known *a priori* not to change while others are subject to change". This procedure is also used for the study of growth curve models. Among the participants at the symposium, pioneers in advocating Bayesian procedures were Dempster and Lindley. Dempster emphasizes the use of "evidence including but not restricted to statistical data". Lindley examines partial likelihood from the Bayesian context and concludes that "although the partial likelihood method is asymptotically sound, the approach to the limit is slow, so that the method can lose much information even with large samples". The Bayesian procedure, because of the subjective nature of the probabilistic assessment of the characteristics of the system, has received some strong criticism. In this regard Kyburg advocates "direct inference" over the Bayesian procedure except when the procedure can be put in the framework of "direct inference"; he titles his paper, "The Basic Bayesian Blunder".

Contributions to the symposium on the foundations of statistics have been made over a broad spectrum of areas. Blyth and Bondar examine Fisher's early papers on fiducial probability and obtain fiducial distributions for discrete random variables using randomization. Plante gives a decision-likelihood solution to the problem of comparing two simple hypotheses; Hedges and Olkin develop a statistical test for the overlap hypothesis. Bondar shows that the risk of a shrinkage estimator is close to that of a minimum risk equivariant estimator except on a small set. Saleh and Sen study the relative dominance of shrinkage and M-estimators of the intercepts for the multisample regression problem. Fraser and Massam examine concave regression where "the set of regression functions is the Cartesian product of a convex cone and a cylinder; the maximum likelihood/least squares solution is characterized in terms of geometrical properties of the cone-cylinder". Tracy and Jinadasa extend Dolby's ultrastructural relationships model to the multivariate case, and they show that the maximum likelihood method is unable to distinguish between the multivariate ultrastructural relationships model and the multivariate functional relationships models. Provost develops a test procedure based on the likelihood ratio criterion for testing the nullity of the multiple correlation coefficient with incomplete data. M. A. Ali gives alternative estimators of regression parameters with missing values and shows that for a wide class of conditions, in the sense of mean square error, the alternative method is better than the method of linear prediction and the method of maximum likelihood. M. M. Ali shows that the tail probability of the t-statistic for samples from a symmetrically truncated normal density is greater than the corresponding tail probability of the usual t-distribution. Lee provides a closed form expresssion for the maximum likelihood estimator of stochastically ordered multinomial populations with fixed and random zeros. Akahira and Takeuchi give a new definition of asymptotic expectation. Nishishato reviews methods of quantification of categorical data and propose two new methods based on the principle of forced classification and the projection of data on a subspace.

Some very interesting contributions deal with philosophical aspects of statistical inference. As mentioned earlier, Kyburg advocates direct inference over the Bayesian procedure. Skyrms studies the concept of "coherence" along with changes in degree of belief resulting from a Bayesian argument in a betting setup. Pollock gives some justification of *nomic* probability in the sense that conclusions reached by *nomic* probability "bear obvious similarities to the conclusions of orthodox statistical reasoning". Seidenfield discusses some results on entropy and uncertainty in the Bayesian context.

We would like to extend our thanks to all of the authors for their valuable contributions. The papers in this volume should stimulate further interest and research in the field of statistical inference.

A. P. Dempster [1]

PROBABILITY AND THE FUTURE OF STATISTICS

ABSTRACT

One definition holds that statistics is the science of using evidence, including but not limited to statistical data, to assess numerical probabilities meant for prospective assessment of uncertainty. Widespread opportunities exist for developing statistics along these lines, through practical attempts to bridge the gap from evidence to probability representations in appropriately selected domains of science. Illustrations are drawn from policy-related fields of statistics, including analysis of economic time series, small area statistics, and the statistics of employment discrimination.

1. PLAN

In this paper I deliver several short sermons and conclude with brief references to a set of Bayesian analyses. While I am critical of what I see as the drastic imbalance of technical research towards old-fashioned frequentist theory, I am both proud of the real accomplishments of the field of statistics in methodology and applications, and bullish about the prospects for future development—especially if we change our ways. The concluding examples are intended to illustrate the main point of the sermons—that we should participate in scientific problem-solving as co-equals with experts in the substantive fields, for the sake of our common ultimate goals of sound methods and trustworthy new knowledge.

2. FOUNDATIONS OF WHAT?

In a symposium on the foundations of statistical inference it is appropriate to begin by asking: foundations of what? The accepted paradigm

[1] Department of Statistics, Harvard University, Cambridge, Massachusetts 02138

I. B. MacNeill and G. J. Umphrey (eds.), Foundations of Statistical Inference, 1–7.

of statistical inference is to draw inferences from random samples to populations, emphasizing that only the information in the sample data is to be used. I argue that this paradigm is too narrow, in fact so narrow that technical statistics risks dismissal as insufficiently relevant to science, even in situations where the main task is to draw uncertain inferences from samples to populations.

There are two main arguments. The first is the familiar one of the Bayesian/non-Bayesian controversies of 25 years ago, where I accept the position, not that Bayes is immutably the correct story in all circumstances, but certainly that frequentist theory is logically inadequate for the task of uncertain estimation (it provides right answers to wrong questions) and thus pays too high a price for achieving the requirement that only sample information be used.

The second argument is less technical, but more fundamental. I believe that there needs to be much more integration of the logic of statistical inference with the reasoning processes used routinely by working scientists. In particular, scientists are required to integrate all the knowledge currently accepted and relevant to a question under study, whence statistical inference must provide formulations which can process such knowledge in the form which it actually takes. Frequentist statistics, whatever its logical merits or demerits, inputs external information only in the form of hypotheses about sampling processes, and its outputs lack formal procedures for combining with other forms of knowledge. Bayesian methods, and, even more, belief function methods do provide limited tools for combining knowledge from disparate sources. But the bridge from statistical technologies to actual working science remains sketchy.

In a recent paper (Dempster, 1983), I identified this gap as belonging to "functional" rather than "technical" statistics and I referred to the problem of where models come from as the "missing half" of statistics. Bridging the gap will require hard thought, and willingness to move closer through changing attitudes, from both scientists and statisticians. For example, while random sampling and randomization of treatments are basic to the statistician's craft, we need to move towards formulations of uncertain inference which can cope with evident failures and absences of randomization. Some thoughts on this topic are outlined by Dempster (1985).

For now, we should perhaps refer to "inference about statistical questions" rather than "statistical inference", to suggest that some broadening of patterns of formal inference is in order. By *statistical questions* I mean a cluster of uncertain inference tasks which often face applied statisticians, including estimating population characteristics and how they have changed, forecasting the future development of observable dynamic processes, and drawing inferences about how such populations and processes would be altered by treatments or policy changes. Statistics might then be defined as the union of those subfields of traditional sciences (e.g., biology or economics) dealing with statistical questions.

3. ON PROBABILISTIC ASSESSMENT OF UNCERTAINTY

Major contributors to the early history of mathematical probability, notably Jacob Bernoulli, Laplace, and Gauss, took it for granted that numerical probabilities were intended to quantify specific instances of uncertainty, such as the uncertainty over the parameters of the orbit of a specific asteroid. Paradoxically, while recent generations of scientists are blessed with revolutionary advances in the mathematical techniques of probability, the original motivation and interpretation is usually suppressed or forgotten, even in a field such as statistics where real world uncertain inference is generally concerned with specific outcomes. I believe we should get back to the original motivation and interpretation, and define a central goal of statistics, perhaps the dominant goal, to be the production of high quality prospectively interpretable probabilities.

Meanwhile, what has been happening in other fields of science? I find two aspects of the situation quite remarkable. The first is that although uncertainty is an obvious fact of life in many scientific or decision-making problem areas, most highly trained scientists and professionals do not use, and indeed have only superficial knowledge of, formal tools for assessing prospective probabilities which have developed in and around the statistics profession over the past 300 years and especially in the past 50 years. For example, medical doctors learn to quote certain statistical frequencies, such as survival rates during and after serious operations of specific kinds, but for weighing complex special features of individual patients it becomes necessary to appeal to clinical judgment as more art than science, and to avoid formal assessments or expressions of uncertainty. Similarly, professional economists, who are well aware of the existence and importance of uncertainty and are important consumers of advanced statistical technologies, also rely heavily on informal judgment, and generally report only point estimates when forecasting macroeconomic phenomena. Meanwhile, the basic model of microeconomics is that actors use expectations to make the decisions which drive an economy. But where are the accounts of how evidence and information are converted into the probability assessments which must underlie forecasts and expectations? It appears that economic scientists and the actors in their theories share the same *lacuna*.

The second remarkable aspect of current scientific and professional handling of uncertainty is the tendency to rediscover the problem and then to set out to rediscover the solution from first principles, often by organizing a small conference, or a session at a meeting, and inviting participation from philosophers, psychologists, statisticians, artificial intelligencers, decision analysts, risk analysts, and others brave enough to express their views. Important issues are raised, but complex issues are confused, and rarely is much clarity or synthesis achieved. A good example is provided by the emerging field of artificial intelligence. The proceedings of a recent workshop on "Uncertainty and Probability in Artificial Intelligence" held at U.C.L.A. on August 14–16, 1985 has 34 contributions of extraordinary diversity but no trace of a consensus which would lead to systematic development of models and techniques.

I have two suggestions which may help to bring order out of confusion. The first is to emphasize uncertainty assessments based on what R. A. Fisher quaintly called Mathematical Probability, since it permits the exact reasoning of mathematics to be applied to the fundamentally inexact topic of uncertainty.

By probability, however, I mean much more. In contrast to some quantitative approaches which use numbers that have no stated verifiability criteria, and more strongly in contrast to approaches which advocate purely verbal characterizations of uncertainty, numerical probabilities are subject to calibration requirements. If you and I share common evidence in a sequence of analyses of related or unrelated problems, and if I compute a probability assessment for each analysis, then I should be required to offer you the choice of sides in any gamble and expect to have the long run ratio of gains to losses approach unity. While there can never be a guarantee that I can achieve this goal, the goal is surely the *sine qua non* of Mathematical Probability, whence strong empirical evidence that my probabilities can be selected against is grounds for reconstruction of the mathematical models which generated the probabilities. Note the critical role of common knowledge bases: if you know more than I do, you *should* be able to successfully win in the long run.

The second suggestion is that statisticians might wish to study the theory of belief functions (Shafer, 1976) which is attracting renewed interest especially in artificial intelligence applications. As one of the originators of the theory, my suggestion is subject to prejudice. The belief function formulation is firmly probability-based, however, and is much more flexible than its specialization to Bayesian theory. I have in mind especially the need noted above to be able to reflect multiple sources of evidence, and the need to tailor inferences to available evidence without the pretence that prior probabilities should be prescribed for all unknown outcomes.

The present barriers to wider experimentation and thence informed study of both Bayesian and belief function methodologies are largely computational. I believe that mathematical understanding of relevant computational processes is rapidly improving, stimulated of course by the hardware revolution, and hence the future of statistical probability assessment technologies is bright.

4. EXAMPLES OF STATISTICAL QUESTIONS

The preceding discussion is conditioned by the types of applications I have recently worked on. I conclude by briefly describing several such Bayesian modelling efforts. Details may be found in the references.

First, suppose random samples are available for each of several subregions of a large region, and the task is to estimate population means or totals of some quantity for each of the subregions. It is by now a familiar story that many statisticians advocate mixing information from the combined sample with information from a particular sample in estimating the mean of the corresponding population. From the Bayesian (or, prospective probabilist) perspective there are two sources of evidence being combined here, whence the task is simply to construct an acceptable probability model from which a corresponding inference is automatic. Dempster and Raghunathan (1985) recently carried out a small modelling exercise for a paper presented at a conference on small area statistics, where we considered the need to incorporate within subregion covariation for a covariate with known population distributions. The old technology of proposing estimators and evaluating their sampling properties is generally inconclusive as well as conceptually flawed. An estimator which is biased from a sampling perspective may easily be Bayesianly unbiased. On the other hand, Bayesian estimators require study beyond the level of our paper, for example, to consider the plausibility and consequences of allowing population distributions with Pareto tails. Here again, Bayesian analysis under alternative models is conceptually straightforward, and the issue of model choice is put where it belongs, in the hands of the analyst who must compare all relevant knowledge with the catalogue of available analyses.

A similar issue is faced by Carlin *et al.* (1985) in the context of modelling monthly economic time series. Again, there is no lack of proposals in the literature for seasonal adjustment or for short term forecasting, some purely *ad hoc,* some model-based, and some even Bayesian in spirit. We chose to study the effects on Bayesian inference of different types of low frequency components in the nonseasonal components, and corresponding models for slowly changing seasonal components. I believe that such problems are intractable from a traditional statistical perspective, because there is no procedure which can cope with important parameters which are simply not well identified in the data. Other forms of evidence will need to be addressed and relied upon before convincing analyses can be devised.

My second pair of examples differ from the first pair because they involve questions of statistical bias in the real sense as contrasted with the technical sense of frequentist statistics. By statistical bias in the real sense I mean systematic deviations between the phenomenon under study and the statistical data sets designed to represent the phenomenon. To study real statistical bias it is necessary to assemble as much evidence as possible about the sampling processes generating the data, and to arrive at uncertain quantitative assessments of real biases, usually without benefit of directly relevant controlled studies.

In this connection, I suggest the paper by Freedman and Navidi (1986) on the possibility of adjusting for census errors using postenumeration survey data, and more especially the sets of comments by Kadane and by myself. Since the postenumeration surveys are deliberately randomized, I strongly believe that they need to be taken seriously, and their flaws subjected to careful quantitative assessment by any and all credible means available.

Finally, I point to Dempster (1984) on the difficulties of assessing causal mechanisms from poorly controlled employee record data. A little analysis of the substantive situation turns up an astonishing range of possible ailments afflicting the data. From all of these examples, and especially perhaps the last, it is clear that statistics provides many questions whose answers require drawing on sources of evidence beyond the usual types of statistical data, and translating all the evidence into usable prospective probabilities.

ACKNOWLEDGMENT

This work was facilitated in part by National Science Foundation Grant DMS-85-4332.

REFERENCES

Carlin, J. B., A. P. Dempster, and A. B. Jonas (1985), "On models and methods for Bayesian time series analysis". *Journal of Econometrics* **30**, 67–90.

Dempster, A. P. (1983), "Purposes and limitations of data analysis". In *Scientific Inference, Data Analysis, and Robustness*, ed. G. E. P. Box, Tom Leonard and Chen-Fu Wu, pp. 117–133. New York: Academic Press.

Dempster, A. P. (1984), "Alternative models for inferring employment discrimination from statistical data". In *W. G. Cochran's Impact on Statistics*, ed. Poduri S. R. S. Rao and Joseph Sedranski, pp. 309–330. New York: Wiley and Sons.

Dempster, A. P. (1985), "Probability, evidence, and judgment". In *Bayesian Statistics 2*, ed. J. M. Bernardo, M. H. DeGroot, D. V. Lindley and A. F. M. Smith, pp. 119–132. New York: North-Holland.

Dempster, A. P., and T. E. Raghunathan (1985). "Using a covariate for small area estimation: a common sense Bayesian approach". Research Report S-101, Department of Statistics, Harvard University.

Freedman, D. A., and W. C. Navidi (1986), "Regression models for adjusting the 1980 census". (With comments by Joseph B. Kadane, Eugene P. Ericksen, A. P. Dempster, Kirk M. Wolter, Albert Madansky, I. P. Felleghi, Lincoln E. Moses, and Gad Nathan) *Statistical Science* **1**, 3–39.

Proceedings of the Workshop on Uncertainty and Probability in Artificial Intelligence, U.C.L.A., August 14–16, 1985, American Association for Artificial Intelligence and RCA Advanced Technology Laboratories.

Shafer, G. (1976). *A Mathematical Theory of Evidence.* Princeton, New Jersey: Princeton University Press.

Colin R. Blyth and James V. Bondar [1]

A NEYMAN-PEARSON-WALD VIEW OF FIDUCIAL PROBABILITY

ABSTRACT

This is an examination of Fisher's early papers on Fiducial probability, from the performance point of view. A Fiducial distribution for a real-valued parameter θ is identified as a distribution-function-valued statistic (each observed value is a probability distribution on the parameter space) such that for each γ between 0 and 1 the γ-quantile of this distribution has probability γ of exceeding θ. The view taken is that there are usually many possible Fiducial distributions, and that the problem is to find one that is in some sense optimal. It is shown that Fiducial distributions can be found for discrete random variables using randomization: this is done for the Binomial (n, p) by inverting the uniformly most powerful one-sided tests. The resulting Fiducial distribution has the corresponding optimum property that its γ-quantile, subject to having probability γ of exceeding the true p, has uniformly minimum probability of exceeding each $p_1 > p$.

1. FIDUCIAL DISTRIBUTIONS

Let X be a real or vector valued random variable, whose family of possible distributions P_θ, $\theta \in \Omega$ is indexed by the real-valued parameter θ, with Ω an interval.

Let H_X be a statistic, each of whose possible values is a cumulative probability function on the real line: If $X = x$ is observed, then the observed value of H_X is $H_x(\cdot)$ which is a cumulative probability function, taken to be continuous and strictly increasing on Ω.

Write $Q_\gamma(x) = H_x^{-1}(\gamma)$ for the γ-quantile of H_x: If W is a random variable with cumulative probability function H_x, then for each possible γ

[1] Department of Mathematics and Statistics, Queen's University, Kingston, Ontario K7L 3N6 (both authors)

I. B. MacNeill and G. J. Umphrey (eds.), Foundations of Statistical Inference, 9–20.

value, $0 < \gamma < 1$ we have

$$P\{W \le Q_\gamma(x)\} = P\{W \le H_x^{-1}(\gamma)\} = \gamma.$$

Fiducial Distribution. The distribution-function-valued statistic H_X is called ***a Fiducial distribution function for θ*** if there is probability exactly γ that its γ-quantile is at least θ, this for every $\theta \in \Omega$ and for all γ values $0 < \gamma < 1$:

$$P_\theta\{\theta \le Q_\gamma(X)\} \equiv_\theta \gamma \qquad \text{for all } \gamma \text{ values } 0 < \gamma < 1,$$

with $\equiv_\theta$ standing for identity in θ. That is, for each γ, the random interval $\{-\infty, Q_\gamma(X)\}$ is a one-sided confidence interval for θ, with coverage probability identically γ for all $\theta \in \Omega$. [And for $\gamma_1 < \gamma_2$ the random interval $\{Q_{\gamma_1}(X), Q_{\gamma_2}(X)\}$ is a confidence interval for θ, with coverage probability identically $\gamma_2 - \gamma_1$ for all $\theta \in \Omega$.]

This is an ordinary probability, with the ordinary correspondence to physical frequencies: In a sequence of trials, no matter what sequence of θ-values may be encountered, the proportion of trials in which the γ-interval covers θ will converge to γ, this for each γ value between 0 and 1. We are taking the classical view that in repeated applications with $\theta_1, \theta_2, \ldots$ the parameter values occurring, either the situation is evolving so that for $A \subset \Omega$ the proportion of occurrences of $(\theta_i \in A)$ does not stabilize to a limit; or if these do stabilize the limits are irrelevant because any losses are borne not by the procedure but by each individual user, so that performance has to be considered separately for each possible value of θ.

A Fiducial distribution for θ is equivalent to a family of one-sided confidence intervals $-\infty$ to $Q_\gamma(X)$ for θ, one for each γ value $0 < \gamma < 1$, for which $P_\theta\{\theta \le Q_\gamma(X)\} \equiv \gamma$, and in which for $\gamma_1 < \gamma_2$ the γ_1-interval is always a proper subset of the γ_2-interval. If you have either a Fiducial distribution or such a family of confidence intervals, then the relation between Q_γ and H_x gives you the other. As noted above, the associated family of confidence intervals has the property of being "nested": $\gamma_1 < \gamma_2$ implies the γ_1-interval lies in the γ_2-interval. Chernoff (1951) and Stein (1951) showed that confidence intervals (even if optimal) need not have this nesting property; this means that existence of a Fiducial distribution is a strictly stronger condition than existence of a family of confidence sets.

Constructing a confidence region for θ amounts to testing every possible θ value and making a list of all the acceptable ones. A confidence region for θ, with confidence coefficient γ, is a random region $R(X) \subset \Omega$ satisfying

$$\inf_{\theta_0 \in \Omega} P_{\theta_0}\{\theta_0 \in R(X)\} = \gamma.$$

This region $R(X)$ is equivalent to the family of tests of

$$H : \theta = \theta_0 \quad \text{against} \quad A : \theta \neq \theta_0,$$

one for each $\theta_0 \in \Omega$,

$$\text{Accept } \theta_0 \text{ for } X \in A(\theta_0), \qquad \text{Reject } \theta_0 \text{ for } X \notin A(\theta_0)$$

given by the equivalence

$$\{X \in A(\theta_0)\} \equiv \{\theta_0 \in R(X)\}$$

and of course satisfying

$$\inf_{\theta_0 \in \Omega} P_{\theta_0}\{X \in A(\theta_0)\}$$

so that each of the tests has size $\leq 1 - \gamma$.

In particular, a confidence region with coverage probability identically γ is equivalent to a family of tests each of size exactly $1 - \gamma$.

Restrictions are usually made on the shape of the regions $R(x)$. Many interesting shapes can be prescribed when θ is vector valued, but for a real valued θ the choice is very limited. To begin with, $R(x)$ is usually restricted to being a connected set: an interval $\{U(x) \leq \theta \leq V(x)\}$. Further restrictions can be made, to intervals of fixed length $\{U(x) \leq \theta \leq V(x)\}$, or to one-sided intervals, either $\{-\infty < \theta \leq V(x)\}$ or $\{U(x) \leq \theta < \infty\}$.

We are expressing Fiducial distributions in terms of upper confidence limits. To restrict values of $R(x)$ to being one-sided intervals $\{-\infty < \theta \leq V(x)\}$ is to make this restriction:

$$\text{If } \theta_0 \in R(x) \text{ then } \theta_1 \in R(x) \text{ whenever } \theta_1 < \theta_0,$$

that is,

$$\text{If } x \in A(\theta_0) \text{ then } x \in A(\theta_1) \text{ whenever } \theta_1 < \theta_0.$$

That is, the acceptance regions $A(\theta_0)$ are decreasing in θ_0, as shown in Figure 1. This implies that $A(\theta_0)$ gives a test of the one-sided hypothesis

$$H : \theta \geq \theta_0 \quad \text{against} \quad A : \theta < \theta_0$$

of size

$$1 - P_{\theta_0}\{X \in A(\theta_0)\}.$$

Also implied: There is a real-valued statistic $T = T(X)$ for which

$$\{X \in A(\theta_0)\} \equiv \{T \geq c(\theta_0)\}$$

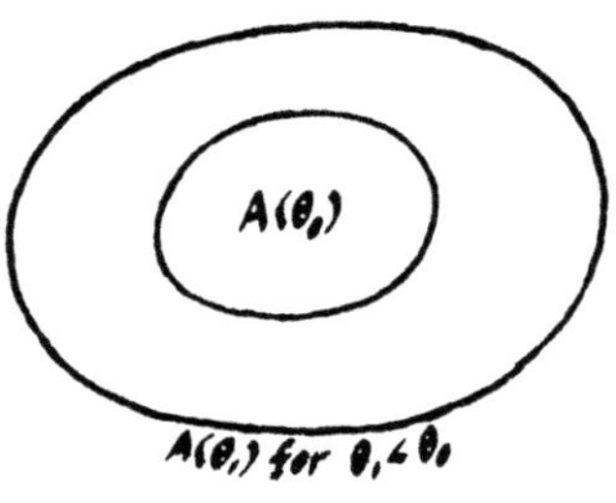

Figure 1. *Acceptance regions $A(\theta_0)$ decreasing in θ_0.*

and since $A(\theta_0)$ is decreasing in θ_0, it follows that $c(\theta_0)$ must be increasing in θ_0. We are requiring

$$P_{\theta_0}\{X \in A(\theta_0)\} = P_{\theta_0}\{T \geq c(\theta_0)\} = \gamma$$

so that $c(\theta_0) = T_{1-\gamma}(\theta_0)$, the $1-\gamma$ quantile of the distribution of T for $\theta = \theta_0$. For T with continuous distribution function strictly increasing between 0 and 1 the picture is as shown in Figure 2. For $\theta_\gamma(T)$ defined by

$$\{T \geq T_{1-\gamma}(\theta_0)\} \equiv \{\theta_0 \leq \theta_\gamma(T)\},$$

we have, for every $\theta_0 \in \Omega$,

$$P_{\theta_0}\{T \geq T_{1-\gamma}(\theta_0)\} = P_{\theta_0}\{\theta_0 \leq \theta_\gamma(T)\} = \gamma,$$

so that $\{\theta \leq \theta_\gamma(T)\}$ is a confidence interval for θ with constant coverage probability γ, and for a Fiducial distribution of θ we have H_T, whose observed value for $T = t$ is H_t, the cumulative probability function whose γ-point is $\theta_\gamma(t)$.

This is the concept of Fiducial distribution introduced by Fisher (1930, p. 532). He takes the statistic T to be the maximum likelihood estimator of θ. Working with $\gamma = 0.05$ as an example, he also writes T for the .95 point of T corresponding to a given θ. For clarity, let us instead write $T_{.95}(\theta_0)$ for the .95 point of T corresponding to $\theta = \theta_0$. In this notation Fisher next says that if as usually (if not always) happens $T_{.95}(\theta_0)$ is increasing in θ_0, then we can use

$$\{\theta \leq \theta_{.05}(T)\} \equiv \{T \geq T_{.95}(\theta)\}$$

to define $\theta_{.05}(T)$, and will then have

$$P\{\theta \leq \theta_{.05}(T)\} = P\{T \geq T_{.95}(\theta)\} = .05.$$

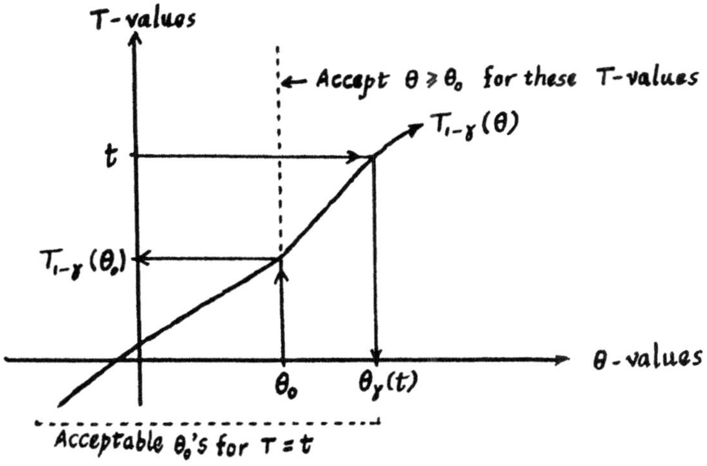

Figure 2. *Acceptance regions vs. confidence limits*

Fisher points out that $P\{\theta \leq \theta_{.05}(T)\} = .05$ is an ordinary probability statement: if we make a table of values of $\theta_{.05}(t)$, then in just 5% of a large number of trials of T will we observe t's for which $\theta \leq \theta_{.05}(t)$; this regardless of the sequence of θ values encountered.

Later on Fisher says that in the same way any other percentile in the Fiducial distribution of θ could be found, identifying $\theta_\gamma(t)$ as the γ-quantile of the observed Fiducial distribution of θ for observed $T = t$. So Fisher's Fiducial distribution is equivalent to a family of confidence intervals $\{\theta \leq \theta_\gamma(T)\}$, one for each γ value, with $\theta_\gamma(T)$ increasing in γ. An observed value of the Fiducial distribution gives observed values for upper confidence limits, simultaneously for all confidence levels.

As an example, Fisher says that if a size 4 sample correlation coefficient $R = .99$ is observed, then the population correlation coefficient ρ can be less than .765 only in the event that R has exceeded its 95 per cent point, an event known to occur just once in 20 times, and that "In this sense ρ has [Fiducial] probability .05 of being less than .765." This sense is exactly that used by Neyman: $P\{\rho \leq \rho_{.05}(R)\} = .05$ for the random R. For a realized value, say $R = .99$, Neyman's language is that the probability .05 refers to the random process for getting an interval, not to the particular realization at hand; and that the probability of $\rho < .99$ is either 0 or 1 and will not be considered.

Fisher takes T to be the maximum likelihood estimator $\hat{\theta}(X)$ of θ. In the Neyman-Pearson-Wald view of statistics only the performance of a procedure matters, so that *any* T giving the probability property would give a Fiducial distribution, whether the maximum likelihood estimator $\hat{\theta}(X)$ does or not. For testing the hypothesis $\theta \geq \theta_0$ the test that accepts when $\hat{\theta}(X) \geq c(\theta_0)$ can be a poor test, so these tests may give poor confidence limits.

Fisher (1933, p. 346), for $X_1, \ldots, X_n$ independent each with normal (μ, σ^2) distribution finds an upper $\gamma = .01$ confidence limit for σ. He bases it on the estimator $S = \sqrt{\sum(X_i - \overline{X})^2/(n-1)}$ for σ. This is not the maximum likelihood estimator of σ, but is a constant multiple of it so it gives the same confidence limit. The condition

$$P_{\sigma_0}\{S \geq c(\sigma_0)\} = \gamma$$

is equivalent to

$$P_{\sigma_0}\left\{\frac{(n-1)S^2}{\sigma_0^2} \geq \frac{(n-1)[c(\sigma_0)]^2}{\sigma_0^2}\right\} = \gamma,$$

and $(n-1)S^2/\sigma_0^2 = \sum(X_i - \overline{X})^2/\sigma_0^2$ has chi-square $(n-1)$ distribution, so writing $a_{1-\gamma}$ for the $(1-\gamma)$ quantile of that distribution we have

$$\frac{(n-1)[c(\sigma_0)]^2}{\sigma_0^2} = a_{1-\gamma}.$$

This gives

$$c(\sigma_0) = \sqrt{\frac{a_{1-\gamma}}{n-1}}\sigma_0$$

so we have

$$P_{\sigma_0}\left\{S \geq \sqrt{\frac{a_{1-\gamma}}{n-1}}\sigma_0\right\} = \gamma$$

which is equivalent to

$$P_{\sigma_0}\left\{\sigma_0 \leq \sqrt{\frac{n-1}{a_{1-\gamma}}}S\right\} = \gamma$$

so that $\{\sigma \leq \sqrt{\frac{(n-1)}{a_{1-\gamma}}}S\}$ is a γ-confidence interval for σ. For example, for $n = 11$, $\gamma = .01$ we have $a_{1-\gamma} = 23.09$, and

$$P\left\{\sigma \leq \frac{S}{\sqrt{2.309}}\right\} \equiv .01.$$

In giving this example, Fisher has 2.309 in error for $\sqrt{2.309}$ throughout.

For observed $S = s$, the observed Fiducial distribution H_s of σ is such that if W has cumulative probability function H_s, then

$$P\left\{W \leq \sqrt{\frac{n-1}{a_{1-\gamma}}}\,s\right\} = \gamma.$$

This is equivalent to

$$P\left\{\frac{(n-1)s^2}{W^2} \leq a_{1-\gamma}\right\} = 1-\gamma$$

so $(n-1)s^2/W^2$ has chi-square $(n-1)$ density, and it follows that the density of W at the point w is

$$\frac{2}{2^{\frac{n-1}{2}}\Gamma\left(\frac{n-1}{2}\right)}(n-1)^{\frac{n-1}{2}}\frac{s^{n-1}}{w^n}\exp[-(n-1)s^2/(2w^2)] \quad \text{for} \quad w > 0.$$

2. A DISCRETE EXAMPLE

For discrete random variables such as Binomial and Poisson, confidence limits with constant coverage probability γ can be constructed for each γ using randomization, and these give a Fiducial distribution for the parameter. Of course, there is no such thing as a randomized procedure. A randomized procedure based on X is an ordinary deterministic procedure based on X and on the outcome Y of a randomization device that must be explicitly described.

Binomial Example. Let X have Binomial (n, p) distribution, $0 \leq p \leq 1$, and let Y have Uniform $(0, 1)$ distribution independent of X. Then $T = X + Y$ is a sufficient statistic for X, Y and has an absolutely continuous distribution. The density of T at the coordinate point t is

$$f_p(t) = \begin{cases} (1-p)^n & \text{for } 0 < t < 1, \\ np(1-p)^{n-1} & \text{for } 1 < t < 2, \ldots \end{cases}$$

and in general

$$f_p(t) = \binom{n}{[t]} p^{[t]}(1-p)^{n-[t]}$$

for all non-integers between 0 and $n+1$, with $[t]$ standing for the greatest integer in t. For testing the hypothesis

$$H: p \geq p_0 \quad \text{against} \quad A: p < p_0,$$

a uniformly most powerful size $(1-\gamma)$ test is given by the acceptance region $T \geq c(p_0)$ with $c(p_0)$ is chosen so that

$$P\{T \geq c(p_0) \mid p = p_0\} = \gamma.$$

The $c(p_0)$'s are easily calculated and are increasing in p_0, so the confidence set corresponding to this family of tests is the confidence interval $\{0, p_\gamma(T)\}$ with $p_\gamma(t)$ defined by

$$\{0 \leq p_0 \leq p_\gamma(t)\} \equiv \{c(p_0) \leq t \leq n+1\}$$

so $p_\gamma(t)$ is the p_0-value for which $c(p_0) = t$. This confidence interval has constant coverage probability

$$P\{p \leq p_\gamma(T)\} \equiv_p P\{T \geq c(p)\} \equiv_p \gamma.$$

Now the equation

$$P(T \leq t \mid p) = 1 - \gamma \tag{1}$$

for fixed γ determines $t = c(p)$ as a function of p, and $p = p_\gamma(t)$ as a function of t; so for fixed t it determines γ as a function of p.

For $T = t$, the observed Fiducial distribution H_t of p is the distribution whose γ-point is $p_\gamma(t)$. If W is a random variable with distribution H_t, then

$$P\{W \leq p_\gamma(t)\} = \gamma.$$

So the cumulative probability function of H_t is given by

$$P\{W \leq p\} = \gamma = 1 - P(T \leq t \mid p),$$

and the probability density of H_t at the coordinate point p is

$$\frac{dP\{W \leq p\}}{dp} = \frac{d}{dp}\{1 - P(T \leq t \mid p)\}.$$

For $0 < t \leq 1$, equation (1) is

$$(1-p)^n t = 1 - \gamma,$$

so the cumulative probability distribution of W is given by

$$P(W \le p) = \gamma = 1 - (1-p)^n t$$

and the probability density of W at the coordinate point p is

$$\frac{dP(W \le p)}{dp} = nt(1-p)^{n-1} \quad \text{for} \quad 0 < p < 1.$$

In particular, the distribution of W

$$\begin{cases} \text{For } t = 0 \text{ is } P(W = 0) = 1, \\ \text{For } t = 1 \text{ is the Beta } (1, n) \text{ density } n(1-p)^{n-1} \text{ for } 0 \le p \le 1, \\ \text{For } 0 < t < 1 \text{ is the } (1-t), t \text{ mixture of these two.} \end{cases}$$

For $1 \le t \le 2$ equation (1) becomes

$$(1-p)^n + np(1-p)^{n-1}(t-1) = 1 - \gamma,$$

so the Fiducial cumulative distribution is given by

$$P(W \le p) = \gamma = 1 - (1-p)^n - np(1-p)^{n-1}(t-1)$$

and the Fiducial density at the coordinate point p is

$$\begin{aligned} \frac{dP(W \le p)}{dp} &= np(1-p)^{n-1} - n(1-p)^{n-1}(t-1) \\ &\quad + n(n-1)p(1-p)^{n-1}(t-1) \\ &= np(1-p)^{n-1}\{1-(t-1)\} + n(n-1)p(1-p)^{n-2}(t-1). \end{aligned}$$

In particular, the Fiducial distribution

$$\begin{cases} \text{For } t = 1 \text{ is the Beta } (1, n) \text{ density } n(1-p)^{n-1} \text{ for } 0 \le p \le 1, \\ \text{For } t = 2 \text{ is the Beta } (2, n-1) \text{ density } n(n-1)p(1-p)^{n-2} \\ \quad \text{for } 0 \le p \le 1, \\ \text{For } 1 < t < 2 \text{ is the } 1-(t-1), (t-1) \text{ mixture of these two.} \end{cases}$$

Continuing in the same way, it is easy to find the Fiducial distribution H_t of p for all possible values t of T:

$$\begin{cases} \text{For } t = 0, \text{ Probability 1 on the point } 0; \\ \text{For } t = i, \text{ the Beta } (i, n-i+1) \text{ distribution, } i = 1, 2, \ldots, n; \\ \text{For } t = n+1, \text{ Probability 1 on the point } 1; \\ \text{For } t \text{ between integers } i \text{ and } i+1, \text{ the} \\ \quad 1-(t-i), (t-i) \text{ mixture of the distributions for } i, i+1. \end{cases}$$

3. A TWO-PARAMETER EXAMPLE

Fisher (1935) considers a two parameter problem: Let $X_1, \ldots, X_n$ be independent, each normal (μ, σ^2); write $\overline{X}$ for $\sum X_i/n$, and S^2 for $\sum(X_i - \overline{X})^2/(n-1)$. Fisher says that for the fiducial argument we must know the exact distribution of an estimator of the unknown parameter, and this distribution must be continuous; but in finding a Fiducial distribution for μ, he uses the exact distribution of $T = (\overline{X} - \mu)/(S/\sqrt{n})$ which is not an estimator of μ and is not even a statistic since it can not be calculated from observations only. For testing the hypothesis

$$H : \mu \geq \mu_0 \quad \text{against} \quad A : \mu < \mu_0,$$

the uniformly most powerful unbiased size $1-\gamma$ test is the one-sided Student test given by the acceptance region

$$\left\{ \frac{\overline{X} - \mu_0}{S/\sqrt{n}} \geq t_{1-\gamma} \right\}$$

with $t_{1-\gamma}$ the $1-\gamma$ point of the Student $(n-1)$ distribution. For these acceptance regions we have

$$P\left\{ \frac{\overline{X} - \mu}{S/\sqrt{n}} \geq t_{1-\gamma} \right\} \equiv \gamma,$$

which is equivalent to

$$P\left\{ \mu \leq \overline{X} - t_{1-\gamma}\frac{S}{\sqrt{n}} \right\} \equiv \gamma,$$

that is, to

$$P\left\{ \mu \leq \overline{X} + t_{\gamma}\frac{S}{\sqrt{n}} \right\} \equiv \gamma, \tag{2}$$

so that $\mu \leq \overline{X} + t_\gamma S/\sqrt{n}$ is a confidence interval for μ with constant coverage probability γ. [Fisher, 1935, page 392, says that since the right hand side of the inequality in (2) can take all real values, we may state the probability that μ is less than any assigned value. But that right hand side is a random variable: its value can not be assigned.]

The observed Fiducial distribution H_x of μ, for observed $X = x$, $\overline{X} = \overline{x}$, $S = s$, is the distribution having $\overline{x} + t_\gamma s/\sqrt{n}$ for its γ-quantile. That is, if W is a random variable with distribution H_x, then

$$P\left\{ W \leq \overline{x} + t_\gamma \frac{s}{\sqrt{n}} \right\} = \gamma.$$

This is equivalent to

$$P\left\{\frac{W-\overline{x}}{s/\sqrt{n}} \le t_\gamma\right\} = \gamma,$$

so that $(W-\overline{x})/(s/\sqrt{n})$ has Student $(n-1)$ distribution. The distribution of W is therefore that of a Student $(n-1)$ random variable multiplied by $s/\sqrt{n}$ and translated by $\overline{x}$ so as to be centered about $\overline{x}$.

4. THE QUESTION OF UNIQUENESS

From the Neyman-Pearson-Wald viewpoint that only a procedure's performance matters, it is the last two paragraphs of Section I of Fisher (1935) that cause difficulties.

Here Fisher emphasizes that the statistic used must contain the whole of the relevant information: the estimator used must be a sufficient statistic. But real-valued sufficient statistics exist only in a few special problems so this requirement rules out Fiducial distributions in most problems, even though there may be distribution-function-valued random variables that have the required probabilities, and that have optimum properties corresponding to optimum properties of the equivalent families of hypothesis tests.

Here Fisher first insists that the Fiducial distribution of a parameter is unique, and not just any distribution-function-valued statistic whose (random) γ-quantile has probability γ of exceeding the parameter value, for each γ between 0 and 1. This view and the Neyman-Pearson-Wald view are compared in the following simple example.

Example. Let $X = X_1, \ldots, X_n$, with the X_i's independent each with Normal $(\mu, 1)$ distribution. Write $\overline{X} = (1/n)\sum X_i$, and for observed values write

$$x = x_i, \ldots, x_n \quad \text{and} \quad \overline{x} = (1/n)\sum x_i.$$

For Fiducial distributions, Fisher's view is that

$$H_X \text{ given by } H_x = \text{Normal }(\overline{x}, 1/n)$$

is the unique Fiducial distribution for μ, and that

$$K_X \text{ given by } K_x = \text{Normal }(x_1, 1/n)$$

is not a Fiducial distribution because it "does not contain all the relevant information about μ."

A Neyman-Pearson-Wald view is that "contain all the relevant information" is meaningless; that K_X is also a Fiducial distribution because it gives the required coverage probabilities; but that K_X is inadmissible because H_X is uniformly better in the sense that its γ-quantile has smaller probability of exceeding μ_1, for all $\mu_1 > \mu$.

In general, Fisher's view has the disadvantage of ruling out the possiblity of a Fiducial distribution whenever a real-valued sufficient statistic fails to exist, as for $Y_1, \ldots, Y_n$ independent, each Uniform $(\theta - \frac{1}{2}, \theta + \frac{1}{2})$. In such problems there can be distribution-function-valued statistics with the performance properties of a Fiducial distribution: for each γ between 0 and 1 its γ-quantile has probability γ of exceeding μ. Moreover, among these there can be one that is uniformly optimal in the sense that each γ-quantile has uniformly minimum probability of exceeding μ_1, for each $\mu_1 > \mu$; or in the sense of having uniformly minimum expected variance, or minimax expected variance.

For Fiducial intervals, Fisher's view is that $\overline{X} + 1.645/\sqrt{n}$ is the unique upper .95 Fiducial limit for μ, and that $X_1 + 1.645$ is not a Fiducial limit for μ because it does not contain all the information, even though it does have coverage probability .95. The Neyman-Pearson-Wald view of course is that if $P\{\mu \leq V(X)\} \equiv \gamma$ then $V(X)$ is an upper γ-confidence limit. Also, in general, although $P\{\theta \leq V_1(X)\} \equiv \gamma_1$ and $P\{\theta \leq V_2(X)\} \equiv \gamma_2 > \gamma_1$ together imply that $V_1(X) \leq \theta \leq V_2(X)$ is a $\gamma_2 - \gamma_1$ confidence interval, there may be better $\gamma_2 - \gamma_1$ intervals not of this form, as happens for the (randomized) Binomial.

For point estimation Fisher's view would lead to the view that $\overline{X}$ is *the* estimated μ value, uniquely. The Neyman-Pearson-Wald view is that every function $T(X)$ is an estimator of θ, most of them of course very poor estimators, and that ordinarily there will be no one T that is uniformly better than all the others.

REFERENCES

Chernoff, H. (1951), "A property of some type A regions". *Annals of Mathematical Statistics* **22**, 472–474.

Fisher, R. A. (1930), "Inverse probability". *Proceedings of the Cambridge Philosophical Society* **26**, 528–535.

Fisher, R. A. (1933), "The concepts of inverse probability and fiducial probability referring to unknown parameters". *Proceedings of the Royal Society of London,* Series A **139**, 343–348.

Fisher, R. A. (1935), "The fiducial argument in statistical inference". *Annals of Eugenics* **6**, 391–398.

Stein, C. M. (1951), "A property of some tests of composite hypotheses". *Annals of Mathematical Statistics* **22**, 475–476.

M. Evans, [1] D. A. S. Fraser [2] and G. Monette [3]

STATISTICAL PRINCIPLES AND TANGENT MODELS

ABSTRACT

Some recent results in foundational theory are surveyed and some general implications noted. Central to this and focussing on the standard principles of sufficiency and conditionality is an underlying intention of separating *error* from *effect*. This manifests itself in the ordinary analysis of variance and also for the generalizations of that analysis for the transformation models and for the exponential models. This paper gives preliminary introduction to the notion of tangent model which in turn allows the partial extension of this error-effect separation to more general statistical models. The emphasis is on finding techniques for determining both significance tests and confidence regions for general models, particularly for complex multiparameter models.

INTRODUCTION

The statistical principles of sufficiency (S), conditionality (C), and likelihood (L) were discussed by Birnbaum (1962, 1972) and were given a formal presentation in terms of equivalence relations on the collection of model-data combinations, called *inference bases*. Central to this work was a proof that sufficiency and conditionality imply likelihood. The result was enigmatic in the sense that sufficiency and conditionality enjoyed wide acceptance while likelihood had little acceptance.

More recently a model construction called cross-embedding has been proposed by Evans *et al.* (1985a,b) and used to prove that conditionality alone implies the likelihood principle. The accompanying discussion focussed

[1] Department of Statistics, University of Toronto, Toronto, Ontario M5S 1A1

[2] Department of Statistics, University of Toronto, Toronto, Ontario M5S 1A1; also University of Waterloo and York University

[3] Department of Mathematics, York University, Downsview, Ontario M3J 1P3

I. B. MacNeill and G. J. Umphrey (eds.), Foundations of Statistical Inference, 21–29.

attention sharply on the mechanism by which this enigmatic result was obtained. The essential step involved a process of "concealing" *given* information in an ancillary variable; as ancillary variables are assumed to contain no information this is the process by which model information is eliminated and thereby the likelihood principle deduced.

The noting of the mechanism by which conditionality implies likelihood provides an enlightening base for reassessing the roles of conditionality and sufficiency. We do not pursue this here, but some brief discussion is given by Evans *et al.* (1985a,b).

An overview of the preceding provides some grounds for the view that the intrinsic role for the basic principles of inference is that of separating *error* from *effect*. These terms, of course, have had consideration and some definition in experimental design contexts. In those contexts the typically referenced model is the ordinary normal linear model.

In a broader context the separation of error and effect has been examined for two classes of general linear models by Fraser (1985), the exponential linear model and the transformation linear model. These two model types are distinct except for the intersection, which is the normal linear model. The analysis involves a separation of the parameters into a sequence of real parameters with corresponding fitting and testing, closely paralleling that for the normal analysis of variance. The two general model types have a separation of error and effect but in quite different ways, conditioning on residuals with the transformation model and on previously fitted parameters with the exponential model. Both cases involve a form of conditionality principle, one that uses implicit structure from the relevant application context. Of these the stronger form of conditionality occurs with the transformation model. Interestingly the two very different analyses collapse to the same procedure, the ordinary procedure, for the normal intersection case. Also it is found that orthogonality as typically considered with an analysis of variance is in essence irrelevant; the appropriate separation occurs without reference to it.

This paper gives preliminary discussion to some procedures for extending this error-effect separation to more general contexts, with primary concern for the many parameter case.

2. ANCILLARITY AND ERROR

The argument from conditionality (C) to likelihood (L) by Evans *et al.* (1985b) uses the cross-embedding structure introduced by Evans *et al.* (1985a). The cross-embedded model has two competing ancillary variables and its use in the argument casts interesting light on the possibilities that can arise in the use of ancillary variables. As these possibilities in general do not conform to the motivating examples typically profferred to support the conditionality principle, we are in a real sense in the position of not having support for the general use of conditionality. The cross embedding argument provides a prime example of the use of ancillarity in clear violation of the supporting arguments for the conditionality principle.

For illustration suppose we have a Poisson (θ) model and a data value 2. We consider the cross embedding of this with the Bernoulli $(\theta^2 e^{-\theta}/2)$ model and data value 1:

$y =$		0	1	2	3	$\cdots$
$z =$	1	$.5e^{-\theta}$	$.5\theta e^{-\theta}$	$.5\theta^2 e^{-\theta}/2$	$.5\theta^3 e^{-\theta}/6$	$\cdots$
	0	$.5\theta^2 e^{-\theta}/2$	0	$.5(1 - \theta^2 e^{-\theta}/2)$	0	

Note the Poisson model probabilities in the first row of the 2-row array of probabilities and the Bernoulli $(\theta^2 e^{-\theta}/2)$ in the third column $(y = 2)$ of the array. (The Bernoulli $(c\theta^2 e^{-\theta}/2)$ can more generally be obtained by a simple adjustment (Evans *et al.*, 1985a), by having probability $c/(1+c)$ in the first row and $1/(1+c)$ in the second row, and with $(1 - \theta^2 e^{-\theta}/2)$ replaced by $(1 - c\theta^2 e^{-\theta}/2)$). The Poisson inference base and the Bernoulli inference base can be compared via the cross embedded model with data value $(z, y) = (1, 2)$.

The indicator function for the first row is ancillary; it has a Bernoulli (1/2) distribution. Thus the inference base (Poisson (θ), data $= 2$) has by C the same inference content as the combined-model inference base.

The indicator function for the third column $(y = 2)$ is ancillary; it has a Bernoulli (1/2) distribution. Thus the inference base (Bernoulli $(\theta^2 e^{-\theta}/2)$, data $=1$) has by C the same inference content as the combined-model inference base.

Thus by the conditionality principle (C) the Poisson inference base and the Bernoulli inference base have the same inference content. This extends trivially to the Bernoulli $(c\theta^2 e^{-\theta}/2)$ and may be found in full detail and with discussion in Evans *et al.* (1985b).

Our concern here is with the mechanism of the argument and the special case just presented suffices for this. The full Poisson model structure is contained in the first row of the array. By contrast if we consider the ancillary indicator for the third column, we have that the two third-column probabilities are lumped at that indicator value 1 and all other probabilities are lumped at 0. The use of this ancillary variable with data value 1 thus suppresses or "conceals" all the Poisson model probabilities save that at the $y = 2$ data point.

The original motivation for the use of an ancillary variable as part of conditioning is that it contains no information and thus in the language pursued later in this paper represents error. Certainly the standard ancillary examples (Cox and Hinkley, 1974; Fraser, 1979) involve ancillaries with "pure" error and no concealed model structure. These standard examples supporting the conditionality principle however do not support the kind of ancillarity and conditioning involved with the cross-embedded models and the argument above.

Two general implications are available. The standard examples for conditionality support only a much more restricted conditionality principle than that commonly considered. This is pursued elsewhere. The result that conditionality (C) implies likelihood (L) only follows by what can be considered an abuse of the conditionality principle and thus it need no longer be considered as an enigmatic result; some detailed discussion is given by Evans *et al.* (1985).

Our direction in this paper is toward the separation of error and effect in general contexts, and the issue of the use of conditionality to suppress model structure should be a serious concern.

3. ANALYSIS OF VARIANCE

The basic manipulations of the analysis of variance involve the decomposition of Euclidean squared length in terms of various orthogonal subspaces that are of interest in relation to the parameters of the problem. This finds theoretical justification in the framework of the normal linear model and from a data analysis viewpoint can be viewed as a linear type separation of apparent effects in accord with different parameter components.

The analysis of variance procedure typically involves fitting a succession of parameters followed by a reverse sequence testing of whether a last added parameter is needed; at any such stage a confidence interval or region is available.

The description just given for the analysis of variance has been used (Fraser, 1985) to develop the analysis of variance for two types of generalization of the normal linear model. The two generalizations are in fact quite distinct as is seen in part from the fact that the intersection of the two classes of model contains just the normal linear model (Heichelheim, 1966).

As a simple illustration consider the transformation model

$$f(y_1 - \theta_1,\ y_2 - \theta_2,\ y_3) \tag{3.1}$$

and the exponential model

$$f(y_1, y_2, y_3)\ \exp\{\theta_1 y_1 + \theta_2 y_2 - \phi(\theta_1, \theta_2)\} \tag{3.2}$$

on R^3 with $f > 0$. Note parenthetically that these coincide for $\theta_1 = \theta_2 = 0$. The models are considered in the context of the forward sequence of hypotheses, $H_0 = \{\theta_1 = \theta_2 = 0\}$, $H_1 = \{\theta_1 = 0,\ \theta_2 \in R\}$, $H_2 = \{\theta_1 \in R,\ \theta_2 \in R\}$ and with testing in the reverse order; test H_1 given H_2; test H_0 given H_1.

Conditionality arguments are used by Fraser (1985) to develop analysis of variance tables. A prime distinguishing feature in each case is the replacement of squared length by an appropriate linear variable. Another distinguishing feature involves the replacement of the product space (based on orthogonalization) structure by a succession of conditional or fibre structures. The details in general involve substantial notation but can be illustrated fairly easily for the simple example involving (3.1) and (3.2).

We record the tables for the two cases but omit the supporting arguments which may be found in *(ibid)*; we write $\theta_1 = \theta_1^0$ and $\theta_2 = \theta_2^0$ for $\theta_1 = 0$ and $\theta_2 = 0$.

Transformation model

Parameter	Effect	Variable	Observed
θ_2^0/θ_1	θ_1 vs. θ_1^0	$y_1 \mid y_2^0, y_3^0$	y_1^0
θ_2/θ_1	θ_2 vs. θ_2^0	$y_2 \mid y_3^0$	y_2^0

Exponential Model

Parameter	Effect	Variable	Observed
θ_2^0/θ_1	θ_1 vs. θ_1^0	y_1	y_1^0
θ_2/θ_1	θ_2 vs. θ_2^0	$y_2 \mid y_1^0$	y_2^0

The relevant conditional distribution for making a test or forming a confidence interval is indicated by the column 'variable' which shows a real variable examined conditionally as designated and marginally as implied by the omission of variables. The structure is that of fibres or foliations and is much simplified by the special coordinate-aligning nature of the parameters in (3.1) and (3.2).

The conditionality arguments that lead to these tables can be recast in terms of a process of separating error and effect. Indeed the construction of the tables can be viewed as providing two very general paradigms for this process. Our interest in this paper is in terms of extending such separation to general statistical models.

4. TRANSFORMATION TANGENT MODELS

Consider a sample $y_1, \ldots, y_n$ from a stochastically increasing model $f(y \mid \theta)$ and suppose the derivative with respect to θ exists and is continuous.

We examine the model $f(y \mid \theta)$ in a 'first-derivative' neighbourhood of θ_0 and derive a *location model* $g\big(t(y) - \theta\big)\,dt$ that coincides with $f(y \mid \theta)$ in the first derivative neighbourhood at θ_0:

$$f\,(y \mid \theta_0) = g\big(t(y) - \theta_0\big) \mid dt(y)/dy \mid ,$$
$$f'\,(y \mid \theta_0) = g'\big(t(y) - \theta_0\big) \mid dt(y)/dy \mid ;$$

the prime denotes differentiation with respect to θ_0. The transformation $t(y)$ is an increasing transformation and the distribution for the transformed variable is of location type in the 'first-derivative' neighbourhood at θ_0.

We outline the derivation in terms of a finite increment θ_0 to $\theta_0 + \delta$. For this let $F(y \mid \theta)$ be the distribution function and $F'(y \mid \theta)$ be its derivative with respect to θ. An increment $h(y)$ in y will correspond to an increment δ in θ if

$$-F'\,(y \mid \theta_0)\,\delta = f\,(y \mid \theta_0)\,h.$$

Let $t(y)$ be a new variable which increments at the same rate as θ:

$$\frac{dy}{dt} = \frac{-F'\,(y \mid \theta_0)}{f\,(y \mid \theta_0)},$$
$$t(y) = \int^{y} \frac{f\,(y \mid \theta_0)}{-F'\,(y \mid \theta_0)}\,dy.$$

Let y_0 be some reference location, say the median, for $f\,(y \mid \theta_0)$; then

$$t(y) = y_0 + \int_{y_0}^{y} \frac{f\,(y \mid \theta_0)}{-F'\,(y \mid \theta_0)}\,dy$$

can be viewed as a topological adjustment of the real line on either side of y_0 to give a distribution that locally (θ close to θ_0) has location form. Let $y = y(t)$ designate the inverse function.

The distribution function for the new variable $t(y)$ is given as

$$F\big(y(t) \mid \theta\big),$$

and the corresponding location model agreeing in the θ_0 neighbourhood has distribution function

$$F\Big(y\big(t-(\theta-\theta_0)\big) \mid \theta_0\Big)$$

and density function

$$f\Big(y\big(t-(\theta-\theta_0)\big) \mid \theta_0\Big)\frac{-F'\Big(y\big(t-(\theta-\theta_0)\big) \mid \theta_0\Big)}{f\Big(y\big(t-(\theta-\theta_0)\big) \mid \theta_0\Big)}$$

$$= -F'\Big(y\big(t-(\theta-\theta_0)\big) \mid \theta_0\Big).$$

For additional details, although from a different viewpoint, see Fraser (1964).

For the original problem, $y_1, \ldots, y_n$ from $f(y \mid \theta)$, there may not be a clear choice for a test of θ_0 or for a confidence region. The preceding development provides a location model approximation at any θ_0 value and makes available procedures that are appropriate and convenient for such models. We do not pursue this here but note that computer implementation is accessible and involves one dimensional integration.

Consider briefly the multiparameter case, say $\mathbf{y}_1, \ldots, \mathbf{y}_n$ a sample from $f(\mathbf{y} \mid \theta)$. Preliminary developments indicate that the monotonicity becomes a more incisive multiparameter requirement and that some conditions may then be required for a unique tangent model at a point θ_0.

5. EXPONENTIAL TANGENT MODEL

Consider a regular statistical model $f(y \mid \theta)$ with a real parameter and suppose that derivatives, as needed, with respect to θ exist.

We examine the model for θ in a first derivative neighbourhood of θ_0 and construct an exponential model

$$g(y \mid \theta) = f(y \mid \theta_0)\exp\{s(y)(\theta-\theta_0) - \phi(\theta-\theta_0)\}$$

that coincides with $f(y \mid \theta)$ in this neighbourhood:

$$f(y \mid \theta_0) = g(y \mid \theta_0),$$
$$f'(y \mid \theta_0) = g'(y \mid \theta_0).$$

It follows easily that

$$s(y) = \frac{\partial}{\partial\theta}\ln f(y \mid \theta)\,|_{\theta_0}$$

is the score at θ_0 and $\phi(\theta - \theta_0)$ is the cumulant type function

$$e^{\phi(\delta)} = \int \exp\{\delta s(y)\} f(y \mid \theta_0)\, dy.$$

Exponential models are available that provide a higher order approximation at θ_0. For this let $\delta = \theta - \theta_0$ and consider the following:

$$g_r(y \mid \theta) = f(y \mid \theta) \exp\left\{\sum_1^r s_i(y)\delta^i/i! - \phi_r(\delta)\right\}.$$

This is an exponential model of degree r and coincides to the rth degree at θ_0 by choice

$$s_i(y) = \frac{\partial^i}{\partial\theta^i} \ln\ f(y \mid \theta)\,|_{\theta_0}$$

with

$$\exp\{\phi(\delta)\} = \int \exp\left\{\sum \delta^i s_i(y)\right\} f(y \mid \theta_0)\, dy.$$

As will be shown elsewhere the use of a higher order approximation can facilitate the determination of tests and confidence regions.

The multiparameter case for the exponential tangent models is available in a straightforward manner, in contrast to the complications that arise for the transformation model as mentioned in the preceding section.

6. CONCLUDING REMARKS

An underlying intent for the material in this paper is that of deriving tests and confidence regions for multiparameter problems. Hypothesis testing theory provides the basic reference material for this task. For implementation, however, the routine success lies closely to models of exponential type, and the principle of sufficiency is the common reduction method.

Location and transformation models (or the corresponding structural models) have a natural (or necessary) preliminary reduction based on an ancillary (or on observed error). This preliminary reduction can simplify the problem to the point where the usual hypotheses testing procedures can be invoked.

Most problems involve testing or the construction of confidence regions in the presence of nuisance parameters. This is exemplified in a broad setting by the standard analysis of variance involving a succession of parameters being introduced and a reverse testing procedure as parameters are dropped. This analysis of variance can be extended to the exponential family and to the transformation family and makes available procedures for testing and forming confidence intervals provided that some sequence is given for examining the parameter components.

For implementation we note that both generalizations lead to well defined tests which typically involve both conditioning and marginalization. The conditioning part is typically routine. It is the marginalization that causes the difficulties; for this the dimension is the number of nuisance parameters in the transformation case and is the number of error coordinates in the exponential case. Various techniques are under development for addressing these difficulties.

The structure of exponential and transformation models provides substantial access to tests and confidence intervals by the analysis of variance procedures in Section 3.

For more general statistical models the available procedures are sharply limited. Likelihood methods are commonly used with asymptotic results available as a partial guide towards tests and confidence regions.

This paper proposes the use of local properties of a statistical model, local on the parameter space. The local properties can be summarized in the form of either a transformation model or an exponential model. Both are tangent models but represent different propagations of the local properties.

The special properties of the tangent and exponential models can be used to develop tests and confidence regions.

REFERENCES

Birnbaum, A. (1962), "On the foundations of statistical inference." *Journal of the American Statistical Association* **57**, 209–332.

Birnbaum, A. (1972), "More on concepts of statistical evidence." *Journal of the American Statistical Association* **67**, 858–886.

Cox, D. R., and D. V. Hinkley (1974), *Theoretical Statistics*. London: Chapman and Hall.

Evans, M., D. A. S. Fraser, and G. Monette (1985a), "Mixtures, embedding, and ancillarity." *Canadian Journal of Statistics* **13**, 1–6.

Evans, M., D. A. S. Fraser, and G. Monette (1985b), "On the sufficiency-conditionality to likelihood argument." *Canadian Journal of Statistics* **14**.

Fraser, D. A. S. (1964), "Local conditional sufficiency." *Journal of the Royal Statistical Society*, Series B **26**, 52–62.

Fraser, D. A. S. (1979), *Inference and Linear Models*. Toronto: DAI, University of Toronto.

Fraser, D. A. S. (1985), "On analysis of variance." University of Toronto, Department of Statistics.

Heichelheim, P. (1966), "A comparison between sufficiency and structural models." Ph.D. thesis, University of Toronto.

V. P. Godambe [1]

DATA BASED CHOICE OF AN ANCILLARY STATISTIC

ABSTRACT

A data based method is presented for choosing from different ancillary statistics, for the purpose of conditioning.

1. INTRODUCTION

Basu (1964) presented some examples of non-unique ancillary statistics which are functions of a minimal sufficient statistic. Barnard and Sprott (1971) presented a criterion based on the considerations of the shape of the likelihood function to choose between the different ancillary statistics. Another criterion for such a choice, based on the dispersion of the conditional Fisher information, was given by Cox (1971). The criterion suggested below, unlike the above two, is data based.

2. THEORY

Godambe (1977, 1980) provided a theory for eliminating a nuisance parameter by conditioning on some statistic called "ancillary" in the sense that its distribution does not contain any information about the parameter of interest, the conditional distribution being free of the nuisance parameter. Similar considerations can be implemented when *apparently* there is no nuisance parameter. We say apparently for the following reason. One may be uncertain about some aspects of the form of the distribution; this then could be said to be the nuisance parameter. It is proposed that we should choose from the competing ancillary statistics, one which renders the inference independent of or less dependent on the part of the distribution one is

[1] Department of Statistics and Actuarial Science, University of Waterloo, Waterloo, Ontario N2L 3G1

I. B. MacNeill and G. J. Umphrey (eds.), Foundations of Statistical Inference, 31–38.

uncertain about. Suppose $f_\theta(x)$ is a given model with θ the unknown parameter and x the minimal sufficient statistic. Let $a(x)$ and $b(x)$ be two ancillary statistics. That is, the distribution functions of a and b, namely $h_1(a)$ and $h_2(b)$, are independent of the parameter θ. To follow our criterion, if one is uncertain *a priori* about the distribution h_1, but not about the distribution h_2, one should condition on the ancillary a. More realistically, the choice can be made *after* the data $x = x_0$ are available. The values $a(x_0) = a_0$ and $b(x_0) = b_0$ may, in some situations, tell which of the distributions h_1 or h_2 is doubtful. For instance, if "a_0" is in the far tail end of the distribution h_1 but "b_0" is a much less extreme value with respect to the distribution h_2, one should condition on a_0 and not on b_0. In other words, we carry out significance tests for the distributions h_1 and h_2 on the basis of the observations a_0 and b_0 respectively to make the choice of distribution when one is doubtful. Such a procedure, of course, would not always be successful. But sometimes it would. If h_1 is found to be the doubtful distribution then, for inference about θ, we would use the conditional distribution,

$$[f_\theta(x)/\{h_1(a_0)\}]. \tag{1}$$

One obvious objection to the above procedure is that it is implicitly based on a statistic which is a function of a and b together and this function in general will *not* be an ancillary statistic. This objection is dealt with as follows. Suppose, using the notation of the preceeding paragraph, the distribution h_1 is chosen on the basis of the data $x = x_0$ or $a(x_0) = a_0$. Then we replace the underlying model $f_\theta(x)$ by the enlarged model

$$U_{\theta,\eta}(x) = \{f_\theta(x)/h_1(a)\}\{h_{1,\theta,\eta}(a)\}, \tag{2}$$

where $h_{1,\theta,\eta}(a)$ is the marginal distribution of "a" containing an additional or nuisance parameter $\eta; \{\eta\} = I$. Note, because of this adjustable parameter η, the distribution of "a", namely $h_{1,\theta,\eta}(a)$, for any value of a, need not be "doubtful". Further assuming that for each fixed θ, the family $\{h_{1,\theta,\eta} : \eta \in I\}$ is complete or satisfies suitable conditions according to Godambe (1977, 1980), the statistic "a" is the *ancillary for the parameter* θ, *in the presence of the nuisance parameter* η, in the distribution given by (2). Hence, inference about θ on the basis of x should be based on the conditional distribution of x, conditional on the observed value of $a = a_0$ in (1); that is, on

$$[f_\theta(x)/\{h_1(a_0)\}], \tag{3}$$

the *same* as in (1).

Note that under suitable conditions the ancillary statistic for θ, in the presence of a nuisance parameter η, is *unique* (Godambe, 1977). If these conditions are satisfied by the model $U_{\theta,\eta}(x)$ in (2), the statistic "b" which was ancillary for the model $f_\theta(x)$ is no more an ancillary in $U_{\theta,\eta}(x)$.

3. ILLUSTRATION

We consider the following multinomial distribution with four cells. If we use the notation of §2 with $x = (n_1, n_2, n_3, n_4)$ we have the following:

$$f_\theta(n_1, n_2, n_3, n_4) = \frac{n!}{n_1!n_2!n_3!n_4!}(1-\theta)^{n_1}(1+\theta)^{n_2}(2-\theta)^{n_3}(2+\theta)^{n_4} \cdot \left(\frac{1}{6}\right)^n, \qquad (4)$$

where n_1, n_2, n_3, n_4 are different cell frequencies with $n = n_1+n_2+n_3+n_4$, and θ being the unknown parameter. Two ancillary statistics are given by

$$a(n_1, n_2, n_3, n_4) = n_1 + n_2 \qquad (5)$$

and

$$b(n_1, n_2, n_3, n_4) = n_1 + n_4. \qquad (6)$$

The distributions $h_1(a)$ and $h_2(b)$ are given by the following:

$$h_1(a) = \frac{n!}{(n_1+n_2)!\,(n_3+n_4)!}\left(\frac{1}{3}\right)^{n_1+n_2}\left(\frac{2}{3}\right)^{n_3+n_4}, \qquad (7)$$

and

$$h_2(b) = \frac{n!}{(n_1+n_4)!\,(n_2+n_3)!}\left(\frac{1}{2}\right)^{n_1+n_4}\left(\frac{1}{2}\right)^{n_2+n_3}. \qquad (8)$$

Now evidently large values of n_1 and n_2, that is small values of n_3 and n_4, tend to throw doubt on the distribution h_1 in (7). However, if in addition to n_1 and n_2 being large, $n_2 + n_3$ and $n_1 + n_4$ are nearly equal, then one may *not* have any reason to doubt the validity of the distribution h_2 in (8). In (4), let $n = 20$ and let the observed values be $n_1 = n_2 = 8$ and $n_3 = n_4 = 2$. This surely would cast doubt on the distribution h_1 but not on h_2. Hence, with such a sample (n_1, n_2, n_3, n_4) we would choose ancillary statistic a and *not* b for conditioning. On the other hand if the observed values are $n_1 = 6$, $n_2 = 1$, $n_3 = 1$ and $n_4 = 12$ one would surely doubt the validity of the distribution h_2 in (8) and *not* h_1 in (7). Hence, here we would choose the ancillary statistic b in preference to a. Thus, it follows from (1), (7) and (8) that the inference about θ will be based on one of the conditional distributions,

$$[f_\theta(x)/\{h_1(a)\}]$$

or

$$[f_\theta(x)/\{h_2(b)\}], \tag{9}$$

according to whether the ancillary statistic a or b is chosen.

If the ancillary statistic a is chosen we may enlarge our model along the lines of (2) by introducing a nuisance parameter for the distribution of $a = n_1 + n_2$:

$$h_{1,\theta,\eta}(a) = h_{1,\eta}(a) = \frac{n!}{(n_1+n_2)!\,(n_3+n_4)!}(\eta)^{n_1+n_2}(1-\eta)^{n_3+n_4} \text{ say} . \tag{10}$$

Now in the enlarged model, namely

$$U_{\theta,\eta}(x) = \{f_\theta(x)/h_1(a)\}\,\{h_{1,\eta}(a)\}, \tag{11}$$

where f_θ, h_1 and $h_{1,\eta}$ are given by (4), (7), and (10) respectively, the statistic $a = n_1 + n_2$, is the ancillary for the parameter θ in presence of the nuisance parameter η, as in (2). Hence, the inference about θ should be based on the conditional distribution:

$$\frac{U_{\theta,\eta}(x)}{h_{1,\eta}(a)} = \frac{\frac{n!}{n_1!n_2!n_3!n_4!}(1-\theta)^{n_1}(1+\theta)^{n_2}(2-\theta)^{n_3}(2+\theta)^{n_4}\left(\frac{1}{6}\right)^n}{\frac{n!}{(n_1+n_2)!(n_3+n_4)!}(2)^{n_1+n_2}(4)^{n_3+n_4}\left(\frac{1}{6}\right)^n}, \tag{12}$$

which is the *same* as f_θ/h_1 in (9).

Note, the statistic $b_1 = \eta_1 + \eta_4$ in (6), which was an ancillary for the model $f_\theta(x)$ in (4), is no more an ancillary for θ in the presence of the nuisance parameter η in the model $U_{\theta,\eta}(x)$ in (11).

4. EXPERIMENTAL ANCILLARY STATISTIC

As in §2, let the model $f_\theta(x)$ admit two ancillary statistics, namely $a(x)$ and $b(x)$. Now the models obtained by conditioning on the observed values, namely $a = a_0$ and $b = b_0$ respectively are:

$$f_\theta(x)/h_1(a_0) \text{ and } f_\theta(x)/h_2(b_0),$$

where as before h_1 and h_2 are the densities of a and b under f respectively. We consider two cases:

(1) The experimenter *knew* that in fact the experiment was carried out holding the ancillary statistic a fixed at $a = a_0$.

(2) The experimenter *suspects* a possibility that in fact that experiment was carried out holding *either* "a" fixed at $a = a_0$ *or* "b" fixed at $b = b_0$, but he does *not* know which.

In case (1), the experimenter's inference would follow from the conditional model $f_\theta(x)/h_1(a_0)$. Note in this conditional model, "b" is no larger an ancillary statistic. Here we call "a" the *experimental ancillary statistic*. Similar terminology, in a different context, is used by Kalbfleisch (1975). In case (2), the theory of §2 provides the following method: if the observed value of $a = a_0$, unlike b_0, belongs to the tail of the distribution of "a" computed under $f_\theta(x)$, call "a" the experimental ancillary statistic and inference should proceed with the model $f_\theta(x)/h_1(a_0)$. That is, in (2) we put $h_{1,\theta,\eta}(a_0) = 1$ and $h_{1,\theta,\eta}(a) = 0$ for $a \neq a_0$ and restrict the original model f_θ to the conditional one, f_θ/h_1. We discount the rare chance that the value a_0 could come from the distribution $h_1(a)$.

The following illustration is taken from Fraser (1973). A population of experimental units $\mathcal{P}$ is divided into two exclusive groups A_1 and A_2 with the ratio of sizes 1:2. A treated unit from A_1, produces a reaction with probability $(1+\theta)/2$ and a treated unit from A_2 produces a reaction with probability $(2+\theta)/4$, with θ being the unknown parameter. Further, the same population $\mathcal{P}$ is alternatively divided in two exclusive groups B_1 and B_2 with ratio of sizes 1:1. A treated unit from B_1 produces a reaction with probability $(1+\theta)/3$ and a treated unit from B_2, with probability $(2+\theta)/3$. For a sample of n units from the population $\mathcal{P}$, let n_1 be from A_1 and n_2 from A_2; similarly m_1 from B_1 and m_2 from B_2. Thus, $n_1 + n_2 = m_1 + m_2 = n$. In the following tables of cell frequencies and corresponding probabilities x_1 and x_2 denote, in the appropriate groups, the number of units treated which showed reaction; $n_1 - x_1$ and $n_2 - x_2$ are the number of treated units without reaction.

Frequencies

	B_1	B_2	
A_1	x_1	$n_1 - x_1$	n_1
A_2	$n_2 - x_2$	x_2	n_2
	m_1	m_2	n

Probabilities

	B_1	B_2	
A_1	$(1+\theta)/6$	$(1-\theta)/6$	1/3
A_2	$(2-\theta)/6$	$(2+\theta)/6$	2/3
	1/2	1/2	1

In the notation of §2 we have $x = (x_1, n_1 - x_1, n_2 - x_2, x_2)$ and $f_\theta(x)$ is the multinomial distribution with cell "Probabilities" given above. The underlying model $f_\theta(x)$ is based on the assumption that *the sample of n units was drawn at random from the population $\mathcal{P}$*. The model $f_\theta(x)$ admits two ancillary statistics $a(x) = n_1$ and $b(x) = m_1$. Suppose now, as in the case (2) above, the experimenter entertains a doubt about the underlying model

$f_\theta(x)$ and suspects that in fact the experiment was carried out holding *either* n_1 fixed *or* m_1 fixed. Thus, following the theory in §2, if the observed value of n_1, unlike that of m_1, belongs to the tail of its distribution computed on the basis of f_θ, we carry out the inference holding the observed value of n_1 fixed; that is, n_1 is called the experimental ancillary statistic. Here the conditional distribution f_θ/h_1 is a mixed binomial and is given by the following:

$$\frac{f_\theta(x)}{h_1(a=n_1)} = \left[\frac{n_1!}{x_1!(n_1-x_1)!}\right]\left[\frac{(1+\theta)}{2}\right]^{x_1}\left[\frac{(1-\theta)}{2}\right]^{n_1-x_1}$$
$$\left[\frac{n_2!}{x_2!\,(n_2-x_2)!}\right]\left[\frac{(2+\theta)}{4}\right]^{x_2}\left[\frac{2-\theta}{4}\right]^{n_2-x_2}$$

This, then, is an answer to the delemma presented by Fraser: "Clearly we could be in an A_1 to A_2 situation or B_1 or B_2 situation ... if we are in the position of not knowing which applies ... then no valid principle of inference can argue that we act as if we are in that situation for which the inference procedures would be most attractive." (Fraser, 1973, p. 47)

Note that for the conditional model f_θ/h_1 given above, m_1 is no more an ancillary statistic.

5. DISCUSSION

In relation to the illustration in §4 and the quotation of Fraser we remark that if the experimenter did not entertain any doubts about the validity of the original model $f_\theta(x)$, a choice between the two ancillary statistics, namely n_1 and m_1, could be made using the criterion of Barnard and Sprott (1971), or of Cox (1971), or something similar (Fraser, 1973).

The above discussion suggests that the uses of the *two types of criteria* for choosing between the ancillary statistics, namely the one given by Barnard and Sprott (1971), and Cox (1971) and the other given in §2 above are somewhat different. If the validity of the underlying model is not in question one should go by the former criteria. But if the validity of the model is in question, one should use the criterion given in §2 to ensure *robust* inference. Again, the two types of criteria and the purposes they serve may not be that distinct. For instance, if the derivative of the score function is either an ancillary statistic or approximately so, Cox's (1971) criterion would suggest inference be made holding this ancillary statistic fixed. But this also provides *robust* inference for large samples, because of the implied asymptotic normality of the corresponding conditional distribution regardless of the underlying distribution (Efron and Hinkley, 1978).

The Fraser example in §4, also brings out the relationship of the theory of §2 above, with Basu's (1964) "performality" of experiment and Kalbfleisch's (1975) "experimental" ancillary statistic.

ACKNOWLEDGMENT

Thanks are due to Harold Mantel and Mary Thompson for useful discussions.

REFERENCES

Barnard, G. A., and D. A. Sprott (1971), "A note on Basu's examples of anomalous ancillary statistics." In *Foundations of Statistical Inference*, ed. V. P. Godambe and D. A. Sprott, pp. 163–170. Toronto, Montreal: Holt, Rinehart, and Winston of Canada.

Basu, D. (1964), "Recovery of ancillary information." *Sankhyā*, Series A **26**, 3–16.

Cox, D. R. (1971), "The choice between alternative ancillary statistics." *Journal of the Royal Statistical Society*, Series B **33**, 251–255.

Efron, B., and D. V. Hinkley (1978), "Assessing the accuracy of the maximum likelihood estimator: observed versus expected Fisher information." *Biometrika* **65**, 457–487.

Fraser, D. A. S. (1973), "The elusive ancillary." In *Multivariate Statistical Inference*, ed. D. G. Kabe and R. P. Gupta, pp. 41–48. Amsterdam: North-Holland.

Godambe, V. P. (1976), "Conditional likelihood and unconditional optimum estimating equations." *Biometrika* **63**, 277–284.

Godambe, V. P. (1980), "On sufficiency and ancillarity in the presence of a nuisance parameter." *Biometrika* **67**, 155–162.

Kalbfleisch, J. D. (1975), "Sufficiency and conditionality." *Biometrika* **62**, 251–259.

Editorial Comment: A referee pointed out the following.

There seem to be two major problems relating to the discussion of the table in Section 4. The way in which the table of probabilities is printed creates the impression that the probabilities are simply those in a 2-way classification, i.e., $(1-\theta)/6$, for example, is the probability that a randomly chosen population unit belongs to the group A_1 and simultaneously to the group B_2. This, however, is false. The case considered by Fraser (1973) is one in which the parent population is either the mixture (A_1, A_2) or the mixture (B_1, B_2), the two hypotheses being mutually exclusive. Thus a unit cannot belong to a group A_i and simultaneously to a group B_j, $i = 1, 2$, $j = 1, 2$. The interpretation of the cell probabilities is as follows. Take, for example, the cell with cell-probability $(1-\theta)/6$. If the parent population has the composition (A_1, A_2), then $(1-\theta)/6$ is the probability that a randomly chosen population unit belongs to group A_1 and shows no reaction when treated. Alternatively, when the parent population has the composition (B_1, B_2), $(1-\theta)/6$ is the probability that a randomly chosen population unit belongs to the group B_2 and shows no reaction when treated.

This source of confusion would be avoided if the marginal headings A_1, A_2, B_1, B_2, are omitted (as in Fraser's paper) from the table of cell probabilities. It may be desirable also to indicate briefly how the probabilities are to be interpreted.

The second difficulty relates to the manner of choosing the sample of size n. The author considers the cases in which the sample is chosen either by holding n_1 fixed or by holding m_1 fixed. But these are not the cases considered by Fraser (1973). The latter are as follows.

Case (I): The statistician knows that the parent population is (A_1, A_2). A random sample of size n from this mixed population is found to contain n_1 units of A_1 (n_1 being a random number). According to Fraser, n_1 is the appropriate conditioning statistic.

Case (II): The statistician knows that the mixed parent population has the composition (B_1, B_2). In a randomly chosen sample of size n, m_1 is the number of units belonging to B_1. According to Fraser (1973) m_1 is the appropriate ancillary for this case.

Case (III): The statistician knows only that the parent population is either (A_1, A_2) or (B_1, B_2), but does not know which of these it is. He has only the numbers x_1, x_3, of units in the two cells in the first row and x_4, x_2 of units in the two cells in the second row. (This implies that the statistician does not know if the x_1 and x_3 units belong to a common subpopulation or whether the x_2 and x_3 units do so.) How is he to make the inference regarding θ in this case?

The above description shows that the problem considered by the author (n_1, m_1 non-random numbers) differ materially from that considered by Fraser. The difficulty in applying the author's criterion to Fraser's problem is that in Cases I and II, it may yield an ancillary different from that accepted by Fraser.

Author's Reply: I am very grateful to the referee for such detailed comments on the 'Fraser example' I discussed in my paper. Yet with all that, I am not very clear about the distinction that the referee is making in my interpretation of the 'example' and Fraser's interpretation. Maybe I will have to meditate more. Anyway 'with *my* interpretation of the Fraser example' I have an illustration of my theory.

Dennis V. Lindley [1]

BERNOULLI PAIRS WITH INVARIANT REVERSALS: AN EXAMPLE OF PARTIAL LIKELIHOOD

ABSTRACT

Bernoulli pairs with invariant reversals provide an inference situation with two parameters, θ and λ ; λ is a nuisance parameter and θ the parameter of interest. The likelihood factorizes and the partial likelihood depends on θ alone. This paper examines the problem of making inferences about θ in detail and concludes that although the partial likelihood method is asymptotically sound the approach to the limit is slow, so that the method can lose a lot of information even with quite large samples. Computations of posterior distributions for 100 pairs illustrate the argument.

1. INTRODUCTION

Consider a situation where data x are dependent on two parameters θ and λ according to a probability density $p(x \mid \theta, \lambda)$, where θ is a parameter of interest and λ is a nuisance parameter. It is required to make inferences about θ alone, λ being of no interest. One method of doing this is to take the likelihood, which is $p(x \mid \theta, \lambda)$ considered as a function of θ and λ for fixed x, and write it in some way as a product $f(\theta)g(\theta, \lambda)$, discard the term $g(\theta, \lambda)$, leaving only $f(\theta)$ and treat this as a likelihood : $f(\theta)$ is sometimes called a partial likelihood. The idea here is that $g(\theta, \lambda)$ contains all the information on the nuisance parameter and some on the parameter of interest. If the latter is heavily confounded with the former it may be of little value so that the greater part of the information on θ is in $f(\theta)$. Clearly the method has most chance of success when this happens and consequently effort has been directed towards determining when $f(\theta)$ is nearly as informative as a full

[1] 2 Periton Lane, Minehead, Somerset, England TA24 8AQ; also Monash University, Australia

I. B. MacNeill and G. J. Umphrey (eds.), Foundations of Statistical Inference, 39–50.

likelihood $f(\theta,\lambda)$. When this happens the method has the great advantage that it is not necessary to consider the nuisance parameter at all.

If sampling-theoretic methods of inference are to be used it is not enough to have some form of likelihood; a sample space has to be constructed as well. This is readily available if the factorization is achieved by means of a statistic $t(x)$. For we may always write

$$p(x \mid \theta,\lambda) = p(t(x) \mid \theta,\lambda)p(x \mid t(x),\theta,\lambda),$$

in terms of the marginal density for $t(x)$ and the density for x conditional on $t(x)$, and if either of these is free of λ we have a factorization and $f(\theta)$ has a well-defined sample space associated with it: either that of $t(x)$, or that of x conditional on the value of $t(x)$ observed, according to which factor is λ-free. The present paper discusses an interesting example of a factorization based on a statistic with a distribution that does not involve λ and which has been claimed to have optimal properties.

2. BERNOULLI PAIRS WITH INVARIANT REVERSALS

The example is given by Lindsay (1980, §9.4). For each $i, 1 \le i \le n$, there is a parameter pair (θ,ϕ_i) such that the data pair (X_i, Y_i) is of independent Bernoulli random quantities with $p(X_i = 1) = p(Y_i = 1) = 1/2 + \theta\phi_i$ (and $p(X_i = 0) = p(Y_i = 0) = 1/2 - \theta\phi_i$). Here θ, $0 \le \theta \le 1/2$, is the parameter of interest: ϕ_i is either $+1$ or -1 and is a nuisance parameter. Given all the parameters the pairs (X_i, Y_i) are independent. For the total data there is therefore a single parameter of interest θ and $\phi = (\phi_1, \phi_2, ..., \phi_n)$ is a vector of n nuisance parameters. Lindsay treats the ϕ_i as random quantities which, given another parameter λ, are independent and identically distributed with probability λ that $\phi_i = +1$ (and hence $1-\lambda$ that $\phi_i = -1$). If this is so the n pairs (X_i, Y_i) are independent and identically distributed conditional on θ and a single nuisance parameter λ. This is often called a mixture model, mixing the original distribution of the pairs with that of the ϕ's. Before proceeding with the model, three points are worth making.

1) To Lindsay's justifications for treating the ϕ_i as i.i.d. given λ the following consideration may be added. In default of any knowledge of the ϕ's it might be reasonable to suppose them exchangeable, in which case de Finetti's theorem shows that they must have the structure just described with the addition of a probability distribution for λ.

2) Writing $p(X_i = 1) = 1/2 \pm \theta$ seems a little strange. An apparently more natural way is to suppose $p(X_i = 1) = \xi$ with probability λ and $1-\xi$ with probability $1-\lambda$, reversing the binomial chances from $(\xi, 1-\xi)$ to

$(1-\xi, \xi)$. However, this parametrization is unidentifiable because replacing ξ by $1-\xi$ and λ by $1-\lambda$ throughout gives the same model, or, in other words, in the unit square of possible values of ξ and λ, the points (ξ, λ) and $(1-\xi, 1-\lambda)$ yield the same probability distributions. Restricting ξ to be not less than 1/2 avoids the difficulty. Hence $\xi = 1/2+\theta$ with $0 \leq \theta \leq 1/2$. The point will have some significance below when we see the effect the reduced range of ξ has.

3) In the mixture form the model is similar to models considered in Empirical Bayes theory. The distribution of the ϕ's plays the role of the prior distribution with λ as the hyperparameter of that distribution. Unlike the usual case, θ is fixed and does not change with each new pair.

We now return to the mixture model. As with any Bernoulli quantities, X_i+Y_i is sufficient for each pair and the values for different i are independent and identically distributed. Attention can therefore be confined to their distribution which is clearly

$$p(X_i + Y_i = 0) = \lambda(1/2-\theta)^2 + (1-\lambda)(1/2+\theta)^2, \tag{2.1}$$

$$p(X_i + Y_i = 1) = 2(1/2-\theta)(1/2+\theta), \quad \text{and} \tag{2.2}$$

$$p(X_i + Y_i = 2) = \lambda(1/2+\theta)^2 + (1-\lambda)(1/2-\theta)^2. \tag{2.3}$$

The data may similarly be condensed to (s, w, t) where these are respectively the numbers of the n pairs having $X_i+Y_i = 0$, 1 and 2: necessarily $s+w+t = n$. (Lindsay also uses the notation w.) Their distribution is trinomial. From (2.2) it is immediately apparent that w has a distribution that does not involve the nuisance parameter λ so that there is a factorization of the type mentioned above, namely

$$p(w \mid \theta, n)p(s \mid w, \theta, \lambda, n)$$

for fixed n. Lindsay defends the use of $p(w \mid \theta, n)$—playing the role of $f(\theta)$ above—and derives an estimate of θ based on it. Throughout the paper it will be supposed that n is fixed and the condition will be omitted from the notation. The constraint will not affect the Bayesian study but is relevant to Lindsay's sample-space argument.

3. THE LIKELIHOOD FUNCTION

From (2.1)–(2.3) it follows that for data (s, w, t) the full likelihood is

$$\begin{aligned}
&(1/2-2\theta^2)^w\{\lambda(1/2-\theta)^2 \\
&\quad + (1-\lambda)(1/2+\theta)^2\}^s\{\lambda(1/2+\theta)^2 + (1-\lambda)(1/2-\theta)^2\}^t \\
&= (1/2-2\theta^2)^w(1/2+2\theta^2)^{s+t}\{1/2-\psi(\lambda-1/2)\}^s \\
&\quad \times \{1/2+\psi(\lambda-1/2)\}^t
\end{aligned} \tag{3.1}$$

after some rearrangement: here

$$\psi = \theta/(1/4 + \theta^2) \tag{3.2}$$

with $0 \leq \psi \leq 1$. In (3.1) the first two terms give a binomial likelihood derived from w with binomial parameter $p = 1/2 - 2\theta^2$. The second two terms (in braces) give another binomial likelihood derived from s, conditional on w, with binomial parameter $p' = 1/2 - \psi(\lambda - 1/2)$. According to the partial likelihood approach the first of these, depending only on θ and not λ, can be used to make inferences about θ. According to conventional wisdom about binomial distributions, w/n is the "best" estimate of the binomial chance $1/2 - 2\theta^2$. Indeed, Lindsay proves that it is the (fully efficient) minimum variance unbiased estimator and is also fully mixture-space efficient asymptotically as $n \to \infty$.

These are high recommendations. Yet the result is unsatisfactory. Suppose $w/n > 1/2$, then a quantity whose value is known to be *less* than 1/2, namely $1/2 - 2\theta^2$, is being estimated as *greater* than 1/2. Such statistical practice is hardly likely to commend the subject to the zoologist within whose field the data arise. What is even worse is the appreciation that the unsound estimate arises not from a quirk of this example but from a piece of respected theory. It is perfectly true that w/n is unbiased: that is $E(w/n \mid \theta) = 1/2 - 2\theta^2$. It is also true that amongst such estimates it is 'best' in several senses. But all this says is that on the average it gives the right answer and is nearer to it than other estimates. So proponents of unbiasedness have to say: yes, the estimate of $1/2 - 2\theta^2$ is 0.6, it may look ridiculous here but on the average all is well.

For this criticism to have any force it is necessary to show that the awkward case $w/n > 1/2$ can reasonably occur. To show this we calculate how large the sample size n has to be to be reasonably certain that w/n will be *less* than half and the difficulty will not arise. Now w/n has mean $p = 1/2 - 2\theta^2$ and variance $p(1-p)/n$. Replacing the binomial with the normal, as is reasonable for large n and p near 1/2, θ near 0, we see that the upper 2 1/2% limit for w/n (at two standard deviations) is $p + 2[p(1-p)/n]^{1/2}$. For this to be less than 1/2, so that we are 97 1/2% sure of a sensible value of w/n less than 1/2, we require $2[p(1-p)/n]^{1/2} < 1/2 - p$ or

$$n > 16p(1-p)/(1-2p)^2.$$

If $\theta = 0.1, p = 0.48$ and n must exceed 2,496. Consequently, the statistician needs a sample of about 2,500 pairs to be reasonably sure that the estimate will not be absurd when $\theta = 0.1$. For $\theta = 0.2$ the corresponding value is 152. We therefore see that the contradictory phenomenon can reasonably occur when θ is near 0, over at least one-fifth of its possible range. Of course, at least according to conventional wisdom, θ is unknown.

The situation calls for re-examination and to do this we use the Bayesian approach which ignores the average, or long-run behaviour and uses the full likelihood together with a distribution for (θ, λ). The former is given by equation (3.1) with ψ given by (3.2).

4. ASYMPTOTIC BAYESIAN RESULTS

To carry through the Bayesian argument it is necessary to evaluate distributions for θ and λ, that for the ϕ's already having been expressed through exchangeability. To do this it is necessary to consider the zoological meaning of the parameters. Lindsay says that the model is adapted from Felsenstein (1973) but neither that paper nor the following one, Farris (1973), helps me to see any connection with zoology. I am therefore most reluctantly forced to "invent" suitable distributions. For most of the calculations we need only the conditional distribution of λ given θ. Firstly, I suppose λ and θ independent. Secondly, following Bayes and others after him, I suppose λ uniform in $(0, 1)$.

In the Bayesian view nuisance parameters are eliminated by integration with respect to them. From the likelihood (3.1) and the uniform distribution of λ, independent of θ, the likelihood for θ, given the data (s, w, t) is

$$(1/2 - 2\theta^2)^w (1/2 + 2\theta^2)^{s+t} \int_0^1 \{1 - \psi(2\lambda - 1)\}^s \{1 + \psi(2\lambda - 1)\}^t d\lambda. \quad (4.1)$$

(Multiplication by 2^{s+t} makes the values of the integrals more nearly of order one.) The partial likelihood approach uses the binomial likelihood before the integral. Hence the difference between that approach and the Bayesian one depends on the integral: the two are only the same when that is constant in θ. It is therefore sensible to study the integral in (4.1).

The logarithm of the integrand is

$$L(\lambda) = s \log\{1 - \psi(2\lambda - 1)\} + t \log\{1 + \psi(2\lambda - 1)\}$$

with derivative

$$L'(\lambda) = \frac{-2s\psi}{1 - \psi(2\lambda - 1)} + \frac{2t\psi}{1 + \psi(2\lambda - 1)}.$$

This vanishes at $\hat{\lambda}$ satisfying

$$2\hat{\lambda} - 1 = (\frac{t - s}{t + s})\frac{1}{\psi}. \quad (4.2)$$

The negative of the second derivative at $\hat{\lambda}$ is

$$-L''(\hat{\lambda}) = \psi^2(t+s)^3/(st).$$

Consequently as $n \to \infty$ (so that s and t do also) $L''(\hat{\lambda})^{-1} \to 0$ and, provided $\hat{\lambda}$ is well within the unit interval of integration, the integral is approximately proportional to

$$e^{L(\hat{\lambda})}\{-L''(\hat{\lambda})\}^{-1/2} \propto \psi^{-1} = (1/2+2\theta^2)/(2\theta). \tag{4.3}$$

In these circumstances the full likelihood (4.1) is approximately

$$(1/2-2\theta^2)^w(1/2+2\theta^2)^{s+t} \cdot (1/2+2\theta^2)/(2\theta)$$

and it is clear that the effect of introducing the additional factor beyond the partial likelihood is negligible. For example, the power of $(1/2+2\theta^2)$ is merely increased from $(s+t)$ to $(s+t+1)$, equivalent to one additional pair of observations. Consequently the asymptotic result is that the loss of information in using only w is negligible. This confirms, in the example of invariant reversals, Lindsay's general result.

However, there is a difficulty. The value of $\hat{\lambda}$ given by (4.2) will often not lie well within the unit interval and when this happens the approximation (4.3) will not be adequate. The condition $0 \leq \lambda \leq 1$ is equivalently

$$\left|\frac{t-s}{t+s}\right| \frac{1}{\psi} \leq 1$$

and this will typically not be true for all ψ (or, equivalently, all θ). The violation will be most serious when ψ, and hence θ, is small. But this, we saw above, is exactly the case when the unsatisfactory feature that w/n can exceed $1/2$ will often occur. It therefore begins to look as though the asymptotic theory is correct but that the limit may take a long time to be effectively reached; this is most likely to occur when θ is small.

There is another feature of (4.3) that is misleading and that is the occurrence of θ^{-1}. For any fixed $\theta > 0$ the argument is sound but the value of n needed for adequate approximation tends to infinity as $\theta \to 0$: a pretty case of non-uniform convergence. For these various reasons a more careful consideration of the integral is needed. Before doing this we take a closer look at the integrand.

5. THE DISTRIBUTION OF s GIVEN w

The integrand in (4.1), most clearly seen in the full likelihood (3.1), is equivalently the distribution of s, given w, and is binomial with index $s+t = n - w$ and parameter $p' = 1/2 - \psi(\lambda - 1/2)$. To study this take first the case where λ is known. One extreme case is $\lambda = 1$ (another, which is smaller, is $\lambda = 0$). The full likelihood from (3.1) is then

$$\begin{aligned}&(1/2 - 2\theta^2)^w(1/2 + 2\theta^2)^{s+t}(1 - \psi)^s(1 + \psi)^t\\ \propto\ &(1/2 - \theta)^{w+2s}(1/2 + \theta)^{w+2t}\end{aligned}$$

on using $\psi = \theta/(1/4 + \theta^2)$ and simplifying. This is quite different from the partial likelihood, having $(w + 2s)/2n$ as the "best" estimate of $(1/2 - \theta)$. If no inversion occurs ($\lambda = 0$) in the Bernoulli trials the full likelihood is available and substantially differs from the partial likelihood. This is no criticism of the approach based on the latter which was not intended to cover this case.

Another extreme case is $\lambda = 1/2$ when the terms in braces in the full likelihood (3.1) reduce to constants having no dependence on ψ (or θ). Here the full and partial likelihoods are the same and the method loses no information at all. These considerations together with those adduced in Section 4 suggest that the partial likelihood approach will experience the greatest difficulties when λ is near 0 or 1, or when θ is near 0. The following analysis supports this.

The distribution of s given w is binomial with parameter $p' = 1/2-\psi(\lambda-1/2)$. As the number of pairs n increases this distribution will effectively determine p' but will not distinguish between values of ψ and λ that give the same value of p'. Figure 1 shows the curves $p' =$ constant in the unit square of possible values for ψ and λ. They are rectangular hyperbolae with axes $\psi = 0$ (or $\theta = 0$) and $\lambda = 1/2$. The distribution of s given w will determine rather well which curve λ, ψ lie on but not where on the curve. If s is nearly equal to t, exact equality being the extreme case, p' is near 1/2 and the curve is like A in the figure: all we know is that either ψ (and hence θ) is small or λ is near 1/2. If s and t are very different p' is near 0 or 1 and the curve is like B in the figure: now we know that both λ is near 0 and ψ near 1 (θ near 1/2).

Our analysis has assumed λ and θ (or ψ) to be independent. Had this not been so then the distribution of s given w could provide a lot of information. For example, suppose $\mid \lambda - 1/2 \mid$ and ψ were positively correlated *a priori*, then they might lie around the dotted curves in the figure which cross the hyperbolae and permit both λ and ψ to be quite well determined.

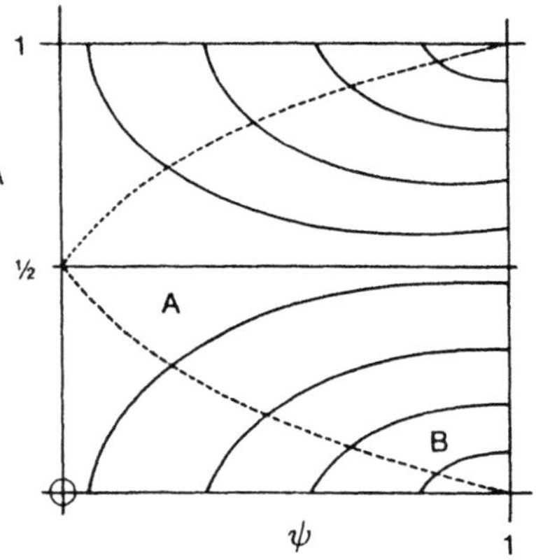

Figure 1. *The solid curves are* $\frac{1}{2} - \psi(\lambda - \frac{1}{2}) =$ *constant. The dotted curves are explained in the text.*

6. NUMERICAL EVALUATION

In this section the results of some numerical work are described. In default of zoological information, it has been supposed that θ is uniform in its range $(0, 1/2)$. Suitably normalized, the likelihood for θ, equation (4.1), is the distribution of θ given all the data (s, w, t). This is written $p(\theta \mid s, w, t)$. Similarly the partial likelihood, again normalized, is the distribution given only w: written $p(\theta \mid w)$. Notice that in the latter case the normalizing factor is not the usual beta-integral associated with the binomial likelihood because of the limitation of the binomial parameter $p = 1/2 - 2\theta^2$ to the interval $(0, 1/2)$ instead of $(0, 1)$.

Figures 2–5 show four cases. In each one there is a sample of $n = 100$ pairs and the curves shown are the densities $p(\theta \mid w)$ and $p(\theta \mid s, w, t,)$ just explained. The partial likelihood method is good whenever these are in close agreement: the more discrepant they are the worse the method is.

Figure 2 is the case $s = 30, w = 54, t = 16$. This could easily have arisen from $\theta = 0.1, \lambda = 0.1$. $w/n = 0.54$ is in excess of $1/2$ and thus the partial likelihood, point estimate is unsatisfactory. $p(\psi \mid w)$ is a beta density with more than half of it truncated, so that $p(\theta \mid w)$ is similar. However, s and t are different so that p' is near 1 and curve B in Figure 1 is more appropriate. This suggests ψ, and hence θ, is not near zero, contrary to what the partial likelihood suggests. In fact, $p(\theta \mid s, w, t)$ has a mode around $\theta = 0.08$ and is markedly different from $p(\theta \mid w)$. The partial likelihood solution performs badly.

Figure 3 has $s = 23, w = 54, t = 23$. Since w has the same value as

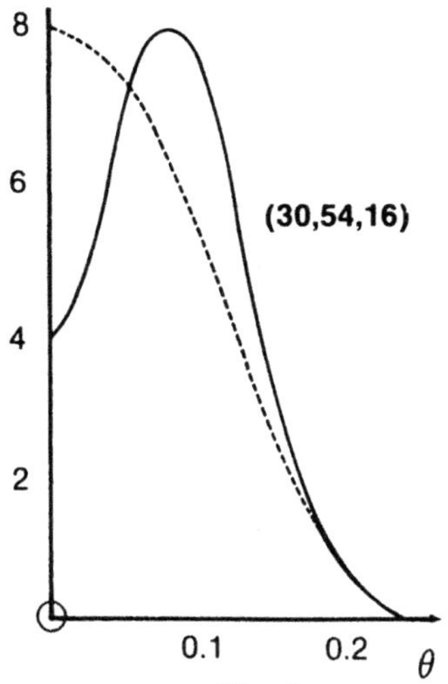

Fig. 2

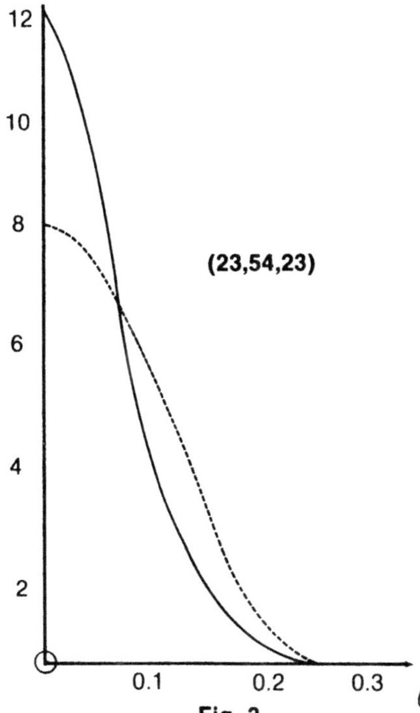

Fig. 3

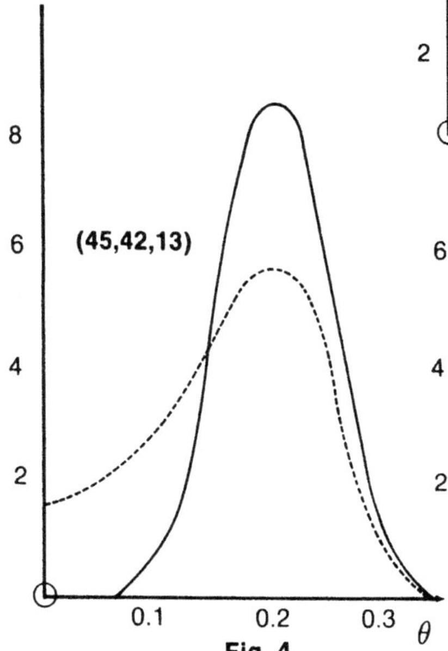

Fig. 4

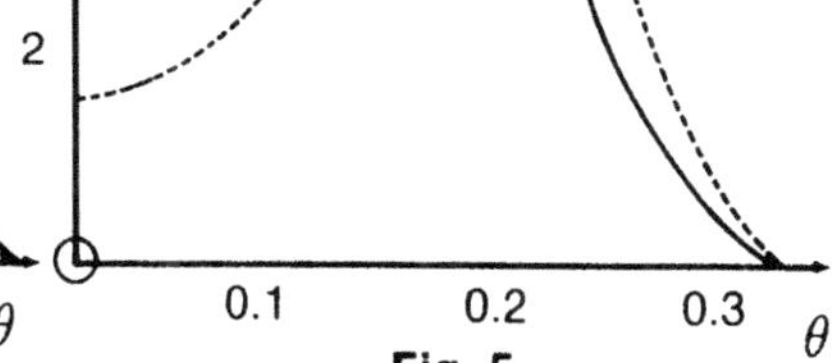

Fig. 5

Figures 2–5. *Each figure gives the densities* $p(\theta \mid s, w, t)$ *(continuous curve) and* $p(\theta \mid w)$ *(broken curve) for the triplet of values* (s, w, t) *indicated.*

before $p(\theta \mid w)$ is unaffected. This case could have arisen from $\theta = 0.1$, as before, but $\lambda = 0.5$. Since $s = t$, values of p' near $1/2$ are suggested and curve A in Figure 1 is appropriate, leading one to think either $\lambda = 0.5$ or θ is small. The density $p(\theta \mid s, w, t)$ has, like $p(\theta \mid w)$, its mode at $\theta = 0$. It is squeezed up against $\theta = 0$ suggesting, more strongly than does the partial likelihood, that θ is small. Again the approach performs badly.

Figure 4 has $s = 45, w = 42, t = 13$. This case could have arisen with $\theta = 0.2, \lambda = 0.1$. Now $w/n < 1/2$, the point estimate makes sense and the truncation effect on $p(\theta \mid w)$ is much smaller, indeed it has a mode at $\theta = 0.2$. As in Figure 2, s and t are unequal and curve B in Figure 1 is appropriate. The overall effect is to make $p(\theta \mid s, w, t)$ have much smaller spread than $p(\theta \mid w)$ though still about the value $\theta = 0.2$. Again the partial approach is disappointing.

Finally Figure 5 has $s = 29, w = 42, t = 29$. This could have arisen with $\theta = 0.2, \lambda = 0.5$. w is as in the previous case so $p(\theta \mid w)$ is as there. But now $s = t$ as with Figure 3 and values of θ near 0 become possible. Indeed, the full density $p(\theta \mid s, w, t)$ is bimodal with modes at $\theta = 0$ and around $\theta = 0.18$. Here the partial likelihood does better but only because we "know" $\theta = 0.2$. In the real world we do not have this knowledge.

In all these cases there are substantial differences between the full information answer $p(\theta \mid s, w, t)$ using all the data and the partial approach $p(\theta \mid w)$ using only w. This is by no means always true. Many other cases have been calculated and the general conclusions are as follows (all with $n = 100$).

(a) For $w \leq 30$, suggesting $1/2 - 2\theta^2 \leq 0.3$ or $\theta \geq 0.32$, there is practically no difference between the two approaches and the densities $p(\theta \mid w)$ and $p(\theta \mid s, w, t)$ are almost indistinguishable. Here the partial likelihood approach does well. (Though it might do less well were θ and λ not independent.)

(b) For $w \geq 30$, suggesting $\theta \leq 0.32$ discrepancies appear, becoming more striking as w increases, being substantial even at $w = 42$ (Figures 4 and 5). When $s = t$ the full density $p(\theta \mid s, w, t)$ is to the left of that based on w above (Figures 3 and 5). If s and t are quite different the shift is in the opposite direction (Figures 2 and 4).

7. CONCLUSIONS

Nearly all the results on partial likelihood known to me are, like Lindsay's, asymptotic (even though w/n is the "best" estimate of p in some sense applicable to finite n). If the case of invariant reversals is typical, we should be wary of asymptotic conclusions: we need also to ask how quickly the asymptotic values are approached. In this case with $\theta = 0.1$, $n = 2,500$

pairs may not be large. 100 is only large if θ exceeds about one-third: and how do we know this?

If we pass from pairs to triplets or higher values the method does even worse. For triplets (X_i, Y_i, Z_i) the probabilities corresponding to (2.1)–(2.3) are

$$\begin{aligned}
p(X_i+Y_i+Z_i=0)&=\lambda(1/2-\theta)^3+(1-\lambda)(1/2+\theta)^3,\\
p(X_i+Y_i+Z_i=1)&=3\lambda(1/2-\theta)^2(1/2+\theta)+3(1-\lambda)(1/2-\theta)(1/2+\theta)^2,\\
p(X_i+Y_i+Z_i=2)&=3\lambda(1/2-\theta)(1/2+\theta)^2+3(1-\lambda)(1/2-\theta)^2(1/2+\theta),\\
p(X_i+Y_i+Z_i=3)&=\lambda(1/2+\theta)^3+(1-\lambda)(1/2-\theta)^3.
\end{aligned}$$

If $w_i = 1$ whenever the sum is 0 or 3, and 0 otherwise, it is easy to see that $p(w=0) = 3(1/4-\theta^2)$, not involving λ. This appears to be the only λ-free factorization that is possible. The distribution of $X_i + Y_i + Z_i$ given w_i is sufficiently rich for λ and θ to be identified (unlike the case with pairs) so that it is clear that restriction to the w's will lose a lot of information.

The following remark is not particularly relevant to the prime issue under discussion here but is worth making. Notice how very informative and immediately comprehensible are the densities plotted in Figures 2–5. They provide much more understandable information than can possibly be obtained by a point estimate and its standard error. The bimodal and truncated forms are particularly revealing. They are only available from the Bayesian approach.

POSTSCRIPT

Subsequent investigation revealed that the example arises in bacterial genetics. Bacteria exist in two forms, A and a, in the ratio $\lambda : 1-\lambda$. They reproduce by splitting into two. In the course of splitting either member of the pair may mutate, changing from A to a, or from a to A. The probability of a mutation of either type is $1/2-\theta$. Assuming independence between bacteria, observations on the new generation only obey the probability model of the paper. Thus an observed AA can either have arisen from A with probability $\lambda(1/2+\theta)^2$ or from a with probability $(1-\lambda)(1/2-\theta)^2$: identifying A with $X = 1$, equations (2.1)–(2.3) follow.

Notice this modelling avoids the lack of identification and the mutation rate is less than 1/2 if θ is non-negative. A uniform distribution for λ is not unreasonable in some circumstances but for θ is typically inappropriate since mutation rates are low; that is, θ is near 1/2. When this happens we see that the partial likelihood method is satisfactory though it ignores the information that the rate is low. Thus the method needs the prior information that θ is near 1/2 for it to work, yet ignores that information in the analysis. More importantly, divorced from its practical context, the method is unsound in samples as large as 2,500 even though it is asymptotically sensible.

ACKNOWLEDGMENTS

I am grateful to an anonymous referee of another paper for drawing my attention to Lindsay's work which, appearing in a journal unfamiliar to statisticians, had escaped my notice; and to Monash University for inviting me to Melbourne and providing hospitality during the writing of the paper.

REFERENCES

Farris, J. S. (1973), "A probability model for inferring evolutionary trees." *Systematic Zoology* **22**, 250–256.

Felsenstein, J. (1973), "Maximum likelihood and minimum-steps methods for estimating evolutionary trees from data on discrete characters." *Systematic Zoology* **22**, 240–249.

Lindsay, B. G. (1980), "Nuisance parameters, mixture models, and the efficiency of partial likelihood estimators." *Philosophical Transactions of the Royal Society of London*, Series A **296**, 639–662.

André Plante [1]

A DECISION-LIKELIHOOD SOLUTION TO THE PROBLEM OF COMPARING TWO SIMPLE HYPOTHESES

ABSTRACT

Hájek's (1967) decision theory version of the likelihood principle is shown to provide a new solution to the problem of comparing two simple hypotheses. The resulting best binary decision criterion is everywhere randomized, the probability of choosing $\theta = 1$ as opposed to $\theta = 0$, being given by

$$\frac{w_1 f_1(x)}{w_0 f_0(x) + w_1 f_1(x)},$$

where w_θ is the loss due to the wrong decision when θ is the state of Nature, where f_θ is a density, and where x is the outcome.

1. INTRODUCTION

Historically, there have been two divergent solutions to the fundamental statistical problem of comparing two simple hypotheses with a fixed sample size, the likelihood ratio, which is expressed on a continuous or multivalued scale, and the Neyman-Pearson fundamental lemma, which uses a binary (0,1) scale. The Neyman-Pearson fundamental lemma has been assimilated within decision theory to the point where it is often regarded as the only possible solution for the statistical comparison of two hypotheses, while it is often assumed that some intrinsic incompatibility exists between Wald's decision theory and Fisher's likelihood function. Both beliefs are untrue. Not only are likelihood and decision approaches compatible for the problem of comparing two simple hypotheses, but the numerical value of the likelihood ratio even provides an optimal decision function.

[1] Département de mathématiques et d'informatique, Université du Québec à Montréal, Montréal, Québec H3C 3P8

I. B. MacNeill and G. J. Umphrey (eds.), Foundations of Statistical Inference, 51–62.

Although decision theory and the likelihood approach evolved for the most part independently, there have been repeated attempts to relate their methods. Barnard (1949) pointed out an analogy between the likelihood approach and Wald's sequential probability ratio test for comparing two simple hypotheses; Lindley (1953) indicated the role of the likelihood function for the solution of a multi-decision problem. Finding that the fundamental lemma provides an unsatisfactory solution for the problem of comparing two simple hypotheses because it does not satisfy the conditionality principle, Birnbaum (1961) justified the likelihood function solution in terms of a multiple decision problem, but later he (Birnbaum, 1962, 1972) used the concept "evidential meaning" which has no decision interpretation. Hájek (1967) gave a decision theory version of both conditionality and likelihood principles of Birnbaum (1962) and proved their equivalence, confirming within decision theory a conjecture of Birnbaum. Robinson (1979a,b), while synthesizing the progress achieved over two decades in Buehler's (1959) betting approach to the conditionality principle, indirectly provided the arguments necessary to complete Birnbaum's criticism of the fundamental lemma solution to the problem of comparing two simple hypotheses, in terms acceptable within a decision theory approach. Indeed, Robinson's systematic set of definitions provides the basis for a strong argument for the use of the conditionality principle or equivalent likelihood principle in decisions based upon a unique sample—the *individual decision problem* described by Plante (1971a,b). The present need for a likelihood solution to the problem of comparing two simple hypotheses within decision theory is now apparent (see Berger and Wolfert, 1984).

In this paper, Hájek's decision theory version of the likelihood principle is shown to solve the problem of comparing two simple hypotheses. The resulting *best binary decision criterion* is shown to be *everywhere randomized* and formally similar to a Bayes posterior distribution when the prior probabilities are replaced by losses.

The main result is stated in Section 3, using the terminology introduced in Section 2. Sections 4, 5, and 6 contain steps in the proof of the main result concluded in Section 7, followed by a brief discussion in Section 8.

2. DEFINITIONS

A *binary statistical* (decision) *situation* is a quintuplet $S = (\mathcal{X}, f_0, f_1, w_0, w_1)$, where $\mathcal{X}$ is a sample space, $f_\theta(\theta = 0, 1)$ are probability densities under two hypotheses, and $w_\theta (> 0)$ is the loss due the wrong decision when the value of the parameter is θ. The situation $S' = (\mathcal{X}, f_1, f_0, w_1, w_0)$ is the *situation alternate to S*.

Given a binary statistical situation $S = (\mathcal{X}, f_0, f_1, w_0, w_1)$, the likelihood ratio is $r(x,S) = f_1(x)/f_0(x)$ where $x \in \mathcal{X}$. The ratio $r(x,S)$ can take the value $+\infty$, and be indeterminate when both f_0 and f_1 are zero. A binary decision problem is solved by determining a *decision function* $\delta : \mathcal{X} \to [0,1]$ where $\delta(x)$ represents the probability of taking the action $\theta = 1$ when x has been observed. The risks associated with δ are $\rho_0(\delta) = w_0 E_0\{\delta(x)\}$ and $\rho_1(\delta) = w_1 E_1\{1 - \delta(x)\}$. Since the solution to a binary decision problem satisfies the condition that

$$\rho_0(\delta) = \rho_1(\delta) \qquad (= \rho(\delta)), \tag{1}$$

as shown by Blackwell and Girshick (1954, p. 195 and p. 198), in formulating a decision-likelihood solution, only the class of decision functions satisfying (1) will be considered. The members of the class will be called *choice functions*. A more general function is necessary in order to associate to each binary statistical situation S a choice function $\delta(\cdot, S)$. This is called a *choice criterion*. In addition, the definition of choice criterion includes the requirement that, if S' is the situation alternate to S, then $\delta(\cdot, S') = 1 - \delta(\cdot, S)$. An example of a choice criterion, based on the fundamental lemma and using the above notation is

$$\delta(x,S) = \begin{cases} 0 & \text{if } r(x,S) < c \\ k & \text{if } r(x,S) = c \\ 1 & \text{if } r(x,S) > c \end{cases}$$

where constants c and k are determined from (1).

A choice criterion δ is *independent of the reference set,* if for every couple of binary statistical situations $S = (\mathcal{X}, f_0, f_1, w_0, w_1)$ and $V = (\mathcal{Y}, g_0, g_1, w_0, w_1)$ with the same losses, the equality $r(x,S) = r(y,V)$, where $x \in \mathcal{X}$ and $y \in \mathcal{Y}$, implies that $\delta(x,S) = \delta(y,V)$. Restricting attention to the choice criteria independent of the reference set is equivalent to looking for a solution which satisfies the decision theory version of the likelihood principle (Hájek, 1967) under condition (1).

3. THEOREM 1.

(a) *The function δ, which associates to a binary statistical situation $S = (\mathcal{X}, f_0, f_1, w_0, w_1)$ the function $\delta(\cdot, S)$ having value*

$$\delta(x,S) = w_1 f_1(x)/\{w_0 f_0(x) + w_1 f_1(x)\} \tag{2}$$

at a point x in $\mathcal{X}$, is a choice criterion independent of the reference set.

(b) *If δ^* is any choice criterion independent of the reference set different from δ, then the relationship*

$$\rho\{\delta(\cdot, S)\} < \rho\{\delta^*(\cdot, S)\}$$

holds between the risks, provided S is such that f_0 and f_1 are essentially different.

With an arbitrary parameter space, equation (2) allows the comparison of any two parameter values θ_0 and θ_1. If each member of the family of densities has the same carrier set, a reference point θ_0 can be chosen so that (2) may be written

$$\delta(x, S_\theta) = 1 - 1/\{1 + (w/w_0)\mathcal{L}(\theta, x)\},$$

where $S_\theta = (\mathcal{X}, f(\cdot, \theta_0),\ f(\cdot, \theta),\ w_0, w)$, where $x \in \mathcal{X}$, and where $\mathcal{L}(\theta, x) = f(x, \theta)/f(x, \theta_0)$ is one version of the likelihood function (see Barnard, 1967). Theorem 1 is proven in Section 7 using a series of lemmas given in Sections 4, 5 and 6.

4. A LINEARLY ORDERED CLASS OF CHOICE CRITERIA INDEPENDENT OF THE REFERENCE SET

Lemma 1 describes a class of choice criteria that are independent of the reference set, and Lemma 2 indicates the unique best choice criterion within that class.

Lemma 1. *Let $\mathcal{S}$ be the class of all binary statistical situations $S = (\mathcal{X}, f_0, f_1, w_0, w_1)$ with fixed losses w_0 and w_1. Then, for each λ in $[0, \min\{w_0, w_1\}]$, the function on $\mathcal{S}$ which associates S in $\mathcal{S}$ with the function $\delta_\lambda(\cdot, S)$ defined by*

$$\delta_\lambda(x, S) = \frac{\lambda f_0(x) + (w_1 - \lambda) f_1(x)}{w_0 f_0(x) + w_1 f_1(x)},$$

where $x \in \mathcal{X}$, is a choice criterion independent of the reference set.

Proof. Suppose without loss of generality that f_0 and f_1 are densities with respect to a σ-finite measure μ, then

$$w_0 \int_{\mathcal{X}} \delta_\lambda(x, S) f_0(x) d\mu(x) + w_1 \int_{\mathcal{X}} \delta_\lambda(x, S) f_1(x) d\mu(x)$$
$$= \int_{\mathcal{X}} \{\lambda f_0(x) + (w_1 - \lambda) f_1(x)\}\, d\mu(x) = w_1$$

from which condition (1) is seen to hold. In $0 \le \delta_\lambda(x, S) \le 1$, the left inequality is evident while the right follows from $\lambda f_1(x) + (w_0 - \lambda) f_0(x) \ge 0$. Also, that $\delta_\lambda(\cdot, S') = 1 - \delta_\lambda(\cdot, S)$, where S' is the situation alternate to S, is easily verified. Thus, δ_λ is a binary choice criterion for every λ in $[0, \min\{w_0, w_1\}]$. That δ_λ is independent of the reference set is evident. The proof of Lemma 1 is now complete. (The class defined by Lemma 1 is the class of all binary choice criteria independent of the reference set, which will be shown in following sections.)

Lemma 2. *For fixed losses w_0 and w_1, consider the class $\mathcal{S}$ and the choice criteria δ_λ defined in Lemma 1. Let $S = (\mathcal{X}, f_0, f_1, w_0, w_1)$ be a member of $\mathcal{S}$ and suppose that f_0 is essentially different from f_1. Then the risk $\rho\{\delta_\lambda(\cdot, S)\}$ is strictly increasing as a function of λ.*

Proof. Let $\rho(\lambda)$ stand for $\rho\{\delta_\lambda(\cdot, S)\}$. Then,

$$\frac{d\rho}{d\lambda} = -w_0 \int_{\mathcal{X}'} \frac{f_1(x)/f_0(x) - 1}{w_0 + w_1\{f_1(x)/f_0(x)\}} f_0(x) d\mu(x),$$

where $\mathcal{X}' = \{x : x \in \mathcal{X},\ f_0(x) > 0\}$ and where μ is a dominating σ-finite measure. The function defined by

$$h(z) = (z - 1)/(w_0 + w_1 z)$$

is strictly concave on $[0, \infty)$. Therefore, by Jensen's inequality,

$$\int_{\mathcal{X}'} h\{f_1(x)/f_0(x)\} f_0(x) d\mu(x) < h \left[\int_{\mathcal{X}'} \{f_1(x)/f_0(x)\} f_0(x) d\mu(x) \right],$$

provided f_0 is not degenerate. The values of $h(z)$ are non-positive in $[0, 1]$, while

$$0 \le \int_{\mathcal{X}'} \{f_1(x)/f_0(x)\} f_0(x) d\mu(x) \le 1.$$

Therefore

$$\frac{d\rho(\lambda)}{d\lambda} > 0,$$

provided f_0 is not degenerate. If $f_0(x_0) = 1$ while $f_0(x) = 0$ for $x \ne x_0$, then

$$\frac{d\rho}{d\lambda} = -w_0 \frac{\{f_1(x_0) - 1\}}{w_0 + w_1 f_1(x_0)},$$

which is positive because f_1 is essentially different from f_0. Therefore, $\rho(\lambda)$ is strictly increasing, which completes the proof of Lemma 2.

5. CHOICE FUNCTIONS FOR SIMPLE BINARY SITUATIONS

Lemmas 3 and 4 concern *simple binary statistical situations,* that is situations where the sample space contains only two points. Given one such situation $S = (\{0,1\}, f_0, f_1, w_0, w_1)$, a relabelling of the sample points will define a new simple situation $S_1 = (\{0,1\}, g_0, g_1, w_0, w_1)$, where $g_\theta(0) = f_\theta(1)$ and $g_\theta(1) = f_\theta(0)$, $(\theta = 0,1)$. Situations S and S_1, and their alternate situations S' and S_1' are all essentially the same as far as choice criteria are concerned. Considering an ordered form of S enables one to avoid having to examine the four possibilities separately.

Let $S = (\{0,1\}, f_0, f_1, w_0, w_1)$ be a simple binary situation; let (y, τ) be a couple in $\{0,1\} \times \{0,1\}$ such that

$$f_\tau(y) = \min\{f_0(0), f_0(1), f_1(0), f_1(1)\}$$

and let $\tau' = | 1 - \tau |$, $y' = | 1 - y |$. *An ordered form of* S is a simple binary situation $S^* = (\{m, M\}, g_m, g_M, v_m, v_M)$ where $g_m(m) = f_\tau(y), g_m(M) = f_\tau(y')$, $g_M(m) = f_{\tau'}(y)$, $g_M(M) = f_{\tau'}(y')$, where $v_m = w_\tau$ and $v_M = w_{\tau'}$. Note that if $f_0 \not\equiv f_1$, then S has exactly one ordered form, if and only if, one does not have the symmetrical case $f_0(0) = f_1(1)$ with $w_0 = w_1$ where it has two ordered forms. To simplify one can write $r^*(z)$ for $r(z, S^*)(z = m, M)$.

Lemma 3. *Let* $S = (\{0,1\}, f_0, f_1, w_0, w_1)$, *a simple binary situation and let* $S^* = (\{m, M\}, g_m, g_M, v_m, v_M)$, *an ordered form of* S. *If* $f_0 \not\equiv f_1$, *then the class of choice functions for the situation* S *can be described by*

$$\delta_\lambda(x) = \frac{\lambda f_0(x) + (w_1 - \lambda) f_1(x)}{w_0 f_0(x) + w_1 f_1(x)}, \tag{3}$$

where $x \in \{0,1\}$ *and where* λ *is a constant satisfying*

$$-v_M \frac{r^*(M)}{1 - r^*(M)},\ -v_m \frac{1}{r^*(m) - 1} \leq \lambda \leq v_m \frac{1}{1 - r^*(M)},\ v_M \frac{r^*(m)}{r^*(m) - 1}, \tag{4}$$

where $r^*(z) = g_M(z)/g_m(z)$ for $z = m, M$.

Proof. Let δ be a choice function associated with the situation $S = (\{0,1\}, f_0, f_1, w_0, w_1)$ with $f_0 \not\equiv f_1$. Correspondingly, let $S^* = (\{m, M\}, g_m, g_M, v_m, v_M)$ be an ordered form of S and let $\delta^*(z) =$ probability of choosing $\theta = M$ when z is observed. Let us define

$$\lambda = \frac{-\delta^*(m)\{v_m g_m(m) + v_M g_M(m)\} + v_M g_M(m)}{g_M(m) - g_m(m)} \tag{5}$$

so that

$$\delta^*(m) = \frac{\lambda g_m(m) + (v_M - \lambda) g_M(m)}{v_m g_m(m) + v_M g_M(m)}. \tag{6}$$

Equations (1) and (6), after some manipulations, give

$$\delta^*(M) = \frac{\lambda g_m(M) + (v_M - \lambda) g_M(M)}{v_m g_m(M) + v_M g_M(M)}, \tag{7}$$

so that λ also satisfies

$$\lambda = \frac{\delta^*(M)\{v_m g_m(M) + v_M g_M(M)\} - v_M g_M(M)}{g_m(M) - g_M(M)}. \tag{8}$$

By using equations (5) and (8), it can be seen that the inequalities $0 \leq \delta^*(m) \leq 1$ and $0 \leq \delta^*(M) \leq 1$, subject to condition (1), are equivalent to (4). Equation (3) follows from equations (6) and (7), which completes the proof.

Lemma 4. *Let S^* be the class of ordered forms of simple binary situations $S^* = (\{m, M\}, g_m, g_M, v_m, v_M)$, where $g_m \not\equiv g_M$, where $g_m(m) > 0$, and where v_m and v_M are fixed. Then, the mapping which sends S^* in S^* into the point $(r^*(m), r^*(M))$, where $r^*(z) = g_M(z)/g_m(z)$ $(z = m, M)$, is one-one and has the range $R^* = \{(r^*(m), r^*(M)) : 0 < 1/r^*(m) \leq r^*(M) < 1\}$.*

Proof. If we let $p_m = g_m(m)$ and $p_M = g_M(m)$, then there is a one-one mapping from S^* onto the plane region $Q = \{(p_m, p_M) : 0 < p_m \leq \min(1 - p_m, p_M, 1 - p_M)\} = \{(p_m, p_M) : 0 < p_m < p_M \leq 1 - p_m\}$. Now, consider the mapping

$$r^*(m) = p_M/p_m, \qquad r^*(M) = (1 - p_M)/(1 - p_m), \tag{9}$$

and suppose that both (p_m, p_M) and (p'_m, p'_M) in Q have the same image. From the equalities $p_M/p_m = p'_M/p'_m$ and $(1-p_M)/(1-p_m) = (1-p'_M)/(1-p'_m)$ we obtain that $p'_M - p'_m = p_M - p_m$, from which $p'_m = p_m$ and $p'_M = p_M$ follow, since $(p_M - p_m)/p_m = (p'_M - p'_m)/p'_m$. Therefore, the mapping (9) is one-one on Q. By inverting (9) for (p_m, p_M) in Q, one obtains

$$\begin{aligned} p_m &= \{1 - r^*(M)\}/\{r^*(m) - r^*(M)\}, \\ p_M &= r^*(m)\{1 - r^*(M)\}/\{r^*(m) - r^*(M)\}. \end{aligned} \tag{10}$$

The denominator $r^*(m) - r^*(M)$ in (10) is clearly positive. Consequently, using (10), the image R^* of Q is given by

$$R^* = \{(r^*(m),\ r^*(M)) :\ 0 < 1 - r^*(M) < r^*(m)[1 - r^*(M)] \leq r^*(m) - 1\}.$$

Because of the equivalence of the sets of inequalities involved, it follows that R^* can also be written $R^* = \{(r^*(m),\ r^*(M)) :\ 0 < 1/r^*(m) \leq r^*(M) < 1\}$, which completes the proof of Lemma 4.

6. CHOICE CRITERIA FOR SIMPLE BINARY SITUATIONS

Having characterized simple situations, and simple choice functions, *simple criteria* can now be examined. A simple criterion associates each simple statistical situation S with a choice function $\delta(\cdot, S)$ such that $\delta(\cdot, S') = 1 - \delta(\cdot, S)$, where S' is the statistical situation alternate to S. Lemmas 5, 6 and 7 deal with reference set independence in the restricted context of simple choice criteria.

Lemma 5. *Let δ be a reference set independent simple choice criterion; let $S = (\{0,1\}, f_0, f_1, w_0, w_1)$, a simple binary statistical situation with $f_0 \not\equiv f_1$, and let $S^* = (\{m, M\}, g_m, g_M, v_m, v_M)$, an ordered form of S. If we use Lemma 3 to write*

$$\delta(x, S) = \frac{\lambda f_0(x) + (w_1 - \lambda) f_1(x)}{w_0 f_0(x) + w_1 f_1(x)} \qquad (x \in \{0,1\})$$

and

$$\delta(z, S^*) = \frac{\lambda^* g_m(z) + (v_M - \lambda^*) g_M(z)}{v_m g_m(z) + v_M g_M(z)} \qquad (z = \{m, M\})$$

for some λ and some λ^*, then $\lambda^* = \lambda$.

Proof. There are four possibilities according to whether $g_m(m) = f_0(0), f_0(1), f_1(0)$ or $f_1(1)$. However, as each proof is similar, suppose that $g_m(m) = f_0(0)$ and that $v_m = w_0$. Then $r^*(m) = r(0, S)$ so that, using the independence of the reference set, $\delta(m, S^*) = \delta(0, S)$. The equality $\lambda^* = \lambda$ follows, and the proof is complete.

Lemma 6. *If a simple binary choice criterion δ is independent of the reference set, then, for given losses, there exists a number λ such that the choice function $\delta(\cdot, S)$ associated with a simple binary situation $S = (\{0,1\}, f_0, f_1, w_0, w_1)$ can be written as*

$$\delta(x, S) = \frac{\lambda f_0(x) + (w_1 - \lambda) f_1(x)}{w_0 f_0(x) + w_1 f_1(x)} \quad (x \in \{0,1\}). \tag{11}$$

Proof. Let $S = (\{0,1\}, f_0, f_1, w_0, w_1)$ and $T = (\{0,1\}, g_0, g_1, w_0, w_1)$ be any two simple binary statistical situations, such that $f_0 \not\equiv f_1$ and $g_0 \not\equiv g_1$, but with identical losses. By Lemma 3, there exist constants λ and μ such that $\delta(\cdot, S)$ and $\delta(\cdot, T)$ can be written in the form (11) using λ and μ respectively. If $S^* = (\{m, M\}, h_m, h_M, s_m, s_M)$ and $T^* = (\{m, M\}, k_m, k_M, v_m, v_M)$ are ordered forms of S and T respectively, then, using Lemma 5,

$$\delta(z, S^*) = \frac{\lambda + (s_M - \lambda) r(z, S^*)}{s_m + s_M r(z, S^*)}$$

and

$$\delta(z, T^*) = \frac{\mu + (v_M - \mu) r(z, T^*)}{v_m + v_M r(z, T^*)},$$

where $z \in \{m, M\}$. Suppose that $r(m, S^*) \leq r(m, T^*)$, then consider in turn the two possibilities which can arise: $v_m = s_m$, and $v_m = s_M$. If $v_m = s_m$, then because of Lemma 4, there exists a simple binary situation U with the same losses as those of S and T, such that $r(m, U^*) = r(m, S^*)$ and $r(M, U^*) = r(M, T^*)$ for some ordered form U^* of U. Lemma 3 and Lemma 5 imply that, for some ν

$$\delta(z, U^*) = \frac{\nu + (s_M - \nu) r(z, U^*)}{s_m + s_M r(z, U^*)} \qquad (z = m, M).$$

Because δ is independent of the reference set, we find at $z = m$ that $\nu = \lambda$, and at $z = M$ that $\nu = \mu$, so that $\mu = \lambda$. If $v_m = s_M$ ($\neq s_m$), then, because of Lemma 4, there exists a symmetrical simple binary situation U, and an ordered form U^* of U with losses identical to those of S^*, such that $r(m, U^*) = r(m, S^*)$. Because U^* is symmetrical with unequal losses, there exists another ordered form U_1^* of U. This form U_1^* has losses identical to those of T^*. Because of Lemma 5, and because U_1^* is an ordered form of U^*, it follows that the representation for $\delta(\cdot, U_1^*)$ stated in Lemma 3 uses the same λ as the representation for $\delta(\cdot, U^*)$. Since U_1^* has losses which are identical to those of T^*, it follows, using the first part of Lemma 6, that the representation for $\delta(\cdot, T^*)$ and for $\delta(\cdot, T)$ use the same λ. Therefore, the representation for $\delta(\cdot, U)$ and for $\delta(\cdot, T)$ use the same λ, and the proof of Lemma 6 is complete.

Lemma 7. *With the assumptions and notation of Lemma 6, the constant* λ *in equation (11) satisfies the condition*

$$0 \leq \lambda \leq \min\{w_0, w_1\}. \tag{12}$$

Proof. Condition (12) is a direct consequence of the following equations which uses Lemma 3 and the notation of the proof of Lemma 6.

$$\sup_{R^*} \left\{ -s_M \frac{r^*(M)}{1 - r^*(M)}, \quad -s_m \frac{1}{r^*(m) - 1} \right\} = 0,$$

and

$$\inf_{R^*} \left\{ s_m \frac{1}{1 - r^*(M)}, \quad s_M \frac{r^*(m)}{r^*(m) - 1} \right\} = \min\{s_m, s_M\}.$$

7. PROOF OF THE MAIN RESULT

Having proven Lemmas 1 to 7, we can now turn to the main result of the paper.

Proof of Theorem 1. For fixed losses w_0 and w_1 let Δ be the class of choice criteria independent of the reference set, and let Λ be the class of choice criteria δ_λ defined in Lemma 1. Part (a) of Theorem 1 follows from Lemma 1. To prove part (b), it is sufficient, because of Lemma 2, to show that $\Delta \subset \Lambda$.

Let $\delta \in \Delta$, let $S = (\mathcal{X}, f_0, f_1, w_0, w_1)$ be a binary situation, and let $x_i \in \mathcal{X}$ $(i = 1, 2)$, and suppose that $f_0(x_i) \neq f_1(x_i)$ $(i = 1, 2)$. Because of Lemma 4, there exist simple binary situations $S_i = (\{0, 1\}, f_{i0}, f_{i1}, w_0, w_1)$ $(i = 1, 2)$, such that $r(x_i, S) = r(0, S_i)$ $(i = 1, 2)$. Consequently, $\delta(x_i, S) = \delta(0, S_i)$ and, because of Lemma 3, for some λ_i $(i = 1, 2)$,

$$\delta(0, S_i) = \frac{\lambda_i + (w_1 - \lambda_i)r(0, S_i)}{w_0 + w_1 r(0, S_i)}.$$

Lemma 6 implies that $\lambda_1 = \lambda_2$, thus it follows trivially that

$$\delta(x, S) = \frac{\lambda + (w_1 - \lambda)r(x, S)}{w_0 + w_1 r(x, S)} \tag{13}$$

in which, because of Lemma 7, $\lambda \in [0, \min\{w_0, w_1\}]$, and x is any point in $\mathcal{X}$ such that $r(x, S) \neq 1$. If, however, $r(x, S) = 1$, then let $T = (\{0, 1\}, g, g, w_0, w_1)$ where $g(0) = 1/2$. Comparing (x, S) with $(0, T)$, gives $\delta(x, S) = \delta(0, T) = w_1/(w_0 + w_1)$, so that (13) still holds. Therefore, $\delta \in \Lambda$, so that $\Delta = \Lambda$. The proof of Theorem 1 is complete.

8. CONCLUDING REMARKS

In my opinion, two quite distinct uses of the fundamental lemma are usually not clearly distinguished one from the other, which is unfortunate, since one is appropriate and the other a misapplication. I find it quite appropriate to use a test of significance based on a boundedly complete statistic which, at any level of significance, has maximum power at a given point θ_1 in the alternative. However, it is inappropriate to use the fundamental lemma to provide a test statistic for testing a simple null hypothesis against a simple alternative. This is because, as Birnbaum (1961) showed, using the fundamental lemma for binary comparison ignores available information which should be taken into account when taking a decision based upon a unique sample. The proper way to use information in this situation is through conditional probabilities. Because of the results of Birnbaum (1962), Hájek (1967) and of this paper, only the numerical value of the likelihood ratio should be quoted as the result when comparing two simple hypotheses. On the other hand, in the first situation, that of comparing two hypotheses using a boundedly complete statistic, one cannot attempt a likelihood ratio of the two hypotheses, since at least one of the hypotheses is composite, and, as Barnard (1967) pointed out, the likelihood of a composite hypothesis is undefined. A decision based upon a test of significance, and a decision based upon the value of a likelihood ratio are complementary procedures, each useful in its appropriate context; they should not be regarded as opposed.

REFERENCES

Barnard, G. A. (1949) "Statistical inference". *Journal of the Royal Statistical Society,* Series B **11,** 115–139.

Barnard, G. A. (1967), "The use of the likelihood function in statistical practice". *Proceedings of the Fifth Berkeley Symposium on Mathematical Statistics and Probability* **1,** 27–40.

Berger, J., and R. Wolfert (1984), *The Likelihood Principle.* IMS Lecture Notes, The Institute of Mathematical Statistics, Hayward, California.

Birnbaum, A. (1961), "On the foundations of statistical inference: binary experiments". *Annals of Mathematical Statistics* **32,** 414–435.

Birnbaum, A. (1962), "On the foundations of statistical inference". *Journal of the American Statistical Association* **57,** 269–326.

Birnbaum, A. (1972), "More on concepts of statistical evidence". *Journal of the American Statistical Association* **67,** 858–861.

Blackwell, D., and M. A. Girshick (1954). *Theory of Games and Statistical Decisions.* New York: Wiley and Sons.

Buehler, R. J. (1959), "Some validity criteria for statistical inferences". *Annals of Mathematical Statistics* **30,** 845–863.

Hájek, J. (1967), "On basic concepts of statistics". *Proceedings of the Fifth Berkeley Symposium on Mathematical Statistics and Probability* **1,** 139–162.

Lindley, D. V. (1953), "Statistical inference". *Journal of the Royal Statistical Society,* Series B **15,** 30–65.

Plante, A. (1971a), "Counter-examples and likelihood", (with discussion). *Foundations of Statistical Inference,* ed. V. P. Godambe and D. A. Sprott, pp. 357–371. Toronto and Montreal: Holt, Rinehart and Winston of Canada.

Plante, A. (1971b), Contribution to the discussion of H. Rubin's "Occam's razor needs new blades". *Foundations of Statistical Inference,* ed. V. P. Godambe and D. A. Sprott, pp. 376–377. Toronto and Montreal: Holt, Rinehart and Winston of Canada.

Robinson, G. K. (1979a), "Conditional properties of statistical procedures". *Annals of Statistics* **7,** 742–755.

Robinson, G. K. (1979b), "Conditional properties of statistical procedures for location and scale parameters". *Annals of Statistics* **7**, 756–771.

Larry V. Hedges [1] and Ingram Olkin [2]

STATISTICAL INFERENCE FOR THE OVERLAP HYPOTHESIS

ABSTRACT

For prototype growth data suppose that X_t represents the height of an individual at time t. The overlap hypothesis asserts that there is no correlation between X_1 and the gain in height, $X_2 - X_1$, or that the correlation between X_1 and X_2 is the same as that between X_1 and $X_1 - X_2$. An estimate of the correlation when either of these hypotheses holds is obtained. A statistical test for the hypothesis is also developed, and indications of a multivariate version are presented.

1. INTRODUCTION

The study of the change of human characteristics has a long and varied history. The fitting of growth curves to data for a single individual dates at least from the 18th century (Scammon, 1927). One tradition in modern studies of individual growth has been to fit empirically various curves to the individual measurements as a function of time. The functions used have included the simple logistic (Robertson, 1915), the Gompertz curve (Winsor, 1932), several forms of generalized logistic functions (Nelder, 1961, 1962) and linear combinations of logistic functions (Bock, Wainer, Peterson, Thissen, Murray and Roche, 1973). An important refinement in this tradition is the use of Bayesian methods to stabilize the estimation of multiparameter growth curves (Bock and Thissen, 1980).

Another tradition in the measurement of growth has been the use of statistical methodology to fit empirically various curves to groups of individuals. The analysis of polynomial growth curves for grouped data was suggested by Wishart (1938) and his results were extended by Box (1950),

[1] Department of Education, University of Chicago, Chicago, Illinois 60637
[2] Department of Statistics, Stanford University, Stanford, California 94305

I. B. MacNeill and G. J. Umphrey (eds.), Foundations of Statistical Inference, 63–72.

Potthoff and Roy (1964), Rao (1959, 1965), and Grizzle and Allen (1969). Yet another possibility is to fit growth curves empirically by making use of a combination of individual and group data using empirical Bayes methodology (Strenio, Weisberg and Bryk, 1983).

An alternative to the traditions of empirical fitting of growth curves is to start with a model of growth and develop the statistical analysis on the basis of that mode. One such model is the overlap hypothesis, introduced by Anderson (1939) in the context of studying stability and change in human characteristics. Suppose that X_1 and X_2 denote heights at times t_1 and t_2, respectively. Assuming an additive model that the height at time t_2 is the sum of the height at the earlier time t_1 plus the height gain from t_1 to t_2, the difference $X_2 - X_1$ then represents the gain in height from times t_1 to t_2. The validity of the overlap hypothesis was examined by Bloom (1964) using data on heights of boys and girls from age 2 to 18.

By a direct computation the relation between the correlation, ρ_{12}, of X_1 and X_2 and the correlation, ρ_0, between X_1 and the gain $X_2 - X_1$ can be computed to be

$$\rho_{12} = \frac{\sigma_1}{\sigma_2} + \rho_0 \left[\left(1 - \rho_{12} \frac{\sigma_1}{\sigma_2} \right)^2 + \frac{\sigma_{11}}{\sigma_{22}} \left(1 - \rho_{12}^2 \right) \right]^{\frac{1}{2}}, \tag{1.1}$$

where $\sigma_{11} \equiv \sigma_1^2$ and $\sigma_{22} \equiv \sigma_2^2$ are the variances of X_1 and X_2, respectively.

The overlap hypothesis has several interpretations. One is that the relationship between the first measure X_1 and the gain $X_2 - X_1$ is zero, namely that $\rho_0 = 0$. From (1.1) this interpretation is equivalent to the hypothesis

$$H_1 : \rho_{12} = \sigma_1/\sigma_2. \tag{1.2}$$

Note that the correlation ρ_{12} is positive, which is to be expected from the context of the model.

A second interpretation is that the relation between X_1 and X_2 remains stable over time, that is it is equal to the relation between X_1 and $X_2 - X_1$. This relation can be interpreted as $\rho_{12} = \rho_0$. The condition $\rho_{12} = \rho_0$ in (1.1) yields a cubic equation in ρ_{12} with solutions: (a) $\rho_{12} = \pm 1$ or (b) $\rho_{12} = \frac{1}{2}\sigma_1/\sigma_2$. Condition (a) means that X_1 and X_2 are perfectly correlated, which is a degenerate case. Thus, we consider only the second alternative

$$H_2 : \rho_{12} = \frac{1}{2} \frac{\sigma_1}{\sigma_2}. \tag{1.3}$$

In this paper we obtain a statistical test for the hypotheses H_1 and H_2, both of which are special cases of

$$H : \rho_{12} = a \frac{\sigma_1}{\sigma_2}. \tag{1.4}$$

We also obtain a confidence interval for ρ_{12} when the hypothesis H is true. From the fact that $\sigma_{12} = \rho_{12}\sigma_1\sigma_2$, the hypothesis H is equivalent to the hypothesis that the covariance matrix Σ of X_1 and X_2 has the form

$$\Sigma = \begin{pmatrix} \sigma_{11} & a\sigma_{11} \\ a\sigma_{11} & \sigma_{22} \end{pmatrix}. \tag{1.5}$$

This implies, of course, that σ_{11} and σ_{22} are restricted so that

$$\sigma_{22}/\sigma_{11} \geq a^2. \tag{1.6}$$

When H holds, we obtain the maximum likelihood estimates (MLE) of the parameters in Section 3, and its asymptotic distribution in Section 4. The likelihood ratio test is described in Section 5, and an extension to p dimensions is given in Section 6. A summary of results and an example involving growth data is provided first in Section 2.

2. SUMMARY OF RESULTS AND AN EXAMPLE

Under the hypothesis H, the maximum likelihood estimate of ρ_{12} is

$$\hat{\rho}_{12} = a\left[s_{11}/(2a^2 s_{11} - 2as_{12} + s_{22})\right]^{\frac{1}{2}}, \tag{2.1}$$

and an approximate $100(1-\alpha)$ percent confidence interval for ρ_{12}^2 is given by

$$\hat{\rho}_{12}^2 \pm z\ \hat{\sigma}_\infty(\hat{\rho}_{12}^2)/\sqrt{n}, \tag{2.2}$$

where z is the critical value obtained from the standard normal distribution, $\sigma_\infty(\hat{\rho}_{12}^2)$ is the asymptotic standard deviation of the estimator $\hat{\rho}_{12}^2$ given in (4.3), and $\hat{\sigma}_\infty(\hat{\rho}_{12}^2)$ is its estimate obtained by replacing the parameters σ_{11}, σ_{22} by their respective sample estimates $\hat{\sigma}_{11}$, $\hat{\sigma}_{22}$ given by (3.6), and σ_{12} is replaced by $\hat{\sigma}_{12} = a\hat{\sigma}_{11}$ under H. A confidence interval for ρ_{12} when H does not hold can be obtained from the usual Fisher's z transformation.

Example. Data from a longitudinal study of the growth of 66 boys and 70 girls from ages 2 through 18 were reported by Tuddenham and Snyder (1954). The sample covariance matrix of heights (in centimeters) of boys at ages 2 and 4 is

$$\begin{pmatrix} 9.73 & 9.63 \\ 9.63 & 13.84 \end{pmatrix}.$$

The assumption that the correlation between height at age 2 and the gain in height between ages 2 and 4 is zero, is equivalent to the overlap hypotheses that $a = 1$. Using (2.1) with $a = 1$ we obtain the maximum likelihood estimator of ρ_{12} to be

$$\hat{\rho}_{12} = 0.83,$$

and a 95% confidence interval for ρ_{12}^2 as

$$0.50 \leq \rho_{12}^2 \leq 0.88.$$

Since $\rho_{12} \geq 0$, we obtain that a 95% confidence interval for ρ_{12} is

$$0.71 \leq \rho_{12} \leq 0.94.$$

To test the hypothesis that height at age 2 is uncorrelated with the gain in height between 2 and 4, that is,

$$H : \rho_{12} = \sigma_1/\sigma_2$$

versus the alternative $A : a \neq 1$, we use the likelihood ratio test statistic, $L(S)$, given in (5.3). The value obtained in the example is $L(S) = 0.513$. Using the approximation

$$-2 \log L(S) \sim \chi_1^2,$$

we see that $-2 \log L(S) = 1.33$. Since this value is less than the critical chi-square value of 3.8 (at the 5 percent level of significance) we do not reject the hypothesis.

3. DEVIATION OF THE MAXIMUM LIKELIHOOD ESTIMATES UNDER THE HYPOTHESIS

Because the hypothesis relates only to variances and covariances, and not means, we take as our starting point the bivariate Wishart distribution, that is, the joint density of the sample covariances s_{11}, s_{22}, s_{12}:

$$f(S; \Sigma) = c(S) \mid \Sigma \mid^{-n/2} \exp -\frac{1}{2}(\text{ tr } \Sigma^{-1}S), \tag{3.1}$$

where $S = (s_{ij})$, $\Sigma = (\sigma_{ij})$, i, $j = 1, 2$, are the sample and population covariances matrices, respectively, $c(S)$ is a constant dependent on S but independent of Σ, and n is the degrees of freedom.

Remark: Note that if the means are known, then S is the sample covariance matrix where the deviations are from the known mean. If the means are unknown, then the deviations are from the sample mean.

To maximize $f(S;\Sigma)$ under H in (1.5) let $\tau = \sigma_{22}/\sigma_{11}$ and write

$$\Sigma = \sigma_{11}\begin{pmatrix} 1 & a \\ a & \tau \end{pmatrix} \equiv \sigma_{11}\Psi.$$

Then (3.1) becomes

$$f(S;\Sigma) = \frac{c(S) \mid \Psi \mid^{-n/2}}{\sigma_{11}^{n}} \exp -\frac{1}{2}\left(\frac{\operatorname{tr} \Psi^{-1}S}{\sigma_{11}}\right). \tag{3.2}$$

Maximization of $f(S;\Sigma)$ in (3.2) with respect to σ_{11} yields

$$\hat{\sigma}_{11} = (\operatorname{tr} \Psi^{-1}S)/(2n), \tag{3.3}$$

so that

$$\max_{\sigma_{11}} f(S;\Sigma) = \frac{c(S)e^{-n}}{\left[\mid \Psi \mid^{\frac{1}{2}} (\operatorname{tr} \Psi^{-1}S)/(2n)\right]^{n}} = \frac{c(S)(2ne^{-1})^{n}}{\left[\mid \Psi \mid^{\frac{1}{2}} \operatorname{tr} \Psi^{-1}S\right]^{n}}.$$

Note that

$$\mid \Psi \mid^{\frac{1}{2}} \operatorname{tr} \Psi^{-1}S = (\tau s_{11} - 2as_{12} + s_{22})/\sqrt{\tau - a^{2}}, \tag{3.4}$$

so that setting the derivative (with respect to τ) of the logarithm of (3.4) equal to zero yields

$$\hat{\tau} = (2a^{2}s_{11} - 2as_{12} + s_{22})/s_{11}. \tag{3.5}$$

Together (3.3) and (3.5) yield the MLE

$$\hat{\sigma}_{11} = s_{11}/n,\ \hat{\sigma}_{22} = (2a^{2}s_{11} - 2as_{12} + s_{22})/n. \tag{3.6}$$

Note that the constraint (1.6) is satisfied, since

$$\hat{\tau} = \hat{\sigma}_{22}/\hat{\sigma}_{11} = a^{2} + \frac{a^{2}s_{11} - 2as_{12} + s_{22}}{s_{11}} = a^{2} + (a,-1)S(a,-1)' \geq a^{2}.$$

Also under H,

$$\hat{\rho}_{12} = a\left[s_{11}/(2a^{2}s_{11} - 2as_{12} + s_{22})\right]^{\frac{1}{2}}. \tag{3.7}$$

4. DISTRIBUTION OF THE MLE

Because $\hat{\sigma}_{11}$ and $\hat{\sigma}_{22}$ are rational functions of moments, the standard delta method (Cramér, 1946, p. 366; Rao, 1973, p. 388) yields joint asymptotic normality:

$$\sqrt{n}\big((\hat{\sigma}_{11},\hat{\sigma}_{22}) - (\sigma_{11},\sigma_{22})\big) \sim \mathcal{N}(0,\Omega_\infty), \tag{4.1}$$

where the asymptotic covariance matrix, Ω_∞, has elements

$$\mathrm{Var}_\infty(\hat{\sigma}_{11}) \equiv \omega_{11} = 2\sigma_{11}^2, \tag{4.2a}$$

$$\begin{aligned}\mathrm{Var}_\infty(\hat{\sigma}_{22}) \equiv \omega_{22} = 8a^4\sigma_{11}^2 - 16a^3\sigma_{11}\sigma_{12} + 4a^2(\sigma_{11}\sigma_{22} + 3\sigma_{12}^2)\\ - 8a\sigma_{12}\sigma_{22} + 2\sigma_{22}^2,\end{aligned} \tag{4.2b}$$

$$\mathrm{Cov}_\infty(\hat{\sigma}_{11},\hat{\sigma}_{22}) \equiv \omega_{12} = 4a^2\sigma_{11}^2 - 4a\sigma_{11}\sigma_{12} + 2\sigma_{12}^2. \tag{4.2c}$$

Under H,

$$\mathrm{Var}(\hat{\sigma}_{22}) = 2\sigma_{22}^2(2\rho_{12}^4 - 2\rho_{12}^2 + 1) = 4a^4\sigma_{11}^2 - 4a^2\sigma_{11}\sigma_{22} + 2\sigma_{22}^2,$$

and

$$\mathrm{Cov}_\infty(\hat{\sigma}_{11},\hat{\sigma}_{22}) = 2\sigma_{11}\sigma_{22}\rho_{12}^2 = 2a^2\sigma_{11}^2.$$

The asymptotic distribution of $\hat{\rho}_{12}^2$ follows directly from (4.1) and (4.2):

$$\sqrt{n}(\hat{\rho}_{12}^2 - \rho_{12}^2) \sim \mathcal{N}\big(0,\sigma_\infty^2(\hat{\rho}_{12}^2)\big),$$

where

$$\begin{aligned}\sigma_\infty^2(\hat{\rho}_{12}^2) &= (a^2/\sigma_{22}, -a^2\sigma_{11}/\sigma_{22}^2)\Omega_\infty(a^2/\sigma_{22}, -a^2\sigma_{11}/\sigma_{22}^2)'\\ &= \frac{a^4}{\sigma_{22}^2}\left[\omega_{11} + \frac{\sigma_{11}^2}{\sigma_{22}^2}\omega_{22} - 2\frac{\sigma_{11}}{\sigma_{22}}\omega_{12}\right].\end{aligned} \tag{4.3}$$

Under H,

$$\mathrm{Var}_\infty(\hat{\rho}_{12}^2) = 4\rho_{12}^4(1-\rho_{12}^2)^2 = 4a^4\frac{\sigma_{11}^2}{\sigma_{22}^2}\left(1 - \frac{a^2\sigma_{11}}{\sigma_{22}}\right)^2.$$

5. THE LIKELIHOOD RATIO TEST OF THE HYPOTHESIS H

To obtain the likelihood ratio test we require the maximum of $f(S;\Sigma)$ given by (3.1), both under the hypothesis (1.4) and for general Σ. The latter is well-known to yield $\hat{\Sigma} = S/n$, so that

$$\max_{\Sigma} f(S;\Sigma) = c(S)\left|\frac{S}{n}\right|^{-n/2} e^{-n}. \tag{5.1}$$

To obtain the maximum of $f(S;\Sigma)$ under H, we have from (3.4) and (3.5) that

$$\max_{\sigma_{11},r} f(S;\Sigma) = c(S)\left[\frac{e^{-1}n}{s_{11}^{\frac{1}{2}}(a^2 s_{11} - 2as_{12} + s_{22})^{\frac{1}{2}}}\right]^n, \tag{5.2}$$

so that the likelihood ratio statistic (LRS) is given by

$$L(S) \equiv (LRS)^{2/n} = \frac{|S|}{s_{11}(a^2 s_{11} - 2as_{12} + s_{22})}. \tag{5.3}$$

A first approximation for the distribution of the LRS under the null hypothesis is

$$-2\log LRS \sim \chi_1^2. \tag{5.4}$$

Under the alternative hypothesis the asymptotic distribution of $L(S)$ is:

$$\sqrt{n}(L(S) - L(\Sigma)) \sim \mathcal{N}\left(0, \sigma_\infty^2(L(S))\right),$$

where

$$\sigma_\infty^2(L(S)) = (d_{11}, d_{22}, d_{12})\Delta_\infty(d_{11}, d_{22}, d_{12})'$$

and the elements of the quadratic form are

$$\Delta_\infty = \begin{pmatrix} 2\sigma_{11}^2 & 2\sigma_{12}^2 & 2\sigma_{11}\sigma_{12} \\ & 2\sigma_{22}^2 & 2\sigma_{22}\sigma_{12} \\ & & \sigma_{11}\sigma_{22} + \sigma_{12}^2 \end{pmatrix},$$

$$d_{11} = \left.\frac{\partial L(S)}{\partial s_{11}}\right|_{S=\Sigma} = \left[-a^2\sigma_{11}\sigma_{22} + \sigma_{12}^2(2a^2\sigma_{11} - 2a\sigma_{12} + \sigma_{22}\right]/\delta^2,$$

$$d_{22} = \left.\frac{\partial L(S)}{\partial s_{22}}\right|_{S=\Sigma} = \sigma_{11}\left[a^2\sigma_{11}^2 - 2a\sigma_{11}\sigma_{12} + \sigma_{12}^2\right]/\delta^2,$$

$$d_{12} = \left.\frac{\partial L(S)}{\partial s_{12}}\right|_{S=\Sigma} = 2\left[-a^2\sigma_{11}^2\sigma_{12} + a\sigma_{11}\sigma_{12}^2 + a\sigma_{11}^2\sigma_{22} = \sigma_{11}\sigma_{22}\sigma_{12}\right]/\delta^2,$$

and

$$\delta = \sigma_{11}(a^2\sigma_{11} - 2a\sigma_{12} + \sigma_{22}).$$

6. THE MULTIVARIATE CASE

Suppose now that $X_1, X_2, \ldots, X_p$ are measurements at p time points, and we consider the overlap hypothesis for every pair (i, j) of points. Then

$$H : \rho_{ij} = a\sigma_i/\sigma_j, \; i, \; j = 1, \ldots, p, \tag{6.1}$$

where $a = 1/2$ or 1 as described in Section 1. Under the hypothesis H, the population covariance matrix Σ has the form

$$\Sigma_H = \begin{pmatrix} \sigma_{11} & a\sigma_{11} & a\sigma_{11} & \cdots & a\sigma_{11} \\ & \sigma_{22} & a\sigma_{22} & \cdots & a\sigma_{22} \\ & & & \cdots & \\ & & & \cdots & \\ & & & \cdots & \\ & & & & \sigma_{pp} \end{pmatrix}.$$

When $a = 1$, the matrix Σ takes a well-known form known as the Green's matrix (see, for example, Gantmacher, 1959, p. 99; Karlin, 1968, p. 110–114). The inverse of Σ is tri-diagonal and is called a Jacobi matrix. However, it is still sufficiently complicated so that explicit MLE are difficult to obtain. Instead, we note that Σ can be written as a linear combination

$$\Sigma_H = \sigma_{11}A_1 + \cdots + \sigma_{pp}A_p, \tag{6.3}$$

where

$$A_1 = \begin{pmatrix} 1 & a & \cdots & a \\ a & 0 & \cdots & 0 \\ \cdot & & \cdots & \\ \cdot & & \cdots & \\ \cdot & & \cdots & \\ a & 0 & \cdots & 0 \end{pmatrix},$$

$$A_2 = \begin{pmatrix} 0 & 0 & 0 & \cdots & 0 \\ 0 & 1 & a & \cdots & a \\ 0 & a & 0 & \cdots & 0 \\ & \cdot & & & \\ & \cdot & & & \\ & \cdot & & & \\ 0 & a & 0 & \cdots & 0 \end{pmatrix}, \ldots, A_p = \begin{pmatrix} 0 & \cdots & 0 & 0 \\ & \cdots & & \\ & \cdots & & \\ & \cdots & & \\ 0 & \cdots & 0 & 0 \\ 0 & \cdots & 0 & 1 \end{pmatrix}.$$

Anderson (1970) has provided an iterative procedure that converges to the MLE. However, there are currently a number of computer programs that provide MLE for patterned matrices. By using a numerical procedure we are able to obtain a likelihood ratio test, namely,

$$(LRT)^{2/n} = \frac{|S/n|}{|\hat{\Sigma}_H|} \exp\left\{p - \frac{1}{2n} \operatorname{tr} \hat{\Sigma}_H^{-1} S\right\},$$

where $\hat{\Sigma}_H$ is the estimate of Σ when the model (4.3) holds.

ACKNOWLEDGMENT

This work was supported in part by the Spencer Foundation and the National Science Foundation.

REFERENCES

Anderson, J. E. (1939), "The limitations of infant and preschool tests in the measurement of intelligence." *The Journal of Psychology* **8**, 351–379.

Anderson, T. W. (1970), "Estimation of covariance matrices which are linear combinations or whose inverses are linear combinations of given matrices." In *Essays in Probability and Statistics*, ed. R. C. Bose, I. M. Chakravarti, P. C. Mahalanobis, C. R. Rao, and K. J. C. Smith, pp. 1–24. Chapel Hill, North Carolina: The University of North Carolina Press.

Bloom, B. S. (1964), *Stability and Change in Human Characteristics*. New York: Wiley and Sons.

Bock, R. D., and D. M. Thissen (1980), "Statistical problems of fitting individual growth curves." In *Human Physical Growth and Maturation: Methodologies and Factors*, ed. F. E. Johnston, A. F. Roche, and C. Susanne, pp. 265–290. New York: Plenum Press.

Bock, R. D., H. Wainer, A. Peterson, D. Thissen, J. Murray, and A. Roche (1973), "A parametrization for individual human growth curves." *Human Biology* **45**, 63–80.

Box, G. E. P. (1950), "Problems in the analysis of growth and wear curves." *Biometrics* **6**, 362–389.

Cramér, H. (1946), *Mathematical Methods of Statistics*. Princeton, New Jersey: Princeton University Press.

Gantmacher, F. R. (1959), *The Theory of Matrices*, Volume II. New York: Chelsea.

Grizzle, J. E., and D. M. Allen (1969), "Analysis of growth and dose response curves." *Biometrics* **25**, 357–381.

Karlin, S. (1969), *Total Positivity*. Stanford: Stanford University Press.

Nelder, J. A. (1961), "The fitting of a generalization of the logistic curve." *Biometrics* **17**, 89–110.

Nelder, J. A. (1962), "An alternative form of a generalized logistic equation." *Biometrics* **18**, 614–616.

Potthoff, R. F., and S. N. Roy (1964), "A generalized multivariate analysis of variance useful especially for growth curve problems." *Biometrika* **51**, 313–326.

Rao, C. R. (1959), "Some problems involving linear hypotheses in multivariate analysis." *Biometrika* **46**, 49–58.

Rao, C. R. (1965), "The theory of least squares when the parameters are stochastic and its application to the analysis of growth curves." *Biometrika* **52**, 447–458.

Rao, C. R. (1973), *Linear Statistical Inferences and Its Applications*. New York: Wiley and Sons.

Robertson, T. B. (1915), "Studies on the growth of man, I, the prenatal and postnatal growth of infants." *American Journal of Physiology* **37**, 1–42.

Scammon, R. E. (1927), "The first seriatim study of human growth." *American Journal of Physical Anthropology* **10**, 329–336.

Strenio, J. F., H. I. Weisberg, and A. S. Bryk (1983), "Empirical Bayes estimation of individual growth-curve parameters and their relationship to covariates." *Biometrics* **39**, 71–86.

Tuddenham, R. D., and M. M. Snyder (1954), "Physical growth of California boys and girls from birth to eighteen years." *University of California Publication in Child Development* **1**, 183–364.

Winsor, C. P. (1932), "The Gompertz curve as a growth curve." *Proceedings of the National Academy of Sciences of the United States of America* **18**, 1–8.

Wishart, J. (1938), "Growth rate determinations in nutrition studies with the bacon pig, and their analysis." *Biometrika* **30**, 16–28.

I. Guttman and M. S. Srivastava [1]

BAYESIAN METHOD OF DETECTING CHANGE POINT IN REGRESSION AND GROWTH CURVE MODELS

ABSTRACT

A Bayesian method is given to detect a change point of a regression model when some of the parameters are known *a priori* not to change, while others are subject to change. The same method is applied to a growth curve model. An illustrative numerical example is given.

1. INTRODUCTION

The problem of detecting departures from constancy of regression relationships over time has been considered many times in statistical literature; see, for example, Hinkley (1972), Brown *et al.* (1975) and Worsley (1983). While Hinkley and Worsley have given likelihood procedures for detecting the change point, Brown *et al.* included in their analysis cusum procedures as well, similar to that developed by Page (1954). A Bayesian method has been given by Broemeling and Chin Choy (1981), and Broemeling and Moen (1984). However, many regression relationships contain covariables which are not subject to change. For example, a covariable such as sex does not change over the time in which observations are obtained. More specifically, suppose in the regression model we have a $(k \times 1)$ vector $\boldsymbol{\beta}$ of common unknown regression parameters, and that the model can be written as

$$\begin{aligned} y_i &= \boldsymbol{\beta}'\mathbf{x}_{1i} + \boldsymbol{\beta}_1'\mathbf{x}_{2i} + \varepsilon_i, & 1 \le i \le t \\ &= \boldsymbol{\beta}'\mathbf{x}_{1i} + \boldsymbol{\beta}_2'\mathbf{x}_{2i} + \varepsilon_i, & t+1 \le i \le N, \end{aligned}$$

where ε_i's are independently normally distributed with mean 0 and variance α^2, and $\boldsymbol{\beta}_1$ and $\boldsymbol{\beta}_2$ are unknown vectors of order $(q_1 \times 1)$ and $(q_2 \times 1)$

[1] Department of Statistics, University of Toronto, Toronto, Ontario M5S 1A1 (both authors)

I. B. MacNeill and G. J. Umphrey (eds.), Foundations of Statistical Inference, 73–91.

respectively. The problem with which we are concerned here is to find the point of change t which is unknown. In this paper we provide a Bayesian method of finding the change point for the multivariate regression model of which the above problem is a special case. We also consider the growth curve model of Rao (1959), which can be described as follows. Suppose

$$\begin{aligned} E(\mathbf{y}_i) &= K\boldsymbol{\theta}_1, \qquad 1 \le i \le t \\ &= K\boldsymbol{\theta}_2, \qquad t+1 \le i \le n, \end{aligned}$$

where $\mathbf{y}_i, \ldots, \mathbf{y}_N$ are independently normally distributed p-vectors with common covariance matrix Σ, and K is a $p \times r$ matrix of known constants. We shall assume that K is of rank r. The problem is to find the change point t. As a special case, suppose $K = (I_r, 0_{r,p-r})'$ where $0_{r,p-r}$ is a matrix of zeroes of order $[r \times (p-r)]$. This situation arises when the mean of $(p-r)$ components of $\mathbf{y}_i$ are known, a very common phenomenon when it is relatively easier to obtain more observations on some components which are not subject to change over time. The general multivariate mean case has been considered by Sen and Srivastava (1973), Srivastava (1981), Guttman and Menzefrike (1982), and Srivastava and Worsley (1986).

2. A CHANGE POINT PROBLEM FOR THE MULTIVARIATE REGRESSION MODEL INVOLVING COMMON PARAMETERS

In this section we concern ourselves with the general multivariate linear model in which it is suspected that a change occurs from one linear model to another where the different models, have some common parameters. Examples of this situation abound, and illustrations are given below.

We suppose first that $\mathbf{y}_i$, $i = 1, \ldots, t$ are t independent $(p \times 1)$ vectors of observations such that

$$Y_1 = \begin{pmatrix} \mathbf{y}_1' \\ \vdots \\ \mathbf{y}_t' \end{pmatrix} = X_{11}A + X_{12}B_1 + E_1, \tag{2.1}$$

where

$$E_1 = \begin{pmatrix} \mathbf{e}_1' \\ \vdots \\ \mathbf{e}_t' \end{pmatrix}$$

is a $(t \times p)$ matrix whose rows are independently and identically distributed (i.i.d.) as multi-normal, that is,

$$E_1 \sim N(0; \Sigma, I_t),$$

so that

$$p(E_1) = (2\pi)^{-pt/2} \mid \Sigma^{-1} \mid^{t/2} \exp -\frac{1}{2} \operatorname{tr} \Sigma^{-1} E_1' E_1 \tag{2.2a}$$

with the $(p \times p)$ matrix Σ assumed positive definite. The symbol "0" denotes a $(t \times p)$ matrix of zeroes. The matrices X_{11} and X_{12} of constants are of order $(t \times k)$ and $(t \times q_1)$ respectively, while the parameter matrices A and B_1 of (2.1) are of order $(k \times p)$ and $(q_1 \times p)$ respectively. We further assume that the data also include Y_2, independent of Y_1, with

$$Y_2 = \begin{pmatrix} \mathbf{y}'_{t+1} \\ \vdots \\ \mathbf{y}'_N \end{pmatrix} = X_{21}A + X_{23}B_2 + E_2, \tag{2.3}$$

where E_2 is distributed as $N(0; \Sigma, I_{N-t})$, X_{21} and X_{23} are of order $[(N - t) \times k]$ and $[(N-t) \times q_2]$ respectively, with the parameter matrix B_2 of order $(q_2 \times p)$. Note that (2.1) and (2.3) have the parameters A in common.

We may combine (2.1) and (2.3) and write, in familiar multivariate linear regression model form,

$$Y = \begin{pmatrix} Y_1 \\ \cdots \\ Y_2 \end{pmatrix} = X_t G + E$$

$$= \begin{pmatrix} X_{11} & X_{12} & 0 \\ & & \\ X_{21} & 0 & X_{23} \end{pmatrix} \begin{pmatrix} A \\ \cdots \\ B_1 \\ \cdots \\ B_2 \end{pmatrix} + \begin{pmatrix} E_1 \\ \cdots \\ E_2 \end{pmatrix}. \tag{2.4}$$

We note that X_t is $[N \times (k+q)]$, $q = q_1 + q_2$, and G is $[(k+q) \times p]$.

With this model, we will now derive the posterior probability that a change point occurred at t, $1 \le t \le N-1$, given that the data were generated in accordance with (2.4). We assume that prior to observing the data, that information about A, B_1, B_2 (that is G) and Σ, or equivalently Σ^{-1}, is well summarized by the conjugate prior

$$p(t; G; \Sigma^{-1}) \propto \frac{1}{N-1} \mid \Sigma^{-1} \mid^{(N_0-p-1)/2}$$
$$\cdot \exp\left\{-\frac{1}{2} \operatorname{tr} \Sigma^{-1} \left[S_0 + (G - G_0)' C (G - G_0)\right]\right\} \tag{2.5}$$

with C a given symmetric positive definite matrix of order $[(k+q) \times (k+q)]$, and G_0 a given $[(k+q) \times p]$ matrix. Hence, when the likelihood dictated by

(2.4) is combined with the prior (2.5), one obtains

$$p(t;G;\Sigma^{-1}\mid) \propto \mid \Sigma^{-1} \mid^{(N_0+N-p-1)/2} \cdot \exp\left\{-\frac{1}{2}\Sigma^{-1}[\tilde{S}_t + (G-\tilde{G})'A_t(G-\tilde{G})]\right\}, \qquad (2.6)$$

where

$$\begin{aligned} \tilde{S}_t &= S_0 + S_t + (G_0-\hat{G})'X_t'X_tA_t^{-1}C(G_0-\hat{G}), \\ A_t &= X_t'X_t + C, \tilde{G} = A_t^{-1}\{CG_0 + X_t'Y\}, \\ S_t &= Y'\left[I - X_tX_t^{-}\right]Y, \end{aligned} \qquad (2.6a)$$

with X_t^- of order $[(k+q)\times N]$ and such that

$$X_t^- = (X_t'X_t)^- X_t', \qquad (2.6b)$$

where $(X_t'X_t)^-$ is a generalized inverse of $X_t'X_t$, and $\hat{G}$ is a solution to the normal equations

$$X_t'X_t\hat{G} = X_t'Y. \qquad (2.6c)$$

It should be noted that since C is assumed positive definite, A_t is non-singular and hence possesses an inverse. Indeed, if X_t is of full rank $(k+q) \leq N$, then

$$\begin{aligned} \hat{G} &= (X_t'X_t)^{-1} X_t'Y, \\ S_t &= Y'[I - X_t(X_t'X_t)^{-1}X_t']Y, \end{aligned} \qquad (2.6d)$$

and

$$\tilde{S}_t = S_0 + S_t + (G_0-\hat{G})[(X_t'X_t)^{-1} + C^{-1}]^{-1}(G_0-\hat{G}). \qquad (2.6e)$$

If one integrates out G from (2.6) using properties of the matric-normal, one obtains:

$$p(t;\Sigma^{-1}\mid Y) \propto \mid \Sigma^{-1} \mid^{(N_0+N-p-1)/2} \mid \Sigma^{-1} \mid^{-(k+q)/2} \mid A_t \mid^{-\frac{p}{2}} \exp\left(-\frac{1}{2}\,\mathrm{tr}\,\Sigma^{-1}\tilde{S}_t\right), \qquad (2.7)$$

or

$$p(t;\Sigma^{-1}\mid Y) \propto \mid \Sigma^{-1} \mid^{(N_0+N-k-q-p)/2} \mid A_t \mid^{-p/2} \exp\left(-\frac{1}{2}\,\mathrm{tr}\,\Sigma^{-1}\tilde{S}_t\right). \qquad (2.7a)$$

Integration with respect to Σ^{-1}, using properties of the Wishart distribution, yields, for $1 \le t \le N-1$,

$$p(t \mid Y) = \tilde{K} \mid A_t \mid^{-p/2} \mid \tilde{S}_t \mid^{-(N+N_0-k-q)/2} \tag{2.8}$$

with

$$\tilde{K}^{-1} = \sum_{t=1}^{N-1} \mid A_t \mid^{-p/2} \mid \tilde{S}_t \mid^{-(N+N_0-k-q)/2} . \tag{2.8a}$$

We may now examine $p(t \mid y)$ for a given set of data; and in particular, one may estimate t by that value at which $p(t \mid Y)$ has its maximum, that is, by the mode of t. It is interesting to note that the form of the solution (2.8a) is the same as that obtained by Broemeling and Choy (1981) and Guttman and Menzefricke (1982), who dealt with a model in which there were no common parameters, so that the terms involving A in (2.1), (2.3) and (2.4) were not present.

To conclude this section, suppose the prior (2.5) is very diffuse $[S_0, C \simeq 0, N_0 \simeq 0]$, and that the $[N \times k]$ matrix $[X'_{11} : X'_{21}]'$ is of full rank k. Then since $X'_t X_t$ is positive definite only if $t \ge q_1$ and $N - t \ge q_2$, one may note that

$$p(t \mid Y) = K \mid X'_t X_t \mid^{-p/2} \mid S_t \mid^{-(N-k-q)/2} \tag{2.8b}$$

if $q_1 \le t \le N - q_2$, and zero otherwise, where

$$K^{-1} = \sum_{t=q_1}^{N-q_2} \mid X'_t X_t \mid^{-p/2} \mid S_t \mid^{-(N-k-q)/2} . \tag{2.8c}$$

We may now estimate t by that value at which $p(t \mid Y)$ has its maximum, that is, by the mode of t. We illustrate the methodology of this section with an example, in Section 4.

3. CHANGE POINT PROBLEMS INVOLVING GROWTH-CURVE MODELS

In this section, we shall discuss a certain change-point problem that involves a switch, at time t say, from one growth-curve model to another. We will see that after a specific transformation, that the growth-curve models are transformed to two general multivariate linear models, with some of the regression coefficients known. Accordingly, we will first analyze these models, and in the latter half of the section, explore the specific growth-curve model problem.

We begin by supposing that the random matrix Y_1, of order $t \times p$ is such that

$$Y_1 = X_1 B_1 + E_1, \tag{3.1}$$

where the rows of Y_1, say $\mathbf{y}_i'$, $i = 1, \ldots, t$ are t independent observations on the $(1 \times p)$ vector $\mathbf{y}'$, X_1 is a $(t \times k_1)$ matrix of given constants, B_1 is a $(k_1 \times p)$ matrix of (regression) coefficients, and E_1 is a $(t \times p)$ random matrix of errors such that for Σ, a $(p \times p)$ positive definite matrix,

$$E_1 \sim N(0; \Sigma, I_t). \tag{3.2}$$

Also, we assume the data set also includes the $[(N-t) \times p]$ random matrix Y_2, independent of Y_1, with

$$Y_2 = X_2 B_2 + E_2, \tag{3.3}$$

where X_2 is a $[(N-t) \times k_2]$ matrix of given constants, B_2 is a $(k_2 \times p)$ matrix of coefficients, and E_2, independent of E_1, is a $[(N-t) \times p]$ random matrix of errors, such that

$$E_2 \sim N(0; \Sigma, I_{N-t}). \tag{3.4}$$

In addition to the assumptions above, we further assume that parts of the labelling is such that if we partition the B_i matrices mentioned above, such that

$$B_i = [B_i^{(1)} : B_i^{(2)}], \tag{3.5}$$

where $B_i^{(1)}$ is of order $(k_i \times r)$, $B_i^{(2)}$ is of order $(k_i \times s)$, $r + s = p$, then the matrices $B_i^{(2)}$ are *known*, while the $B_i^{(1)}$ are unknown. Given that Y_1 and Y_2 are generated according to (3.1) and (3.3), with the $B_i^{(2)}$ of (3.5) known, our problem is to determine the posterior probability that a change-point occurs at time t. The answer is supplied in the following theorem proved in the Appendix of this paper.

Theorem 3.1. Suppose it is intended that the $(N \times p)$ matrix Y is to be generated from the process

$$Y \sim N(XB; \Sigma; I_N), \tag{3.6}$$

but instead it is the case that for some t, and

$$Y = \begin{pmatrix} Y_1 \\ \cdots \\ Y_2 \end{pmatrix},$$

that Y_1 and Y_2 are generated according to (3.1)–(3.2) and (3.3)–(3.4) respectively, where $1 \leq t \leq N-1$, and where $B_i^{(2)}$, $i = 1, 2$ are known and of order $(k_i \times s)$.

Suppose further that the prior of $(B_1^{(1)}, B_2^{(1)}, \Sigma^{-1})$, given $B_1^{(2)}$, $B_2^{(2)}$, is very diffuse, and is such that it is appropriate to use the following noninformative prior:

$$p\left(B_1^{(1)}, B_2^{(1)}, \Sigma^{-1} \mid B_1^{(2)}, B_2^{(2)}\right) \propto \mid \Sigma^{-1} \mid^{-(p+1)/2} . \tag{3.7}$$

Let the $(p \times p)$ matrix $S^{(t)}$ be defined as follows:

$$\begin{aligned} S^{(t)} &= \sum_{i=1}^{2} Y_i' \left[I - X_i \left(X_i' X_i\right)^{-1} X_i'\right] Y_i \\ &= \begin{pmatrix} S_{11}^{(t)} & S_{12}^{(t)} \\ S_{21}^{(t)} & S_{22}^{(t)} \end{pmatrix}, \end{aligned} \tag{3.8}$$

where $S_{11}^{(t)}$ is $(r \times r)$, $S_{22}^{(t)}$ is $(s \times s)$, and $r + s = p$. Then posterior of t given $Y = (Y_1' : Y_2')'$ and $X = (X_1' : X_2')'$, is,

$$\begin{aligned} p\left(t \mid Y, X; B_1^{(2)}, B_2^{(2)}\right) =& K \mid S_{22}^{(t)} \mid^{-r/2} \mid C_1 \mid^{-r/2} \mid C_2 \mid^{-r/2} \\ & \times \mid S_{22}^{(t)} + Q_1 + Q_2 \mid^{-(N-r)/2} \mid S_{1.2}^{(t)} \mid^{-(N-k)/2} \end{aligned} \tag{3.9}$$

if $k_1 \leq t \leq N - k_2$, and zero otherwise, where

$$k = k_1 + k_2$$

$$\begin{aligned} S_{1.2}^{(t)} &= S_{11}^{(t)} - S_{12}^{(t)} \left(S_{22}^{(t)}\right)^{-1} S_{21}^{(t)}, \\ C_i &= X_i' X_i, \\ Q_i &= [B_i^{(2)} - \hat{B}_i^{(2)}]' C_i [B_i^{(2)} - \hat{B}_i^{(2)}]. \end{aligned} \tag{3.9a}$$

and

$$\hat{B}_i = \left(X_i' X_i\right)^{-1} X_i' Y_i = [\hat{B}_i^{(1)} : \hat{B}_i^{(2)}]; \tag{3.10}$$

note that $\hat{B}_i$ is $(k_i \times p)$, $\hat{B}_i^{(1)}$ is $(k_i \times r)$, $\hat{B}_i^{(2)}$ is $(k_i \times s)$, and $r + s = p$.

As mentioned earlier, the proof of the above theorem is given in the Appendix. It is interesting to note that if $B_i^{(2)}$ is equal to 0, a $(k_i \times s)$ matrix of zeroes, then it can be shown that

$$S_{22}^{(t)} + Q_1 + Q_2 = Y_1^{(2)'} Y_1^{(2)} + Y_2^{(2)'} Y_2^{(2)}, \tag{3.11}$$

where

$$Y_i = [Y_i^{(1)} : Y_i^{(2)}] \tag{3.11a}$$

and $Y_1^{(2)}$ is of order $(t \times s)$, $Y_2^{(2)}$ is of order $[(N-t) \times s]$, etc.

We remark that Theorem 3.1 possesses a version for which (3.7) is replaced by the conjugate prior (see (2.5)) appropriate for the model of this section. The resultant posterior is of the form (3.9), with the relevant matrices suitably altered (see, for example, the results of Section 2) to take into account sample information (supplied by the likelihood) and the prior information (supplied by the conjugate prior). The proof in the Appendix must then be extended, but the extension requires only some additional straightforward and tedious algebra. As the result would require many new definitions of the matrices involved, we omit this case.

One other important point consequent to the work in Section 2, is that it is easy to show for the (unconditional) model (3.1)–(3.4) with all of B_1 and B_2 unknown, that a change-point occurs at t with posterior probability as follows:

$$\begin{aligned} p(t \mid Y; X_1, X_2) &= K \mid X_t' X_t \mid^{-p/2} \mid S_t \mid^{-(N-k)/2}, \qquad k_1 \le t \le N - k_2 \\ &= 0, \qquad \text{otherwise.} \end{aligned} \tag{3.12}$$

Diffuse priors for B_1, B_2 and Σ^{-1} have been assumed, and the $(N \times k)$ matrix X_t is given by

$$X_t = \begin{pmatrix} X_1 & 0 \\ 0 & X_2 \end{pmatrix}, \quad k = k_1 + k_2; \tag{3.12a}$$

hence, since $\mid X_t' X_t \mid = \prod_{i=1}^{2} \mid X_i' X_i \mid$, one may write

$$p(t \mid Y, X_1, X_2) = K \mid X_1' X_1 \mid^{-p/2} \mid X_2' X_2 \mid^{-p/2} \mid S_t \mid^{-(N-k)/2} \tag{3.13}$$

if $k_1 \le t \le N - k_2$, and zero otherwise.

Here $X_1' X_1$ is positive definite only if $t \ge k_1$, and $X_2' X_2$ is positive definite only if $N - t \ge k_2$, which accounts for the stated restriction, $k_1 \le t \le N - k_2$, of (3.12) and (3.13). Hence

$$K^{-1} = \sum_{t=k_1}^{N-k_2} \mid X_t' X_t \mid^{-p/2} \mid S_t \mid^{-(N-k)/2}. \tag{3.13a}$$

We turn now to the central of concern of this section, namely, change point problem for growth-curve models. Very often, applications in the multivariate setting lead to models involving

$$\begin{aligned} E(\mathbf{y}_i) &= K\boldsymbol{\theta}_1, \qquad i = 1, \ldots, t \\ &= K\boldsymbol{\theta}_2, \qquad i = t+1, \ldots, N, \end{aligned} \tag{3.14}$$

where $1 \le t \le N-1$ and K is a $(p \times r)$ matrix of rank r. By a known orthogonal transformation of the vectors $\mathbf{y}_i$ to $\mathbf{w}_i$ we can write the model as follows:

$$\begin{aligned} E(\mathbf{w}_i) &= G\boldsymbol{\theta}_1, \qquad 1 \le i \le t \\ &= G\boldsymbol{\theta}_2, \qquad t+1 \le i \le N, \end{aligned} \tag{3.15}$$

where $G' = ((K'K)^{\frac{1}{2}}; 0)$, $\mathbf{w}_i = \Gamma\mathbf{y}_i$ and $\Gamma' = (K(K'K)^{-\frac{1}{2}}, C)$ with $C'K = 0$ and $C'C = I_{p-r}$. Under the normality assumption for $\mathbf{y}_i$ with covariance matrix Σ, we have, for $1 \le i \le t$,

$$\mathbf{w}_i \sim N_p(G\boldsymbol{\theta}_1, \Gamma\Sigma\Gamma'), \tag{3.16}$$

and for $t+1 \le i \le N$,

$$\mathbf{w}_i \sim N_p(G\boldsymbol{\theta}_2, \Gamma\Sigma\Gamma'). \tag{3.17}$$

We may summarize (3.16)–(3.17) as follows. If we set

$$W_1 = \begin{pmatrix} \mathbf{w}_1' \\ \vdots \\ \mathbf{w}_t' \end{pmatrix}, \text{ and } W_2 = \begin{pmatrix} \mathbf{w}_{t+1}' \\ \vdots \\ \mathbf{w}_N' \end{pmatrix}, \tag{3.18}$$

then

$$W_1 \sim N\{\mathbf{1}_t[\boldsymbol{\theta}_1^{*\prime} \ : \ \mathbf{0}']; \Gamma\Sigma\Gamma'; I_t\}$$

is independent of

$$W_2 \sim N\{\mathbf{1}_{N-t}[\boldsymbol{\theta}_2^{*\prime} \ : \ \mathbf{0}']; \Gamma\Sigma\Gamma'; I_{N-t}\}, \tag{3.19}$$

where the $1 \times r$ vectors $\boldsymbol{\theta}_i^{*\prime}$ are given by

$$\boldsymbol{\theta}_i^{*\prime} = \left(\boldsymbol{\theta}_i'(K'K)^{\frac{1}{2}}\right), \tag{3.20}$$

$\mathbf{0}'$ denotes a $(1 \times s)$ vector of zeroes, $s = p - r$ and in general $\mathbf{1}_m$ is a $(m \times 1)$ vector of ones. We are precisely in the format of the model used to develop Theorem 3.1, with the W_i replacing Y_i of the theorem, with $\mathbf{1}_t$ and $\mathbf{1}_{N-t}$ replacing X_1 and X_2 respectively so that $k_i = 1$, $k = 2$, $C_1 = t$, $C_2 = N - t$, and with $\Lambda = \Gamma\Sigma\Gamma'$ replacing Σ, and with

$$(B_i^{(1)} \ : \ B_i^{(2)}) = (\boldsymbol{\theta}_i^{*\prime} \ : \ \mathbf{0}'). \tag{3.21}$$

Hence, we may write down the posterior of t directly, and we have the following theorem.

Theorem 3.2. Suppose the transformation $\mathbf{w}_i = \Gamma \mathbf{y}_i$ yields to W_1 and W_2 of (3.18) with distributions as specified by (3.19). Then the posterior of t, given W_1 and W_2 is

$$\begin{aligned} p(t \mid \text{ data }) = K[t(N-t)]^{-r/2} \mid S_{22}^{(t)} \mid^{-r/2} \\ \times \mid (W_1^{(2)'} W_1^{(2)}) + (W_2^{(2)'} W_2^{(2)}) \mid^{-(N-r)/2} \mid S_{1.2}^{(t)} \mid^{-(N-2)/2}, \\ 1 \leq t \leq N-1, \end{aligned} \tag{3.22}$$

where

$$S^{(t)} = W_1' \left(I_t - \frac{1}{t} \mathbf{1}_t \mathbf{1}_t' \right) W_1 + W_2' \left(I_{N-t} - \frac{1}{N-t} \mathbf{1}_{N-t} \mathbf{1}_{N-t}' \right) W_2 \tag{3.22a}$$

is partitioned as $(p = r + s)$

$$S^{(t)} = \begin{pmatrix} S_{11}^{(t)} & S_{12}^{(t)} \\ S_{21}^{(t)} & S_{22}^{(t)} \end{pmatrix}. \tag{3.22b}$$

$S_{22}^{(t)}$ is of order $s \times s$, $S_{1.2}^{(t)} = S_{11}^{(t)} - S_{12}^{(t)} S_{22}^{(t)} S_{21}^{(t)}$, and

$$\begin{aligned} K^{-1} = \sum_{t=1}^{N-1} \Big\{ [t(N-t)]^{-r/2} \mid S_{22}^{(t)} \mid^{-r/2} \mid S_{1.2}^{(t)} \mid^{-(N-2)/2} \\ \times \mid W_1'^{(2)} W_1^{(2)} + W_2'^{(2)} W_2^{(2)} \mid^{-(N-r/2)} \Big\}. \end{aligned}$$

In the above, we have partitioned

$$W_i = [W_i^{(1)} : W_i^{(2)}] \tag{3.22c}$$

with $W_1^{(2)}$ of order $(t \times s)$, and $W_2^{(2)}$ of order $[(N-t) \times s]$.

4. AN EXAMPLE

We now illustrate the results of Section 2 with an example. Madalla (1977) has given data which we reproduce in Table 1. The notation of Table 1 is as follows.

$$\begin{aligned} \tilde{y} &= \text{ gross domestic product} \\ \tilde{x}_1 &= \text{ labor input; } \tilde{x}_2 = \text{ capital input.} \end{aligned} \tag{4.1}$$

Table 1

Year	$\tilde{y}$	$\tilde{x}_1$	$\tilde{x}_2$
1929	189.8	173.3	87.8
30	172.1	165.4	87.8
31	159.1	158.2	84.0
32	135.6	141.7	78.3
33	132.0	141.6	76.6
34	141.8	148.0	76.0
35	153.9	154.4	77.7
36	171.5	163.5	79.1
37	183.0	172.0	80.0
38	173.2	161.5	77.6
39	188.5	168.6	81.4
40	205.5	176.5	87.0
41	236.0	192.4	96.2
42	257.8	205.1	104.4
43	277.5	210.1	110.0
44	291.1	208.8	107.8
45	284.5	202.1	102.1
46	274.0	213.4	97.2
47	279.9	223.6	105.9
48	297.6	228.2	113.0
49	297.7	221.9	114.9
50	328.9	228.8	124.1
51	351.4	239.0	134.5
52	360.4	241.7	139.7
53	378.9	245.2	147.4
54	375.8	237.4	148.9
55	406.7	245.9	158.6
56	416.3	251.6	167.1
57	422.8	251.5	171.9
58	418.4	245.1	173.1
59	445.7	254.9	182.5
60	457.3	259.6	189.0
61	466.3	258.1	194.1
62	495.3	264.6	202.3
63	515.5	268.5	205.4
64	544.1	275.4	215.9
65	579.2	285.3	225.0
66	615.6	297.4	236.2
67	631.1	305.0	247.9

The discussion of Madalla indicated that the effect of capital inputs remain stable over the years 1929–1967, while effects of labor inputs may undergo change at some (time) point. Madalla used the Cobb-Douglas function which employs logarithms of the variables in (4.1); hence we let

$$y = \ln \tilde{y}, \; x_j = \ln \tilde{x}_j. \tag{4.1a}$$

In view of the stability of the effect of capital inputs, there is at this point a choice of two descriptions for the the generating process for the y's, and Madalla has given no direction at this point as to which should be explored. The first description for the expected value of y would reflect not only a stable effect for capital input, but a common intercept for both the pre-change process and the after change process; that is,

$$\mathbf{y} = \begin{matrix} t \\ \\ N-t \end{matrix} \begin{pmatrix} \mathbf{y}_1 \\ \cdots \\ \mathbf{y}_2 \end{pmatrix}$$

$$= \begin{bmatrix} \mathbf{1}_t & \mathbf{x}_2^{(t)} & \vdots & \mathbf{x}_1^{(t)} & \vdots & \mathbf{0}_t \\ & & \vdots & & \vdots & \\ \mathbf{1}_{N-t} & \mathbf{x}_2^{(N-t)} & \vdots & \mathbf{0}_{N-t} & \vdots & \mathbf{x}_1^{(N-t)} \end{bmatrix} \begin{pmatrix} a_0 \\ a_1 \\ \cdots \\ \beta_1 \\ \cdots \\ \beta_2 \end{pmatrix} + \begin{pmatrix} \boldsymbol{\varepsilon}_1 \\ \cdots \\ \boldsymbol{\varepsilon}_2 \end{pmatrix}. \tag{4.2}$$

In general, $\mathbf{1}_t$ is a vector of t ones, $\mathbf{0}_t$ is a vector of t zeroes, $\mathbf{x}_1^{(t)}$ is the $(t \times 1)$ vector of the first t-components of $\mathbf{x}_1$, while $\mathbf{x}_1^{(N-t)}$ is the $[(N-t) \times 1]$ vector of the last $(N-t)$ components of $\mathbf{x}_1$, etc. $(N = 39)$. Using the notation of Section 2, $p = 1$, $k = 2$, $q_1 = 1$, $q_2 = 1$, with $A = \mathbf{a} = (a_0, a_1)'$, $B_1 = \beta_1$, $B_2 = \beta_2$, etc. (see (2.4)). A second description is also possible, viz,

$$\mathbf{y} = \begin{pmatrix} \mathbf{y}_1 \\ \cdots \\ \mathbf{y}_2 \end{pmatrix} \tag{4.3}$$

$$= \begin{bmatrix} \mathbf{x}_2^{(t)} & \vdots & \mathbf{1}_t & \mathbf{x}_1^{(t)} & \vdots & \mathbf{0}_t & \mathbf{0}_t \\ & \vdots & & & \vdots & & \\ \mathbf{x}_2^{(N-t)} & \vdots & \mathbf{0}_{N-t} & \mathbf{0}_{N-t} & \vdots & \mathbf{1}_{N-t} & \mathbf{x}_1^{(N-t)} \end{bmatrix} \begin{pmatrix} a \\ \cdots \\ \beta_{01} \\ \beta_{11} \\ \cdots \\ \beta_{02} \\ \beta_{12} \end{pmatrix} + \begin{pmatrix} \boldsymbol{\varepsilon}_1 \\ \cdots \\ \boldsymbol{\varepsilon}_2 \end{pmatrix}$$

In model (4.3), the intercept is allowed to change and the effect due to capital input is assumed stable, and we have $p = 1$, $k = 1$, $q_1 = 2$, $q_2 = 2$.

As a matter of fact, had Madalla not stated that the effect of capital input remains stable, but a change point is suspected, the model would be that mentioned in Section 3, specifically:

$$\mathbf{y} = \begin{pmatrix} \mathbf{y}_1 \\ \cdots \\ \mathbf{y}_2 \end{pmatrix}$$

$$= \begin{bmatrix} \mathbf{1}_t & \mathbf{x}_1^{(t)} & \mathbf{x}_2^{(t)} & \vdots & \mathbf{0}_t & \mathbf{0}_t & \mathbf{0}_t \\ \mathbf{0}_{N-t} & \mathbf{0}_{N-t} & \mathbf{0}_{N-t} & \vdots & \mathbf{1}_{N-t} & \mathbf{x}_1^{(N-t)} & \mathbf{x}_2^{(N-t)} \end{bmatrix} \begin{pmatrix} \beta_{01} \\ \beta_{11} \\ \beta_{21} \\ \cdots \\ \beta_{02} \\ \beta_{12} \\ \beta_{21} \end{pmatrix}$$

$$+ \begin{pmatrix} \boldsymbol{\varepsilon}_1 \\ \cdots \\ \boldsymbol{\varepsilon}_2 \end{pmatrix} \tag{4.4}$$

(see (3.1)–(3.4) and (3.12a)). Using the notation of Section 3, we have here that $p = 1$, $k_1 = 3$, $k_2 = 3$, X_1 is $(t \times 3)$, X_2 is $[(N-t) \times 3]$, etc.

The tentative entertaining of the 3 models (4.2), (4.3), and (4.4) raises the question of "which model, given the data, should be chosen"? We advocate the following *ad hoc* procedure. Let the posterior distributions of the change point derived using models (4.2), (4.3) and (4.4) be denoted, respectively by $p_1(t)$, $p_2(t)$ and $p_3(t)$. Find the mode of these distributions, say $t_1^{(m)}$, $t_2^{(m)}$ and $t_3^{(m)}$, respectively. Determine the values $S_{t_i(m)}$ where

$$S_{t_i(m)} = \mathbf{y}'[I - X_t^{(i)}(X_t^{(i)\prime} X_t^{(i)})^{-1} X_t^{(i)\prime}]\mathbf{y} \tag{4.5}$$

with the matrices $X_t^{(i)}$, $i = 1, 2, 3$, defined in the obvious way by (4.2), (4.3) and (4.4). Compute

$$S_{t_1(m)}/(N-4), \ S_{t_2(m)}/(N-5), \ S_{t_3(m)}/(N-6). \tag{4.6}$$

Determine the minimum of the quantities in (4.6) and select that model which has this minimum.

Table 2 gives the posterior probabilities that a change point occurs at t using (4.2), (4.3) and (4.4). Note that the maxima for the various $p_i =$

Table 2. *Posterior Probabilities* of a Change-point at t*

Year	t	$p_1(t \mid \text{data})$	$p_2(t \mid \text{data})$	$p_3(t \mid \text{data})$	$S_{t_j}(m)$
1933	5	.013	.003	.001	
1934	6	.086	.025	.003	
1935	7	.405	.202	.030	$S_{t_1}(m) = .029427$
1936	8	.157	.194	.068	
1937	9	.094	.388	.175	$S_{t_2}(m) = .026448$
1938	10	.009	.042	.118	
1939	11	.001	.011	.058	
1940	12	.001	.010	.033	
1941	13	.000	.008	.042	
1942	14	.000	.003	.130	
1943	15	.000	.002	.176	$S_{t_3}(m) = .022591$
1944	16	.001	.003	.030	
1945	17	.157	.026	.052	
1946	18	.155	.075	.079	
1947	19	.003	.001	.000	
1948	20	.001	.000	.000	

$$t_1^{(m)} = 7;\ t_2^{(m)} = 9;\ t_m^{(m)} = 15$$

[*Rounded off from calculations with nine significant figures in output]

$p_i(t \mid \text{ data })$ occur at $t_1^{(m)} = 7$ (1935), $t_2^{(m)} = 9$ (1937), and $t_3^{(m)} = 15$ (1943). The $S_{t_i(m)}$'s are also given in Table 2, leading to

$$S_{t_1(m)}/35 = .000871, \quad S_{t_2(m)}/34 = .000778, \quad S_{t_3(m)}/33 = .000685.$$
$$(t_1^{(m)} = 7) \qquad\qquad (t_2^{(m)} = 9) \qquad\qquad (t_3^{(m)} = 15)$$

The three quantities $S_{t_j(m)}$ are virtually the same. Accordingly, we would select that model which has largest posterior probability of change. Hence, we would select model (4.2) as being most compatible with the data. (Omitted posterior probabilities are less than .001). It is interesting to note that Worsley (1983), who assumes that both effects of labor and capital input change at some point t, estimates this change point by maximum likelihood to be at $t = 14$ (1942). This model is that of (4.4) for which the maximum of $p_3(t \mid \text{ data })$ is at $t_3^{(m)} = 15$ (1943).

ACKNOWLEDGMENT

This research has been supported by NSERC of Canada, and was completed while I. Guttman was visiting Professor at the Graduate School of Business, University of Chicago, Spring 1985.

APPENDIX

In this Appendix, we give a proof of Theorem 3.1. Using (3.2) and (3.4) for the likelihood, and assuming that the Jeffrey's vague prior is appropriate, that is, the prior is

$$p\left(t, \Sigma^{-1}, B_1, B_2\right) \propto \mid \Sigma^{-1} \mid^{-(p+1)/2}, \tag{A.1}$$

we easily see that the posterior, given Y_1, Y_2, X_1, X_2 is such that

$$p\left(t, \Sigma^{-1}, B_1, B_2, Y_1, Y_2; X_1, X_2\right) \propto \mid \Sigma^{-1} \mid^{(N-p-1)/2} \exp\left(-\frac{1}{2} \operatorname{tr} \Sigma^{-1} Q\right), \tag{A.2}$$

where the $(p \times p)$ matrix Q is given by

$$Q = \sum_{i=1}^{2} \{S_i^{(t)} + (B_i - \hat{B}_i)' C_i (B_i - \hat{B}_i)\}, \tag{A.3}$$

and where

$$C_i = X_i' X_i, \ \hat{B}_i = C_i^{-1} X_i' Y_i,$$
$$S_i^{(t)} = Y_i' [I - X_i (X_i' X_i)^{-1} X_i'] Y_i. \tag{A.3a}$$

Now, in conformity to the partitioning of (3.5), partition the $k_i \times p$ matrices $\hat{B}_i$ as follows:

$$\hat{B}_i = [\hat{B}_i^{(1)} \; : \; \hat{B}_i^{(2)}], \tag{A.4}$$

where $\hat{B}_i^{(1)}$ is $(k_i \times r)$, $\hat{B}_i^{(2)}$ is $(k_i \times s)$, with $r + s = p$. Now write

$$\Sigma^{-1} = \Omega = \begin{pmatrix} \Omega_{11} & \Omega_{12} \\ \Omega_{21} & \Omega_{22} \end{pmatrix}, \tag{A.5}$$

where Ω_{11} is $(r \times r)$, Ω_{12} is $(r \times s)$, $\Omega_{12}' = \Omega_{21}$, and Ω_{22} is $(s \times s)$. Since $f(Z_1, Z_2) = f_1(Z_1 \mid Z_2) f_2(Z_2)$, it follows from a property of the normal distribution that

$$\begin{aligned} \text{tr } \Sigma^{-1} Q = & \text{tr } \Omega S^{(t)} + \sum_{i=1}^{2} \text{tr } \Omega_{11} [B_i^{(1)} - M_i]' C_i [B_i^{(1)} - M_i] \\ & + \sum_{i=1}^{2} \text{tr } \Omega_{2.1} [B_i^{(2)} - \hat{B}_i^{(2)}]' C_i [B_i^{(2)} - \hat{B}_i^{(1)}], \end{aligned} \tag{A.6}$$

where

$$S^{(t)} = S_1^{(t)} + S_2^{(t)}$$

and

$$\Omega_{2.1} = \Omega_{22} - \Omega_{21} \Omega_{11}^{-1} \Omega_{12}. \tag{A.6a}$$

We note in passing that $\Omega_{2.1}^{-1} = \Sigma_{22}$, the lower right hand $(s \times s)$ partition of Σ, and

$$M_i = \hat{B}_i^{(1)} - (B_i^{(2)} - \hat{B}_i^{(2)}) \Omega_{21} \Omega_{11}^{-1}. \tag{A.6b}$$

Now in Section 3, the model assumed presupposes that $B_i^{(2)}$ are known, so that we have,

$$\begin{aligned} & p\left(t, \Sigma^{-1}, B_1^{(1)}, B_2^{(1)} \mid Y_1, Y_2; X_1, X_2; B_1^{(2)} B_2^{(2)}\right) \\ & \propto \mid \Sigma^{-1} \mid^{(N-p-1)/2} \exp\left(-\frac{1}{2} \text{ tr } \Sigma^{-1} Q\right), \end{aligned} \tag{A.7}$$

where tr $\Sigma^{-1} Q$ is given by (A.6). Hence, on integrating out the $B_i^{(1)}$, we find, using properties of the $(k_i \times r)$ matric-normal density that

$$\begin{aligned} p\left(t, \Sigma^{-1} \mid \text{ data }\right) \propto & \prod_{i=1}^{2} \mid \Omega_{11} \mid^{-k/2} \mid C_i \mid^{-r/2} \mid \Sigma^{-1} \mid^{(N-p-1)/2} \\ & \cdot \exp\left(-\frac{1}{2} \text{ tr } \Omega S^{(t)}\right) \exp\left(-\frac{1}{2} \sum_{i=1}^{2} \text{ tr } \Omega_{2.1} Q_i\right), \end{aligned} \tag{A.8}$$

where

$$Q_i = [B_i^{(2)} - \hat{B}_i^{(2)}]' C_i [B_i^{(2)} - \hat{B}_i^{(2)}]. \tag{A.8a}$$

It is understood that "data" stands for $[Y_1, Y_2; X_1, X_2\colon B_1^{(2)}, B_2^{(2)}]$.

Before proceeding to integrate out Σ^{-1}, we note that

$$\begin{aligned}\Omega &= \begin{pmatrix} \Omega_{11} & \Omega_{12} \\ \Omega_{21} & \Omega_{22} \end{pmatrix} \\ &= \begin{pmatrix} 0 & 0 \\ 0 & \Omega_{2.1} \end{pmatrix} + \begin{pmatrix} I \\ \Omega_{21}\Omega_{11}^{-1} \end{pmatrix} \Omega_{11} \left(I, \Omega_{11}^{-1}\Omega_{12}\right),\end{aligned}$$

so that

$$\begin{aligned}\Sigma^{-1} &= \begin{pmatrix} \Sigma_{11} & \Sigma_{12} \\ \Sigma_{21} & \Sigma_{22} \end{pmatrix}^{-1} \\ &= \begin{pmatrix} 0 & 0 \\ 0 & \Sigma_{22}^{-1} \end{pmatrix} + \begin{pmatrix} I \\ -\Sigma_{22}^{-1}\Sigma_{21} \end{pmatrix} \Sigma_{1.2}^{-1} \left(I, -\Sigma_{12}\Sigma_{22}^{-1}\right),\end{aligned} \tag{A.9}$$

where

$$\Sigma_{1.2} = \Sigma_{11} - \Sigma_{12}(\Sigma_{22})^{-1}\Sigma_{21} = \Omega_{11}^{-1}.$$

If one makes the one-to-one transformation

$$\begin{aligned}\Lambda_{21} &= \Omega_{21}\Omega_{11}^{-1}, \\ \eta_{11} &= \Omega_{11}, \\ H_{2.1} &= \Omega_{2.1},\end{aligned} \tag{A.10}$$

one finds that the Jacobian of the transformation is $|\ \Omega_{11}\ |^{s} = |\ \eta_{11}\ |^{s}$ (see Srivastava and Khatri, 1979, page 8). In reference to the exponent in (A.8), we may show, using (A.9), that,

$$\begin{aligned}&\operatorname{tr} \Omega S^{(t)} + \operatorname{tr} \Omega_{2.1}[Q_1 + Q_2] \\ &= \operatorname{tr} H_{2.1}[S_{22}^{(t)} + Q_1 + Q_2] \\ &\quad + \operatorname{tr} \eta_{11}\{S_{1.2}^{(t)} + (\Lambda_{21} + S_{22}^{(t)-1}S_{21}^{(t)})' S_{22}^{(t)} (\Lambda_{21} + S_{22}^{(t)-1}S_{21}^{(t)})\},\end{aligned} \tag{A.11}$$

where

$$S_{1.2}^{(t)} = S_{11}^{(t)} - S_{12}^{(t)}(S_{22}^{(t)})^{-1}S_{21}^{(t)}. \tag{A.11a}$$

The $(p \times p)$ matrix $S^{(t)}$ has been partitioned as follows:

$$S^{(t)} = \begin{pmatrix} S_{11}^{(t)} & \vdots & S_{12}^{(t)} \\ \cdots & \vdots & \cdots \\ S_{21}^{(t)} & \vdots & S_{22}^{(t)} \end{pmatrix}, \tag{A.11b}$$

where $S_{21}^{(t)\prime} = S_{12}^{(t)}$, $S_{11}^{(t)}$ is $(r \times r)$, $S_{12}^{(t)}$ is $r \times s$, $r + s = p$, etc. If one notes that

$$| \Sigma^{-1} | = | \Omega | = | \Omega_{11} | \times | \Omega_{2.1} | = | \eta_{11} | \times | H_{2.1} |, \tag{A.12}$$

one may show that the posterior of the new variables defined in (A.10) and if t is:

$$\begin{aligned} & p(t, \Lambda_{21}, \eta_{11}, H_{2.1} \mid \text{ data }) \\ & \propto \ | \eta_{11} \ |^{(N-k+2s-p-1)/2} | \ H_{2.1} \ |^{(N-p-1)/2} \\ & \times | \ C_1 \ |^{-r/2} | \ C_2 \ |^{-r/2} \exp -\frac{1}{2} \text{ tr } H_{2.1}[S_{22}^{(t)} + Q_1 + Q_2] \\ & \times \exp -\frac{1}{2} \text{ tr } \eta_{11}[S_{1.2} + (\Lambda_{21} + S_{22}^{(t)-1} S_{21}^{(t)})' S_{22}^{(t)} (\Lambda_{21} + S_{22}^{(t)-1} S_{21})]. \end{aligned}$$

If one integrates out Λ_{21}, η_{11} and $H_{2.1}$ in that order, one obtains

$$\begin{aligned} p(t \mid \text{ data }) \propto & \ | \ C_1 \ |^{-r/2} | \ C_2 \ |^{-r/2} | \ S_{22}^{(t)} \ |^{-r/2} \\ & \times | \ S_{22}^{(t)} + Q_1 + Q_1 \ |^{-(N-r)/2} | \ S_{1.2}^{(t)} \ |^{-(N-k)/2}, \end{aligned}$$

which completes the proof.

REFERENCES

Broemeling, L. D., and J. H. Chin Choy (1981), "Detecting structural change in linear models." *Communications in Statistics A, Theory and Methods* **10**, 2551–2561.

Broemeling, L. D., and D. H. Moen (1984), "Testing for a change in the regression matrix of a multivariate linear model." *Communications in Statistics A, Theory and Methods* **13**, 1521–1532.

Brown, R. L., J. Durbin, and J. M. Evans (1975), "Techniques for testing the constancy of regression relationships over time." *Journal of the Royal Statistical Society*, Series B **37**, 149–192.

Guttman, I., and U. Menzetricke (1982), "On the use of loss functions in the change point problem." *Annals of the Institute of Statistical Mathematics* **34**, 319–326.

Hinkley, D. V. (1972), "Time-ordered classification." *Biometrika* **59**, 509–523.

Madalla, G. S. (1977), *Econometrics*. New York: McGraw-Hill.

Page, E. S. (1954), "Continuous inspection schemes." *Biometrika* **41**, 100–114.

Potthoff, R. F., and S. N. Roy (1964), "A generalized multivariate analysis of variance model useful especially for growth curve problems." *Biometrika* **51**, 313–326.

Rao, C. R. (1959), "Some problems involving linear hypotheses in multivariate analysis." *Biometrika* **46**, 49–58.

Sen, A. K., and M. S. Srivastava (1973), "On multivariate tests for detecting change in mean." *Sankhya*, Series A **35**, 173–186.

Srivastava, M. S. (1981), "On tests for detecting change in the multivariate mean, in statistical distributions in scientific work." *Nato Advanced Study Institute Series*, ed. C. Taille, G. P. Patil and B. Baldersari, pp. 181–190. Dordrecht, Holland: Reidel.

Srivastava, M. S., and C. G. Khatri (1979), *An Introduction to Multivariate Statistics*. New York: North-Holland.

Srivastava, M. S., and K. J. Worsley (1986), "Likelihood ratio tests for a change in the multivariate normal mean." *Journal of the American Statistical Association* **81**, 199–204.

Worsley, K. J. (1983), "Testing for a two-phase multiple regression." *Technometrics* **25**, 35–42.

James V. Bondar [1]

HOW MUCH IMPROVEMENT CAN A SHRINKAGE ESTIMATOR GIVE?

ABSTRACT

Estimation problems which are invariant under an amenable group are considered, under regularity conditions and the assumption that a unique minimum risk equivariant estimator δ_E exists. We show that for any shrinkage estimator, δ_S, whose risk is everywhere better than that of δ_E, $\delta_S(x)$ is close to $\delta_E(x)$ except on a set of x-values which is small in a sense which we make precise. Further, the values of the risk functions of δ_E and δ_S are close except on a small set.

As a corollary we show that when independent estimation problems are combined, large Stein effects occur only on small sets. In Appendix I, we show that a similar result holds if the parameter space is compact, whether the model is invariant or not.

1. INTRODUCTION

A number of authors have examined shrinkage estimators and their risk functions for invariant models; in typical examples the values of all shrinkage estimators turn out to be close to those of the classical estimator (which is often a minimum risk equivariant estimator) for large parameter values. We generalize some of these results in Section 3: the value of a shrinkage estimator must be close to that of the minimum risk equivariant estimator except on a small set of points in the sample space. This places in a general context a number of isolated observations which have appeared scattered through the literature; these observations have the moral that shrinkage estimators are unlikely to significantly improve a statistician's estimates unless he has some sort of prior information which can be used to tell what values he

[1] Department of Mathematics and Statistics, Carleton University, Ottawa, Ontario K1S 5B6

I. B. MacNeill and G. J. Umphrey (eds.), Foundations of Statistical Inference, 93–103.

should shrink toward.

An example of a situation in which such prior information might be supposed to exist is an analysis of variance where past experience with similar problems indicates that the differences between treatment means are often of the same magnitude as, or less than the standard deviations of the treatment estimates. It is then not unreasonable to shrink the treatment means toward a common mean, provided the loss function of interest is the sum of the losses for the individual treatment means. If such a plausible reason for shrinking is not present, then shrinkage must be accompanied by great good luck or there will be no substantial improvement over the classical estimator.

Theorem 1 of Section 2 shows that the values of the risk function of a shrinkage estimator are close to those of the classical estimator save on a small set. A similar result is given by Heath and Sudderth (1978), proved by the little-known and unintuitive methods of integration over non-σ-additive measures. Since both this author and some of his colleagues have had misgivings about Heath and Sudderth's proofs, we have thought it useful to give a proof using the more classical and easy to verify results of Stone (1970), which are presented in Appendix II.

The present work concerns models invariant under an amenable group; this includes most classical parametric statistical problems, except for those multivariate problems invariant under the general linear group of all nonsingular $p \times p$ matrices (this group is not amenable for $p \geq 2$). See Bondar and Milnes (1981, Section 2).

A well-known discovery of Stein is that combining several independent invariant estimation problems may allow the existence of a (non-equivariant) estimator of the parameters of the combined problem, whose risk is everywhere lower than that of the estimator obtained by simply using the best equivariant estimators for the individual estimation problems. A corollary of the result in our Section 2 is that such lowering of risk is small except on a small subset of the parameter space. Moreover, in Section 3 we show that the values of these improved estimators are close to the equivariant estimators, except on small sets.

These results on the Stein effect are related to (but weaker than) the very striking work of Professor V. M. Joshi (1979) which demonstrates that there is no Stein effect when combining problems in sampling finite populations. The analysis required for Joshi's theorems is considerably more delicate than ours, e.g., he uses prior distributions which must be chosen with much more care than our own. Furthermore, his context is different: in finite population sampling, the group is not transitive on the parameter space.

In this connection it is interesting to note the paper of Gutmann (1982) which shows that no Stein effect can occur in estimation problems (invariant or not) whose sample spaces have finitely many points. It may also be shown that no Stein effect can occur in problems whose parameter spaces have finitely many points or for which admissible implies Bayes; a proof is given in Appendix I of this paper. Now, finiteness implies compactness. Also, when the results of the present paper are specialised to compact groups, it follows from Corollary 2 that no Stein effect can occur (an already known result). When compactness is replaced by something weaker (e.g., by amenability as in the present paper) the best we can prove is a weaker result: Stein effects are small in some sense. All this suggests that compactness is a key to such results. This makes Joshi's results appear all the more striking since he doesn't have compactness. It appears that we still do not understand what are the circumstances under which Stein effects can occur, and why.

There is one more point of contact between the present work and that of Professor Joshi: in the mid-sixties Joshi (1967) and Brown (1966) discovered the existence of shrinkage-type set estimators for multivariate problems (for a brief discussion of Joshi's work on these estimators, see Hwang and Casella, 1982, page 869). Our methods can be extended to handle set estimators, and show that the improvement given by shrinking is small except on small sets.

2. LARGE STEIN EFFECTS ARE LIMITED TO SMALL SETS

We consider an invariant estimation problem (for definitions and discussion, see Lehmann, 1983, chapter 3); i.e., an invariant model (Lehmann, 1959, chapter 6) with an equivariant loss function. A brief summary of the relevant definitions follows: we are given an open set $\mathcal{X}$ in Euclidean space E^q, and a set $\{P_\theta : \theta \in \Omega\}$ of probabilities on $\mathcal{X}$. Let a locally compact topological group G act on $\mathcal{X}$ (i.e. there is a continuous multiplication $(g,x) \to gx$ from $G \times \mathcal{X}$ to $\mathcal{X}$ such that $g_1(g_2x) = (g_1g_2)x$ and $ex = x$ for all $g_i \in G$ and $x \in \mathcal{X}$ where e is the identity of G. We define an action of G on Ω by defining $g\theta : P_{g\theta}(E) = P_\theta(g^{-1}E)$ for all $g \in G$, $\theta \in \Omega$ and measurable $E \subset G$; we require $g\theta$ to lie in Ω for all $\theta \in \Omega$ and $g \in G$. For all θ and d in Ω, let the loss $L(\theta, d)$ be the loss resulting from getting the estimate d when the true value of the parameter is θ. Let L be equivariant, i.e. $L(\theta, d) = L(g\theta, gd)$ for all θ, $d \in \Omega$ and $g \in G$. L is also required to be non-negative and Borel measurable on $\Omega \times \Omega$. This completes the definition of the *invariant estimation problem*.

We shall restrict ourselves to the case in which G acts exactly transitively on $\mathcal{X}$ and Ω, i.e., that given $x_1, x_2 \in \mathcal{X}$ there exists precisely one $g \in G$ such that $gx_1 = x_2$, and given $\theta_1, \theta_2 \in \Omega$ there exists precisely one $g' \in G$ such that $g'\theta_1 = \theta_2$. This is equivalent to saying that $G = \mathcal{X} = \Omega$. The exact transitivity of G on $\mathcal{X}$ can be dropped at the expense of adding some technical topological conditions that are annoyingly difficult to state. The requirement of exact transitivity on Ω is essential to our results. We shall also assume that each P_θ is dominated by a right invariant measure ν on G.

An *estimator*, δ, will be any Borel-measurable function from $\mathcal{X}$ to Ω. Its risk at θ is

$$R(\delta,\theta) = \int L(\theta,\delta(x))dP_\theta(x). \tag{2.1}$$

The estimator δ is *equivariant* if $\delta(gx) = g\delta(x)$ for all $x \in \mathcal{X}$ and $g \in G$; it is well-known that the risk of such an estimator does not depend on θ. We shall assume that our estimation problem has a unique equivariant estimator δ_E of lowest risk. In addition to the standard definitions above, we make one not-so-standard (but useful, we believe) definition: A *shrinkage estimator* is an estimator δ_S such that

$$R(\delta_S,\theta) \leq R(\delta_E,\theta) \ \text{ for all } \ \theta \in \Omega, \tag{2.2}$$

with strict inequality for at least one value of θ. Of course, a shrinkage estimator cannot be equivariant.

Let G be an *amenable* group, i.e., there exists a sequence $\{G_n : n = 1,2,\ldots\}$ of compact measurable subsets of G such that $\bigcup_{n=1}^{\infty} G_n = G$, $G_n \subset G_{n+1}$ (all n) and for each compact $K \subset G$,

$$\frac{\nu(G_nK \cap G_n)}{\nu(G_n)} \to 1 \ \text{ as } \ n \to \infty. \tag{2.3}$$

(G_nK is defined as $\{gk \mid g \in G_n,\ k \in K\}$; ν is a right invariant measure on G. Such a sequence $\{G_n\}$ is called a *summing sequence*).

If F is any real-valued measurable function on Ω, if $\{G_n\}$ is a summing sequence and the limit as $n \to \infty$ of the sequence

$$\frac{1}{\nu(G_n)} \int_{G_n} F(\theta)d\nu(\theta) \ \ (n = 1,2,3,\ldots) \tag{2.4}$$

exists, then (2.4) will be called the *average* of F. (This average is known in the harmonic analysis literature as the value of an invariant mean at F). A set $E \subset G$ will be said to be of *zero density* if the average of its indicator function is zero. Any compact set is of zero density, but the converse is not true. Nevertheless, for a set to be of zero density means that it is small in a certain sense. (To speak fancifully, if one could imagine oneself standing back infinitely far from G and E, E is of zero density if and only if it is too small to be visible at such a distance).

Example 1 (Estimating a multivariate normal mean with known covariance): Let X be an observation with distribution $N_p(\boldsymbol{\mu}, I)$ with unknown $\boldsymbol{\mu}$. If we wish to estimate μ and our loss $L(d, \mu)$ is a function of $d - \mu$, then we have an invariant estimation problem with $\mathcal{X} = \Omega = E^p$, and $G = E^p$ with addition as the "multiplication" in G. A convenient summing sequence for G is given by choosing G_n to be the parallelepiped centered at the origin with sides of length $2n$.

Example 2 (Estimating a univariate mean in a location and scale model): Let (X, S) be the mean and standard deviation respectively of a random sample from a univariate distribution with probability density function $f(x_i) = \frac{1}{\sigma} f\left(\frac{x_i - \mu}{\sigma}\right)$ where f is itself a density; μ and σ are unknown.

Now we have an invariant statistical model with $X = \Omega = G$ where $G = \{(\mu, \sigma) \mid \sigma > 0, \ -\infty < \mu < \infty\}$ and the action of G on X is given by: $(\mu, \sigma)(x, s) = (\mu + \sigma x, \sigma s)$ and the multiplication in G is $(\mu_1, \sigma_1)(\mu_2, \sigma_2) = (\mu_1 + \sigma_1 \mu_2, \sigma_1 \sigma_2)$. A convenient choice of G_n is the convex hull of the points $(n, n^2), (n, 1), (1, 1/n), (-n, n^2), (-n, 1)$ and $(-1, 1/n)$ in the μ-σ plane.

If we attempt to estimate μ with a loss function of the form $L(\mu, \sigma, d) = L((d - \mu)/\sigma)$, an estimator of μ takes values in R, not G, hence we don't have an estimation problem of the type discussed above. We can use the following device: if δ is an estimator of μ, we consider the associated estimator δ' with $\delta'(x) = (\delta(x), s)$ which *does* take values in G. This device can be used in problems of estimating one co-ordinate of a parameter θ.

For this problem, we are looking at the case in which there is a two-dimensional sufficient statistic. We can examine the general case, but this means that $G \neq X$, and as we remarked earlier we will not give this generalization here.

Example 3: Given two independent invariant estimation problems $(\mathcal{X}_i, \Omega_i, G_i, L_i)$, $\mathcal{X}_i = \Omega_i = G_i$ $(i = 1, 2)$; if we wish to combine them and estimate $(\theta_1, \theta_2) \in \Omega_1 \times \Omega_2$ with loss function $L((\theta_1, \theta_2), (d_1, d_2)) = L(\theta_1, d_1) + L(\theta_2, d_2)$, then the combined problem is also an invariant estimation problem with group $G = G_1 \times G_2$ and action $(g_1, g_2)(x_1, x_2) = (g_1 x_1, g_2 x_2)$ on $\mathcal{X} = \mathcal{X}_1 \times \mathcal{X}_2$. If G_1 and G_2 are amenable, then so is G; one can use $G_n = G_{1n} \times G_{2n}$, to use the obvious notation. It is clear that if each individual problem has a minimum risk equivariant estimator δ_{iE}, then the minimum risk equivariant estimator for the combined problem is $(\delta_{1E}, \delta_{2E})$ with risk $R(\delta_{1E}) + R(\delta_{2E})$.

Theorem 1. If an invariant estimation problem satisfies the assumptions at the beginning of this section, if G is amenable, such that $G = \mathcal{X} = \Omega$, and if a minimum risk equivariant estimator δ_E exists, *then* the average risk of any shrinkage estimator is equal to the risk of δ_E.

Proof. First we consider the case of bounded L. Pick a summing sequence

$\{G_n\}$ for G and define the probability π_n on G as follows:

$$\pi_n(E) = \frac{\nu(E \cap G_n)}{\nu(G_n)}, \quad E \subset G.$$

Let δ_S be any shrinkage estimator. We have

$$\int R(\theta, \delta_S) d\pi_n(\theta) = \int \int L(\theta, \delta_S(x)) dP_\theta(x) d\pi_n(\theta);$$

by Fubini this becomes the (well-known) expression

$$\int_x \int_\theta L(\theta, \delta_S(x)) dB_n(\theta \mid x) dP_n^X(x), \tag{2.5}$$

where $B_n(\cdot \mid x)$ is the Bayes posteror for θ given observation x and prior π_n; P_n^X is the marginal distribution of X derived from the Bayesian joint distribution for (X, θ) when θ is given the prior distribution π_n. Of course, the inner integral is with respect to θ.

By Theorem 3 of Appendix II, we see that (2.5) converges to

$$\int_x \int_\theta L(\theta, \delta_S(x)) dB_\nu(\theta \mid x) dP_n^X(x), \tag{2.6}$$

where $B_\nu(\cdot \mid x)$ is the Bayes posterior given observation x and the (perhaps improper) prior ν.

Now, the inner integral in (2.6) (the one with respect to θ) is greater than $\int_\theta L(\theta, \delta_E(x)) dB_\nu(\theta \mid x)$, since $\delta_E(x)$ is precisely that value of d which minimizes this last integral (Lehmann, 1983, chapter 3) and this last integral is equal to the risk of δ_E ($R(\delta_E)$, say). Integrating this inequality with respect to P_n^X gives us that (2.6) is at least equal to $R(\delta_E)$. Hence we have proved

$$\int R(\theta, \delta_S) d\pi_n \to (2.6) \geq R(\delta_E).$$

However, by the definition of δ_S, we know that $R(\theta, \delta_S) \leq R(\delta_E)$, hence $\int R(\theta, \delta_S) d\pi_n(\theta) \leq R(\theta_E)$. This proves that $\int R(\theta, \delta_S) d\pi_n(\theta)$ has a limit and this is $R(\delta_E)$, which proves the theorem for bounded L.

For unbounded L, we replace L by the truncated loss function $L_B(\theta, d) = \min\{L(\theta, d), B\}$, apply the result above and let $B \to \infty$. (Note that $R_B(\theta, \delta) \to R(\theta, \delta)$ by the monotone convergence theorem, and that by Theorem 3 of the appendix, the difference between $\int R_B(\theta, \delta_S) d\pi_n(\theta)$ and (2.6) is $\leq B\epsilon$ uniformly in L.) Q.E.D.

Corollary 1. Under the conditions of Theorem 1, for any $\epsilon > 0$, $\{\theta : R(\delta_E) \geq R(\theta, \delta_S) + \epsilon\}$ is a set of zero density.

Proof. If the set in question has density D, the average of $R(\theta, \delta_S)$ is $\leq R(\delta_E) - \epsilon D$, which contradicts Theorem 1. Q.E.D.

If Corollary 1 is applied to Example 3, the following corollary is obtained.

Corollary 2. In the situation of Example 3, if G_1 and G_2 are amenable, then combining the two estimation problems and using the loss function $L(\theta_1, d_1) + L(\theta_2, d_2)$, gives the best equivariant estimator $(\delta_{1E}, \delta_{2E})$. Any other estimator $(\delta_{1S}, \delta_{2S})$ which improves on this one will have a risk less than $R(\delta_{1e}) + R(\delta_{2E}) - \epsilon$ only on a set of zero density in $\Omega \times \Omega$.

It is worth noting that Corollary 2 may be generalized to the combination of any finite number of invariant estimation problems.

3. SHRINKAGE AND CLASSICAL ESTIMATORS ARE CLOSE EXCEPT ON SMALL SETS

In this section we consider an invariant estimation problem with the same assumptions as in Section 2, plus the following additional assumptions:

A. For some compact subset K in G, $P_{\theta=e}[K] = 1$. (This assumption may be weakened, which we hope to do in a future paper.)

B. If $c < \sup_{\theta,d} L(\theta, d)$, then $\{\theta^{-1}d : L(\theta, d) \leq c\}$ is a compact set in G.

We also assume as before, that there is a minimum risk equivariant estimator δ_E (which is unique since G is the sample space). Since estimators δ take values in G, such expressions as $(\delta(x))^{-1}\delta_E(x)$ are meaningful, and measure the distance between the values $\delta(x)$ and $\delta_E(x)$ (if the expression is close to e, then $\delta(x)$ and $\delta_E(x)$ are close).

Theorem 2. For an invariant estimation problem satisfying the assumptions of Theorem 1 and also the assumptions A and B above, if δ_S is any shrinkage estimator, then for any open neighbourhood U of e, the set of x such that $(\delta_S(x))^{-1}\delta_E(x) \notin U$ is a set of zero density.

Proof. By Theorem 1.1 of Bondar and Milnes (1981), G has property H_1, that is, there exists a sequence $\{G_n\}$ of closed sets such that

$$\frac{\nu(\cap_{k\in K} G_n k)}{\nu(G_n)} \longrightarrow 1 \tag{3.1}$$

for every compact subset K of G. We shall denote $\cap_{k\in K} G_n k$ by $G_n[K]$, and note that $G_n[K] = \{h : hK^{-1} \subset G_n\}$. We pick such a sequence $\{G_n\}$ and define the probability π_n by: $\pi_n(E) = \nu(E \cap G_n)/\nu(G_n)$ for $E \subset G$.

Let $B_n(\cdot \mid x)$ and $B_\nu(\cdot \mid x)$ be the posterior probabilities and P_n^X be the marginal distribution of X when π_n is used as a prior. Then the posterior risk given x when d is used as an estimate and ν as a prior is

$$\int_\theta L(\theta, d) dB_\nu(\theta \mid x) = R_\nu(d \mid x), \text{ say.} \tag{3.2}$$

It is well-known that (3.2) is minimized and equals $R(\delta_E)$ when $d = \delta_E(x)$, and that this is the unique value of d which gives a minimum. Because of this uniqueness and assumption B, it follows that for any open set U containing e, there is a $\gamma > 0$ such that $d^{-1}\delta_E(e) \notin U$ implies that $\int L(\theta, d) dB_\nu(\theta \mid x = e) > R(\delta_E) + \gamma$. Since $dB_\nu(g\theta \mid gx) = dB_\nu(\theta \mid x)$ for all x, θ and g (see Stone, 1970, p. 1349) and since L is also equivariant, the value of (3.2) depends only on $d^{-1}\delta_E(x)$. Therefore, for any open set U containing e, there is a $\gamma > 0$ such that $d^{-1}\delta_E(x) \notin U$ implies that (3.2) is greater than $R(\delta_E) + \gamma$, for all x.

Now, if δ_S is any shrinkage estimator, then

$$\int R(\theta, \delta_S) d\pi_n(\theta) = \int_G \int_G L(\theta, \delta_S(x)) dB_n(\theta \mid x) dP_n^X(x)$$

(by (2.5))

$$\begin{aligned} &\geq \int_{x \in G_n[K]} \int_G L(\theta, \delta_S(x)) dB_n(\theta \mid x) dP_n^X(x) \\ &= \int_{x \in G_n[K]} R_\nu(\delta_S(x) \mid x) dP_n^X(x) \end{aligned} \tag{3.3}$$

since $B_n(\cdot \mid x) = B_\nu(\cdot \mid x)$ for $x \in G_n[K]$.

Next we assume that this theorem is incorrect, i.e., that there is an open set U containing e such that $S = \{x : (\delta_S(x))^{-1}\delta_E(x) \notin U\}$ does not have density zero. This means that the sequence $\nu(G_n \cap S)/\nu(G_n)$, $n \geq 1$ has a subsequence converging to a value $D > 0$.

Note that the probability measure P_n^X gives no probability to sets outside $G_n K$, and has the uniform density $\nu(G_n)^{-1}$ (with respect to ν) over the set $G_n[K]$. It then follows that $P_n^X[G_n[K]] = \nu(G_n[K])\nu(G_n)^{-1} \to 1$ by display (3.1). Similarly, $P_n^X[S] \geq \nu(G_n \cap S)\nu(G_n)^{-1}$ which has a subsequence converging to $D > 0$.

Then $R_\nu(\delta_S(x) \mid x) > \gamma$ for $x \in S \cap G_n[K]$, so (3.3) is

$$\begin{aligned} &\geq \int_{S \cap G_n[K]} R_\nu(\delta_S(x) \mid x) dP_n^X(x) + \int_{G_n[K]-S} R_\nu(\delta_S(x) \mid x) dP_n^X(x) \\ &\geq (R(\delta_E) + \gamma) P_n^X[S \cap G_n[K]] + R(\delta_E) P_n^X[G_n[K] - S]. \end{aligned}$$

There is a subsequence of these numbers which, for sufficiently large n, becomes

$$\geq (R(\delta_E) + \gamma)(1 - D) + R(\delta_E)D$$
$$= R(\delta_E) + \gamma D$$

which contradicts the fact that δ_S is a shrinkage estimator, for which $\int R(\theta, \delta_S)d\pi_n(\theta) \leq R(\delta_E)$ for every $n \geq 1$. Q.E.D.

APPENDIX I: IF ADMISSIBLE IMPLIES BAYES, THEN STEIN EFFECTS ARE LIMITED

Suppose we have two estimation problems $(X_i, \Omega_i, \{P_{i\theta}\}, L_i)$, $i = 1, 2$; (our notation is explained in Section 2, especially in Example 3), and we wish to combine them to estimate (θ_1, θ_2) in the combined model $(X_1 \times X_2, \Omega_1 \times \Omega_2, \{P_{1\theta} \times P_{2\theta}\})$ using the loss function $L_1(\delta_1(x_1), \theta_1) + L_2(\delta_2(x_2), \theta_2)$. Let each δ_i be admissible for estimating θ_i in the ith model. If we are in a context in which all admissible estimators are proper Bayes, then δ_i is Bayes for some prior π_i. Therefore the posterior risk given x, namely

$$R_i(d_i \mid x_i) = \int L_i(d_i, \theta_i) dB_{\pi i}(\theta_i \mid x_i)$$
$$= \int L(d_i, \theta_i) \frac{f_i(x_i; \theta_i)}{\int f_i d\pi_i} d\pi_i(\theta_i)$$

is minimized by setting $d_i = \delta_i(x_i)$ a.e. (m) (f_i is the density of $P_{i\theta}$ with respect to a dominating measure m). Next, put the prior $\pi_1 \times \pi_2$ on the combined model; the posterior risk given (x_1, x_2), namely $R(d_1, d_2 \mid x_1, x_2)$, simplifies to give $R_1(d_1 \mid x_1) + R_2(d_2 \mid x_2)$. Clearly this sum is minimized by setting $d_1 = \delta_1(x_1)$ and $d_2 = \delta_2(x_2)$.

Since the value of the combined risk function integrated with respect to $d\pi_1(\theta_1)d\pi_2(\theta_2)$ is equal to $\int R(\delta_1(x_1), \delta_2(x_2) \mid x_1, x_2)dP_1^X(x_1)dP_2^X(x_2)$, where P_i^X is the (Bayesian) marginal distribution of x_i for model i, therefore there cannot be any estimator for the combined problem whose risk function is everywhere strictly less than that of (δ_1, δ_2). In particular, by Wald (1950, Theorem 5.2), our assumption that admissible estimators are proper Bayes is satisfied whenever the parameter space is finite, and Wald's regularity conditions 3.1 to 3.7 hold. The same argument holds for combining more than two problems.

APPENDIX II: THE THEOREM OF STONE

We repeat here a result of Stone (1970) and derive some consequences needed for Section 2. We repeat part of Stone's paper since his original statement of his results was in the form of a necessary and sufficient condition; he introduced conditions which are unnecessary for the result we need, but are required to make its converse hold. We shall state the result in the form needed, without these extra conditions, and derive a consequence of Stone's main theorem.

We require an invariant statistical model satisfying the assumptions spelled out in Section 2, except that the condition that G be amenable is here replaced by the more general condition that G be σ-compact. Let $P_\nu(A \mid x)$ denote the posterior probability of $A \subset \Omega$ given the observation x, when the right invariant measure ν is used as a prior distribution (such a posterior exists a.e. (P_θ) for every θ). A sequence $\{\pi_n : n = 1, 2, \ldots\}$ of probability measures on G is said to be *asymptotically right invariant* if for every measurable subset $E \subset G$, $\mid \pi_n(E) - \pi_n(Eg) \mid \to 0$ as $n \to \infty$, uniformly in g on each compact set in G. Let $P_n(\cdot \mid \cdot)$ be the posterior probability measure corresponding to the prior π_n.

Define

$$d_n(x) = \sup_{A \subset \Omega} \mid P_n(A \mid x) - P_\nu(A \mid s) \mid$$

which measures the closeness of P_n and P_ν. Now consider the observation X to have the marginal distribution obtained by giving x the distribution P_θ, giving θ the distribution π_n (x and θ independent), and then integrating out θ, i.e., we are looking at the marginal Bayes distribution P_n^X for x. We will say that $\{\pi_n\}$ induces *convergence in probability to* P_ν if $\text{plim}_{n\to\infty} d_n(x) = 0$, i.e., for all $\epsilon > 0$, $P_n^X[d_n(x) \leq \epsilon] \to 1$.

Stone's Theorem: given an invariant statistical model satisfying the conditions of Section 2, if $\{\pi_n : n = 1, 2, \ldots\}$ is asymptotically right invariant and $\pi_n << \nu$ for all n, then $\{\pi_n\}$ induces convergence in probability to P_ν.

The proof of this theorem consists of Lemma 3 of Stone's paper. Note that his conditions (i) and (ii) are needed only for his lemmas 1 and 2, which are not a part of the proof of the theorem above.

It is worth noting that a sequence $\{\pi_n\}$ satisfying Stone's theorem will exist if and only if G is amenable (the existence of such a sequence is called condition P in Bondar and Milnes (1981, p. 108); by their Theorem 1.1 this is equivalent to amenability).

Theorem 3. If $\{\pi_n\}$ satisfies the conditions of Stone's theorem, and F :

$G \times G \to R$ is a non-negative bounded measurable function, then

$$| \int_x \int_\theta F(\theta, x) dP_n(\theta \mid x) dP_n^X(x) - \int_x \int_\theta F(\theta, x) dP_\nu(\theta \mid x) dP_n^X(x) | \quad (A.1)$$

converges to zero as $n \to \infty$. Furthermore, if the supremum of $| F |$ is B, say, then for every $\epsilon > 0$ there is an N (independent of F) such that the value of (A.1) is $\leq B\epsilon$.

Proof.

$$| \int_\theta F(\theta, x) dP_n(\theta \mid x) - \int_\theta F(\theta, x) dP_\nu(\theta \mid x) |$$
$$\leq B d_n(x) \quad \text{for each } x.$$

Therefore, the expression (A.1) is

$$\leq B \int d_n(x) dP_n^X(x)$$
$$= B\epsilon P_n^X[d_n(X) \leq \epsilon] + 2BP_n^X[d_n(X) > \epsilon]$$

(since $d_n(x) \leq 2$ for all x). Now, the second term in this last expression tends to zero by Stone's theorem. Q.E.D.

REFERENCES

Bondar, J., and P. Milnes (1981), "Amenability: a survey for statistical applications of Hunt-Stein and related conditions on groups." *Zeitschrift für Wahrscheinlichkeitstheorie und Verwandte Gebiete* **57**, 103–128.

Brown, L. D. (1966), "On the admissibility of invariant estimators of one or more location parameters." *Annals of Mathematical Statistics* **37**, 1087–1136.

Gutmann, S. (1982), "Stein's paradox is impossible with finite sample space." *Annals of Statistics* **10**, 1017–1020.

Heath, D., and W. Sudderth (1978), "On finitely additive priors, coherence, and extended admissibility." *Annals of Statistics* **6**, 333–345.

Hwang, J. T., and G. Casella (1982), "Minimax confidence sets for the mean of a multivariate normal distribution." *Annals of Statistics* **10**, 868–881.

Joshi, V. M. (1967), "Inadmissibility of the usual confidence sets for the mean of a multivariate normal distribution." *Annals of Mathematical Statistics* **38**, 1868–1875.

Joshi, V. M. (1979), "Joint admissibility of the sample means as estimators of the means of finite populations." *Annals of Statistics* **7**, 995–1002.

Lehmann, E. (1959), *Testing Statistical Hypotheses*. New York: Wiley and Sons.

Lehmann, E. (1983), *Theory of Point Estimation*. New York: Wiley and Sons.

Stone, M. (1970), "Necessary and sufficient condition for convergence in probability to invariant posterior distributions." *Annals of Mathematical Statistics* **41**, 1349–1353.

Wald, A. (1950), *Statistical Decision Functions*. Reprinted by Chelsea.

A. K. Md. Ehsanes Saleh [1] and Pranab Kumar Sen [2]

ON SHRINKAGE AND PRELIMINARY TEST M-ESTIMATION IN A PARALLELISM PROBLEM

ABSTRACT

In a multi-sample simple regression model for which homogeneity of the regression slopes is plausible, shrinkage as well as preliminary test versions of M-estimators of the intercepts are considered. In the light of the asymptotic distributional risk, the relative dominance pictures of these estimators are studied.

1. INTRODUCTION

Consider p (≥ 2) independent samples. For each i $(= 1, \ldots, p)$, let $Y_{i1}, \ldots, Y_{in_i}$ be independent random variables (r.v.) with distribution functions (d.f.) $F_{i1}, \ldots, F_{in_i}$, respectively, all defined on the real line R. It is assumed that

$$F_{ij}(x) = F\left(x - \theta_i - \beta_i c_{ij}\right), \quad x \in R, \; 1 \leq j \leq n_i, \; 1 \leq i \leq p, \tag{1.1}$$

where the c_{ij} are known (regression) constants, F is an (unknown) d.f., and $\boldsymbol{\theta} = (\theta_1, \ldots, \theta_p)'$ and $\boldsymbol{\beta} = (\beta_1, \ldots, \beta_p)'$ are unknown vectors (parameters) of *intercepts* and *slopes*, respectively. We are primarily interested in the estimation of $\boldsymbol{\theta}$ when it is suspected that the β_i are all equal (i.e., the regression lines are all parallel). Such a problem arises in a parallel line bio-assay model (Sen, 1971). For this problem, based on the classical *least squares* and *rank* $(R-)$ *estimators*, the theory of *preliminary test estimation* (PTE) has been

[1] Department of Mathematics and Statistics, Carleton University, Ottawa, Ontario K1S 5B6

[2] Department of Biostatistics, Rosenau Hall 201 H, The University of North Carolina at Chapel Hill, Chapel Hill, North Carolina 27514

I. B. MacNeill and G. J. Umphrey (eds.), Foundations of Statistical Inference, 105–119.

studied by Lambert *et al.* (1985) and Akritas *et al.* (1985). Parallel theory of *shrinkage least squares* and *R-estimation* of $\boldsymbol{\theta}$ has also been developed by Saleh and Sen (1985a,b). Earlier, for multivariate location and general estimable parameters, *shrinkage U-statistics* and *R-estimators* were studied by Sen (1984) and Sen and Saleh (1984, 1985). The object of the current study is to focus on the PTE and shrinkage versions of *M-estimators* of $\boldsymbol{\theta}$. Along with the preliminary notions, the proposed estimators are considered in Section 2. The relevant asymptotic theory, developed in Section 3, is incorporated in Section 4 in the formulation of *asymptotic distributional risk* (ADR) results. In the light of the ADR, in the concluding Section, the *asymptotic dominance* and *relative efficiency* of these M-estimators are studied. It is shown that none of the PTE and shrinkage M-estimators may dominate over the other, though each of them fares well relative to the classical M-estimator.

2. PROPOSED SHRINKAGE AND PTE VERSIONS OF M-ESTIMATORS

To define the M-estimators, first, we need to introduce the allied *score functions*. Let $\psi : R \rightarrow R$ be defined by

$$\psi(x) = \psi_1(x) + \psi_2(x), \qquad x \in R, \tag{2.1}$$

where ψ_1 and ψ_2 are both nondecreasing and skew-symmetric; ψ_1 is *absolutely continuous* on any bounded interval in R (and is square integrable with respect to F), while ψ_2 is a *step-function* having finitely many jumps. That is, for some m (≥ 0), there exist open intervals $J_r = (a_r,\ a_{r+1})$, $0 \leq r \leq m$ and real (and finite) numbers d_r, $0 \leq r \leq m$, such that $\psi_2(x) = d_r$, $\forall x \in J_r$, $0 \leq r \leq m$ (where $-\infty = a_0 < \cdots < a_m < a_{m+1} = +\infty$, and conventionally, we let $\psi_2(a_2) = (d_r + d_{r+1})/2$, $r = 1, \ldots, m$; for $m = 0$, $\psi_2 \equiv 0$). Also, we assume that the d.f. F in (1.1) is symmetric about the origin, so that

$$\overline{\psi} = \int_R \psi(x) dF(x) = 0. \tag{2.2}$$

Actually, we assume that the d.f. F has an absolutely continuous probability density function (pdf) f with a finite *Fisher information*

$$I(f) = \int_R \left(f'(x)/f(x)\right)^2 dF(x), \tag{2.3}$$

where $f'(x) = (d/dx)f(x)$ exists almost everywhere (a.e.).

Now, for each i $(= 1, \ldots, p)$ and real (a, b), we define

$$M_{1,n_i}^{(i)}(a,b) = \sum_{j=1}^{n_i} \psi(Y_{ij} - a - bc_{ij}),$$
$$M_{2,n_i}^{(i)}(a,b) = \sum_{j=1}^{n_i} c_{ij}\psi(Y_{ij} - a - bc_{ij}). \tag{2.4}$$

Then, for each i $(= 1, \ldots, p)$, $\left(\tilde{\theta}_{n_i}^{(i)}, \tilde{\beta}_{n_i}^{(i)}\right)$, the M-estimator of (θ_i, β_i) in (1.1), based on the score function ψ, is given by the solution of the equations:

$$M_{1,n_i}^{(i)}(a,b) = 0 = M_{2,n_i}^{(i)}(a,b). \tag{2.5}$$

For this purpose, it is equivalent to consider

$$\tilde{M}_{1,n_i}^{(i)}(a,b) = \sum_{j=1}^{n_i} (c_{ij} - \bar{c}_i)\psi(Y_{ij} - a - bc_{ij}), \tag{2.6}$$

where $\bar{c}_i = n_i^{-1}\sum_{j=1}^{n_i} c_{ij}$, $1 \le i \le p$. (Note that $M_{1,n_i}^{(i)}(a,b) + \mathbf{M}_{2,n_i}^{(i)}(a,b) = M_{2,n_i}^{(i)}(a,b)$, $\forall a \in R,\ b \in R,\ i = 1, \ldots, p$). Thus, (2.5) may also be written as:

$$M_{1,n_i}^{(i)}(a,b) = 0 = \tilde{M}_{2,n_i}^{(i)}(a,b), \qquad 1 \le i \le p. \tag{2.7}$$

These *unrestrained M-estimators* are denoted by

$$\tilde{\boldsymbol{\theta}}_n = (\tilde{\theta}_{n_1}^{(1)}, \ldots, \tilde{\theta}_{n_p}^{(p)})' \text{ and } \tilde{\boldsymbol{\beta}}_n = (\tilde{\beta}_{n_1}^{(1)}, \ldots, \tilde{\beta}_{n_p}^{(p)}), \tag{2.8}$$

where $n = n_1 + \cdots + n_p$ is the total sample size.

Next, assuming that $\beta_1 = \cdots = \beta_p = \beta$ (unknown), to estimate β, we consider the statistic

$$\tilde{M}_{2,n}^{*}(\mathbf{a}, b) = \sum_{i=1}^{p} \tilde{M}_{2,n_i}(a_i, b)$$
$$= \sum_{i=1}^{p}\sum_{j=1}^{n_i} (c_{ij} - \bar{c}_i)\psi(Y_{ij} - a_i - bc_{ij}), \tag{2.9}$$

where $\mathbf{a} \in R^p$ and $b \in R$. Then, $\hat{\boldsymbol{\theta}}_n, \hat{\beta}_n$, the *restrained M-estimator* of $\boldsymbol{\theta}$ and β (assuming that $\beta_1 = \cdots = \beta_p = \beta$), is given by the solution of

$$M_{1,n_i}^{(i)}(a_i, b) = 0, \qquad 1 \le i \le p,$$
$$\tilde{M}_{2,n_i}^{*}(\mathbf{a}, b) = 0. \tag{2.10}$$

To introduce the PTE and shrinkage versions, we need to consider the *residual mean square* (score) *error*:

$$s_n^2 = (n-2p)^{-1} \sum_{i=1}^{p} \sum_{j=1}^{n_i} \psi^2 \left(Y_{ij} - \hat{\theta}_{n_i}^{(i)} - \hat{\beta}_{n_i}^{(i)} c_{ij} \right). \tag{2.11}$$

An *M-test* for $H_0 : \beta_1 = \cdots = \beta_p = \beta$ (against the set of alternatives that the β_i are not all equal) may be framed along the lines of Sen (1982) and Singer and Sen (1985). Consider the aligned statistics

$$\tilde{M}_{2,n_i}^{(i)} \left(\tilde{\theta}_{n_i}^{(i)}, \hat{\beta}_n \right) = \hat{M}_{2,n_i}^{(i)}, \quad \text{say}, \quad 1 \le i \le p. \tag{2.12}$$

Also, let

$$C_{i,n_i}^2 = \sum_{j=1}^{n_i} \left(c_{ij} - \bar{c}_i \right)^2, \qquad 1 \le i \le p. \tag{2.13}$$

Then, we may consider the test-statistic (along the lines of Sen, 1969):

$$L_n = S_n^{-2} \left(\sum_{i=1}^{p} C_{i,n_i}^{-2} \left(\hat{M}_{1,n_i}^{(i)} \right)^2 \right). \tag{2.14}$$

For the time being, we denote the upper $100\alpha\%$ point of the null hypothesis distribution of L_n by $\ell_{n,\alpha} (0 < \alpha < 1)$; we shall make more comments on $\ell_{n,\alpha}$ later on. Then, corresponding to a *level of significance* α (of the preliminary test), the PTE version of the M-estimator of $\boldsymbol{\theta}$ (i.e., *PTME* of $\boldsymbol{\theta}$) may be defined as

$$\hat{\boldsymbol{\theta}}_n^{PT} = I\left(L_n < \ell_{n,\alpha}\right) \hat{\boldsymbol{\theta}}_n + I\left(L_n \ge \ell_{n,\alpha}\right) \tilde{\boldsymbol{\theta}}_n, \tag{2.15}$$

where $I(A)$ denotes the indicator function of the set A. Generally, the $\tilde{\boldsymbol{\theta}}_n$ have larger variability than $\hat{\boldsymbol{\theta}}_n$, and the PTE version attempts to control this by having the sample information on the homogeneity of the β_i through L_n.

To present the *shrinkage M-estimator* (i.e., *SME*) of $\boldsymbol{\theta}$, first, we consider a James-Stein (1961) type M-estimator of $\boldsymbol{\beta}$:

$$\begin{aligned} \hat{\boldsymbol{\beta}}_n^{JS} &= (\hat{\beta}_{1,n}^{JS}, \ldots, \hat{\beta}_{p,n}^{JS})' \\ &= \tilde{\boldsymbol{\beta}}_n - (p-3) L_n^{-1} (\tilde{\boldsymbol{\beta}}_n - \hat{\beta}_n \mathbf{1}_p) \\ &= \hat{\beta}_n \mathbf{1}_p + \{1 - (p-3) L_n^{-1}\} (\tilde{\boldsymbol{\beta}}_n - \hat{\beta}_n \mathbf{1}_p), \end{aligned} \tag{2.16}$$

where $\mathbf{1}_p = (1, \ldots, 1)'$. Note that for L_n close to 0, $\hat{\boldsymbol{\beta}}_n^{JS}$ may not behave very smoothly. To avoid this shortcoming, we consider a truncated version

$$\hat{\boldsymbol{\beta}}_n^{S} = (\hat{\beta}_{1,n}^{S}, \ldots, \hat{\beta}_{p,n}^{S}) = \hat{\beta}_n \mathbf{1}_p I\left(L_n < \varepsilon\right) + \hat{\boldsymbol{\beta}}_n^{JS} I\left(L_n \ge \varepsilon\right) \tag{2.17}$$

where ε is an arbitrary positive number, usually to be taken small. Then, the proposed SME of $\boldsymbol{\theta}$ is defined as the solution of

$$M_{1,n_i}^{(i)}(a_i, \hat{\beta}_{i,n}^{s}) = 0, \qquad 1 \le i \le p. \tag{2.18}$$

We denote the SME of $\boldsymbol{\theta}$ by $\hat{\boldsymbol{\theta}}_n^S = (\hat{\theta}_{1,n}^S, \ldots, \hat{\theta}_{p,n}^S)$. Our basic goal is to study the properties of $\tilde{\boldsymbol{\theta}}_n, \hat{\boldsymbol{\theta}}_n, \hat{\boldsymbol{\theta}}_n^{PT}$, and $\hat{\boldsymbol{\theta}}_n^S$ and to comment on their relative (asymptotic) dominance pictures when H_0 may or may not hold.

3. ASYMPTOTIC DISTRIBUTIONAL RESULTS

Let us define

$$\sigma_\psi^2 = \int_R \psi^2(x)dF(x), \tag{3.1}$$

$$\gamma = \gamma(\psi, F) = \int_R \psi(x)\left\{-f'(x)/f(x)\right\}dF(x). \tag{3.2}$$

Note that by (2.3), γ is finite. It is also possible to rewrite γ as

$$\gamma = \int_R \psi_1'(x)dF(x) + \sum_{r=1}^{m}(d_r - d_{r-1})\, f(a_r), \tag{3.3}$$

(see Sen, 1981, Section 8.2), where for $m = 0$, the second term drops out), so that by the nondecreasing character of ψ_1, ψ_2, γ is a positive number.

Next, borrowing the basic ideas from Sen and Salah (1985), we may argue that if $H_0 : \beta_1 = \cdots = \beta_p = \beta$ does not hold, and we have a fixed alternative hypothesis for which $\beta_i \neq \beta_{i'}$ for at least one pair $(i, i') : 1 \le i < i' \le p$, then as $n \to \infty$, $\hat{\boldsymbol{\theta}}_n^{PT}, \hat{\boldsymbol{\theta}}_n^S$ and $\tilde{\boldsymbol{\theta}}_n$ all become equivalent (up to the order $O(n^{-1/2})$) in probability, so that in this asymptotic setup there is not much to study. To obtain meaningful results, we shall, therefore, confine ourselves to local (Pitman-type) alternatives, for which these estimators are not all asymptotically equivalent (up to the order $O\left(n^{-1/2}\right)$) in probability. Specifically, we consider a class $\{K_n\}$ of alternatives defined by

$$K_n : \boldsymbol{\beta} = \boldsymbol{\beta}_{(n)} = \beta_0 \mathbf{1}_p + n^{-1/2}\boldsymbol{\tau};\ \boldsymbol{\tau} = (\tau_1, \ldots, \tau_p)', \tag{3.4}$$

where β_0 is arbitrary and $\boldsymbol{\tau}$ ($\in R^P$) is some fixed point. In this context, it is further assumed that the following holds.

(i) For each i $(= 1, \ldots, p)$, as $n \to \infty$,

$$\lambda_{ni} = n^{-1}n_i \to \lambda_i : 0 < \lambda_i < 1 \quad \left(\sum_{i=1}^{p}\lambda_i = 1\right), \tag{3.5}$$

and the $\bar{c}_i$, defined after (2.6), (depending on the n_i), all converge, i.e.,

$$\bar{c}_i \rightarrow c_i : 0 \leq | c_i | < \infty, \qquad 1 \leq i \leq p. \tag{3.6}$$

(ii) Define the C^2_{i,n_i} as in (2.13). Then, there exists some real positive numbers $Q_1, \ldots, Q_p$, such that as $n \rightarrow \infty$,

$$n_i^{-1} C^2_{i,n_i} \rightarrow Q_i \, (0 < Q_i < \infty), \qquad i = 1, \ldots, p. \tag{3.7}$$

This ensures that defining

$$C^2_{0,n} = \sum_{i=1}^{p} C^2_{i,n_i}, n^{-1} C^2_{0,n} \rightarrow \overline{Q} = \sum_{i=1}^{p} \lambda_i Q_i. \tag{3.8}$$

(iii) The regression constants satisfy the Noether condition:

$$\max_{1 \leq i \leq p} \left\{ \max_{1 \leq j \leq n_i} (c_{ij} - \bar{c}_i)^2 \Big/ C^2_{i,n_i} \right\} \rightarrow 0, \text{ as } n \rightarrow \infty. \tag{3.9}$$

Note that (3.5)–(3.9) are all standard (design) conditions, and are also required for the asymptotic theory of LSE and R-estimators. By virtue of these assumptions, in (3.4), we may take without any loss of generality, that

$$\sum_{i=1}^{p} \lambda_i Q_i \tau_i = 0. \tag{3.10}$$

Note that unlike the LSE, the M-estimators (of $\boldsymbol{\theta}, \boldsymbol{\beta}$) are not linear functions of the observations (or the scores). To facilitate the study of the asymptotic theory, we therefore make use of the Jurečková linearity results, as further extended by Singer and Sen (1985). We state these results (without proofs) for our subsequent use. Let $H_0^* : \boldsymbol{\theta} = \boldsymbol{\beta} = \mathbf{0}$ and let K $(0 < K < \infty)$ be an arbitrary constant. Then, under H_0^*, as $n \rightarrow \infty$,

$$\sup\{n^{-1/2} \mid M^{(i)}_{1,n_i}\left(\frac{a}{\sqrt{n}}, \frac{b}{\sqrt{n}}\right) - M^{(i)}_{1,n_i}(0,0) + \lambda_i n^{1/2} (a + b\bar{c}_i) \gamma \mid : \mid a \mid \leq K, \mid b \mid \leq K\} \xrightarrow{P} 0, \tag{3.11}$$

$$\sup\{n^{-1/2} \mid \tilde{M}^{(i)}_{2,n_i}\left(\frac{a}{\sqrt{n}}, \frac{b}{\sqrt{n}}\right) - \tilde{M}^{(i)}_{2,n_i}(0,0) + \gamma b n^{-1/2} C^2_{i,\,n_i} \mid : \mid a \mid \leq K, \mid b \mid \leq K\} \xrightarrow{P} 0, \tag{3.12}$$

for $i = 1, \ldots, p$, and

$$\sup\{n^{-1/2} \mid \tilde{M}^*_{2,n}\left(\frac{1}{\sqrt{n}}\mathbf{a}, \frac{1}{\sqrt{n}}b\right) - \tilde{M}^*_{2,n}(\mathbf{0}, 0) + \gamma b n^{-1/2} C^2_{0,n} \mid :$$
$$\mid a_i \mid \leq K, \ 1 \leq i \leq p, \ \mid b \mid \leq K\} \xrightarrow{P} 0. \tag{3.13}$$

Using then the basic arguments of Jurečková (1977), it follows from the above results that as $n \to \infty$,

$$n^{-1/2}(\tilde{\beta}^{(i)}_{n_i} - \beta_i) - \gamma^{-1}\lambda_i^{-1}Q_i^{-1}n^{-1/2}\tilde{M}^{(i)}_{2,n_i}(\theta_i, \beta_i) \xrightarrow{P} 0, \tag{3.14}$$

$$n^{1/2}\left(\tilde{\theta}^{(i)}_{n_i} - \theta_i\right) - \gamma^{-1}\lambda_i^{-1}$$
$$\cdot \left[n^{-1/2}M^{(i)}_{1,n_i}(\theta_i, \beta_i) - \bar{c}_i Q_i^{-1} n^{-1/2}\tilde{M}^{(i)}_{2,n_i}(\theta_i, \beta_i)\right] \xrightarrow{P} 0, \tag{3.15}$$

for $i = 1, \ldots, p$, so that

$$n^{1/2}\begin{pmatrix} \tilde{\boldsymbol{\theta}}_n - \boldsymbol{\theta} \\ \tilde{\boldsymbol{\beta}}_n - \boldsymbol{\beta} \end{pmatrix} \xrightarrow{D} N_{2p}\left(\mathbf{0}, \nu^2 \begin{pmatrix} \mathbf{D}_{11} & \mathbf{D}_{12} \\ \mathbf{D}_{21} & \mathbf{D}_{22} \end{pmatrix}\right), \tag{3.16}$$

where $\nu^2 = \sigma^2_\psi/\gamma^2$ and

$$\mathbf{D}_{11} = \mathrm{Diag}\left(\lambda_i^{-1}\left(1 + Q_i^{-1}c_i^2\right), \quad 1 \leq i \leq p\right), \tag{3.17}$$

$$\mathbf{D}_{12} = \mathbf{D}'_{21} = \mathrm{Diag}\left(-c_i\lambda_i^{-1}Q_i^{-1}, \quad 1 \leq i \leq p\right), \tag{3.18}$$

$$\mathbf{D}_{22} = \mathrm{Diag}\left(\lambda_i^{-1}Q_i^{-1}, \quad 1 \leq i \leq p\right). \tag{3.19}$$

[Note that the $(\tilde{\theta}^{(i)}_{n_i}, \tilde{\beta}^{(i)}_{n_i})$, $i = 1, \ldots, p$, are independent vectors.] For later use, we define here

$$\mathbf{C}^* = \mathrm{Diag}(c_1, \ldots, c_p), \ \mathbf{J} = \mathbf{I}_p - \mathbf{1}_p\mathbf{1}'_p\mathbf{D}_{22}^{-1},$$
$$\mathbf{c} = (c_1, \ldots, c_p), \tag{3.20}$$

$$\boldsymbol{\Lambda}_0 = \mathrm{Diag}\left(\lambda_1^{-1}, \ldots, \lambda_p^{-1}\right), \ \mathbf{D}^*_{11} = \boldsymbol{\Lambda}_0 + \overline{Q}^{-1}\mathbf{c}\mathbf{c}',$$
$$\mathbf{D}^*_{12} = \mathbf{D}^{*'}_{21} = -\overline{Q}^{-1}\mathbf{C}^*\mathbf{1}_p. \tag{3.21}$$

Also, we note that (3.16) remains true for $\boldsymbol{\beta} = \boldsymbol{\beta}_{(n)}$ (i.e., under $\{K_n\}$) as well. By (3.12) and (3.13), we claim that

$$\gamma n^{1/2}(\hat{\beta}_n - \beta_0) \overset{P}{\sim} \left(\sum_{i=1}^{p}\tilde{M}^{(i)}_{2,n_i}(\theta_i, \beta_0)\right) \Big/ \sqrt{n}, \tag{3.22}$$

under H_0, as well as $\{K_n\}$, where the later result follows by an appeal to the contiguity of probability measures under $\{K_n\}$ to those under H_0. As a result, from (3.11) to (3.22), we have

$$n^{1/2}(\hat{\theta}_{n_i}^{(i)} - \theta_i) - \gamma^{-1}\lambda_i^{-1}\{n^{-1/2}M_{1,n_i}^{(i)}(\theta_i, \beta_0) - \bar{c}_i\overline{Q}^{-1}n^{-1/2}\sum_{j=1}^{p}\tilde{M}_{2,n_j}^{(i)}(\theta_j, \beta_0)\} \xrightarrow{P} 0, \tag{3.23}$$

under H_0 as well as $\{K_n\}$. Therefore, under $\{K_n\}$,

$$n^{1/2}\begin{pmatrix} \tilde{\theta}_n - \theta \\ \hat{\beta}_n - \beta_0 \end{pmatrix} \xrightarrow{D} N_{p+1}\left(\begin{pmatrix} \mathbf{C}^*\mathbf{J}\tau \\ 0 \end{pmatrix}, \nu^2\begin{pmatrix} \mathbf{D}_{11}^* & \mathbf{D}_{12}^* \\ \mathbf{D}_{21}^* & D_{22}^* \end{pmatrix}\right), \tag{3.24}$$

where $D_{22}^* = \overline{Q}^{-1}$. Next, we note that by (2.14), (3.12), (3.22) and some standard steps, under $\{K_n\}$,

$$L_n = \nu^{-2}(\mathbf{V}_n + \tau)'\mathbf{J}'\mathbf{D}_{22}^{-1}\mathbf{J}(\mathbf{V}_n + \tau) + o_p(1), \tag{3.25}$$

where $V_n = n^{1/2}\left(\tilde{\beta} - \beta_{(n)}\right)\left(\sim N_p\left(\mathbf{0}, \mathbf{D}_{22}\nu^2\right)\right)$, and where we have made use of the fact [viz. Sen (1982)] that under $\{K_n\}$,

$$s_n^2 \xrightarrow{P} \nu^2 = \sigma_\psi^2/\gamma^2, \qquad \text{as } n \to \infty. \tag{3.26}$$

Thus, under $\{K_n\}$, L_n has asymptotically a non-central chi-square distribution with $p-1$ degrees of freedom (DF) and non-centrality parameter

$$\Delta = \nu^{-2}\tau'\mathbf{J}'\mathbf{D}_{22}^{-1}\mathbf{J}\tau = \nu^{-2}\left(\tau'\mathbf{D}_{22}^{-1}\tau\right), \tag{3.27}$$

as by (3.10), $\tau'\mathbf{D}_{22}^{-1}\mathbf{1}_p = 0$. Under H_0, L_n has asymptotically the central chi-square d.f. with $p-1$ DF, so that in (2.15), $\ell_{n,\alpha}$ may be replaced by $\chi_{p-1,\alpha}^2$, the upper $100\alpha\%$ point of the central chi-square d.f. with $p-1$ DF, when n is large. Beta distributional approximations may work out better for smaller sample sizes. Having (3.24) through (3.27), we may proceed as by Lambert *et al.* (1985) and obtain that under $\{K_n\}$, as $n \to \infty$,

$$\begin{aligned} G^{PT}(\mathbf{x}) &= \lim_{n\to\infty} P\left\{n^{1/2}\left(\hat{\theta}_n^{PT} - \theta\right) \le \mathbf{x} \mid K_n\right\} \\ &= \Phi_p\left(\mathbf{x} - \mathbf{C}^*\mathbf{J}\tau; \mathbf{Q}, \nu^2\mathbf{D}_{11}\right) H_{p-1}\left(\chi_{p-1,\alpha}^2; \Delta\right) \\ &\quad + \int_{E(\tau)} \Phi_p\left(\mathbf{x} + \mathbf{C}^*\mathbf{Z}; \mathbf{0}, \nu^2\mathbf{\Lambda}_0\right) d\Phi_p\left(\mathbf{Z}; \mathbf{0}, \nu^2\mathbf{D}_{22}\right), \end{aligned} \tag{3.28}$$

where $\Phi_p(\mathbf{x}; \mathbf{a}, \mathbf{B})$ stands for the p-variate normal d.f. with mean vector $\mathbf{a}$ and covariance matrix $\mathbf{B}$, $H_m(x; \delta)$ is the non-central chi-square d.f. with m DF and non-centrality parameter δ,

$$E(\boldsymbol{\tau}) = \left\{ \mathbf{Z} : (\mathbf{Z} + \boldsymbol{\tau})' \mathbf{J}' \mathbf{D}_{22}^{-1} \mathbf{J} (\mathbf{Z} + \boldsymbol{\tau}) \geq \nu^2 \chi^2_{p-1,\alpha} \right\},$$

and we have used the fact that $\mathbf{D}_{11} - \mathbf{D}_{12}\mathbf{D}_{22}^{-1}\mathbf{D}_{21} = \mathbf{\Lambda}_0$.

Now, by virtue of (2.16), (3.16) and (3.24),

$$n^{1/2} \parallel \tilde{\boldsymbol{\beta}}_n - \boldsymbol{\beta}_{(n)} \parallel, \; n^{1/2} \mid \hat{\beta}_n - \beta_0 \mid \quad \text{and} \quad n^{1/2} \parallel \hat{\boldsymbol{\beta}}_n^{JS} - \boldsymbol{\beta}_{(n)} \parallel$$

are all $O_p(1)$. Hence, by (2.18) and (3.11), under $\{K_n\}$ and the assumed regularity conditions as $n \to \infty$,

$$\begin{aligned} &n^{1/2} \left(\hat{\theta}^S_{i,n} - \theta_i \right) + \bar{c}_i \left\{ n^{1/2} \left(\hat{\beta}^S_{i,n} - \beta_0 \right) - \tau_i \right\} \\ &\quad - \gamma^{-1} \lambda_i^{-1} n^{1/2} M^{(i)}_{i,n_i} \left(\theta_i, \beta_0 + \frac{1}{\sqrt{n}} \tau_i \right) \xrightarrow{P} 0 \end{aligned} \tag{3.29}$$

for every i $(= 1, \ldots, p)$. Also, note that

$$\begin{aligned} &n^{1/2} \left(\hat{\boldsymbol{\beta}}^S_n - \beta_0 \mathbf{1}_p - n^{1/2} \boldsymbol{\tau} \right) \\ &= n^{1/2} \left(\hat{\beta}_n - \beta_0 \right) \mathbf{1}_p - \boldsymbol{\tau} \\ &\quad + I(L_n \geq \varepsilon) \left\{ 1 - (p-3) L_n^{-1} \right\} n^{1/2} \left(\tilde{\boldsymbol{\beta}}_n - \hat{\beta}_n \mathbf{1}_p \right) \\ &= \left\{ n^{1/2} \left(\hat{\beta}_n - \beta_0 \right) \mathbf{1}_p - \boldsymbol{\tau} \right\} I(L_n < \varepsilon) \\ &\quad + I(L_n \geq \varepsilon) \left[n^{1/2} \left(\tilde{\boldsymbol{\beta}}_n - \beta_0 \mathbf{1}_p \right) - \boldsymbol{\tau} \right] \\ &\quad - (p-3) L_n^{-1} n^{1/2} \left(\tilde{\boldsymbol{\beta}}_n - \hat{\beta}_n \mathbf{1}_p \right). \end{aligned} \tag{3.30}$$

Hence, by (3.14), (3.29), (3.30) and some standard steps, it follows that under $\{K_n\}$ and the assumed regularity conditions, as $n \to \infty$,

$$\begin{aligned} n^{1/2} \left(\hat{\theta}^S_{i,n} - \theta_i \right) &= \gamma^{-1} \lambda_i^{-1} M^{(i)}_{1,n_i} (\theta_i, \beta_i) \\ &\quad - \bar{c}_i I(L_n < \varepsilon) \{ n^{1/2} (\hat{\beta}_n - \beta_0) - \tau_i \} \\ &\quad - \bar{c}_i I(L_n \geq \varepsilon) \{ \gamma^{-1} \lambda_i^{-1} Q_i^{-1} n^{-1/2} \tilde{M}^{(i)}_{2,n_i} (\theta_i, \beta_i) \} \\ &\quad + (p-3) L_n^{-1} \bar{c}_i I(L_n \geq \varepsilon) \gamma^{-1} n^{-1/2} \{ \lambda_i^{-1} Q_i^{-1} \\ &\quad \tilde{M}^{(i)}_{2,n_i} (\theta_i, \beta_i) - \overline{Q}^{-1} \sum_{j=1}^{p} \tilde{M}^{(i)}_{2,n_j} (\theta_j, \beta_0) \} + o_p(1), \end{aligned} \tag{3.31}$$

where we take $\beta_i = \beta_0 + n^{1/2}\tau_i$, $1 \le i \le p$, and note that by (3.10), in (3.20), we may, as well, replace the $\tilde{M}_{2,n_i}^{(i)}(\theta_i, \beta_0)$ by $\tilde{M}_{2,n_i}^{(i)}(\theta_i, \beta_i)$. Thus, on letting $\mathbf{U}_n^* = (U_{n1}^*, \ldots, U_{np}^*)'$ with $U_{nj}^* = n^{-1/2} M_{2,n_j}^{(j)}(\theta_j, \beta_j)$, $1 \le j \le p$, and $\mathbf{V}_n^* = (V_{n1}^*, \ldots, V_{np}^*)$ with $V_{nj}^* = n^{-1/2}\tilde{M}_{2,n_j}^{(j)}(\theta_j, \beta_j)$, $1 \le j \le p$, we have from (3.31) that under $\{K_n\}$ and the assumed regularity conditions, as $n \to \infty$,

$$\begin{aligned} n^{1/2}(\hat{\boldsymbol{\theta}}_n^S - \boldsymbol{\theta}) &= \gamma^{-1}\boldsymbol{\Lambda}_0\mathbf{U}_n^* - I(L_n \ge \varepsilon)\gamma^{-1}\mathbf{C}^*\mathbf{D}_{22}\mathbf{V}_n^* \\ &\quad + (p-3)L_N^{-1}\gamma^{-1}I(L_n \ge \varepsilon)\mathbf{C}^*\mathbf{J}\mathbf{D}_{22}\left(\mathbf{V}_n^* + \mathbf{D}_{22}^{-1}\boldsymbol{\tau}\right) \\ &\quad - I(L_n < \varepsilon)\mathbf{C}^*\{n^{1/2}(\hat{\beta}_n - \beta_0)\mathbf{1}_p + \boldsymbol{\tau}\} + \mathbf{o}_p(1); \end{aligned} \tag{3.32}$$

$$L_n = \sigma_\psi^{-2}\left(\mathbf{V}_n^* + \mathbf{D}_{22}^{-1}\boldsymbol{\tau}\right)'\mathbf{D}_{22}\mathbf{J}'\mathbf{D}_{22}^{-1}\mathbf{J}\mathbf{D}_{22}\left(\mathbf{V}_n^* + \mathbf{D}_{22}^{-1}\boldsymbol{\tau}\right) + \mathbf{o}_p(1), \tag{3.33}$$

where

$$(\mathbf{U}_n^*, \mathbf{V}_n^*) \xrightarrow{\mathcal{D}} N_{2p}\left(\mathbf{0}, \sigma_\psi^2 \cdot \mathrm{Diag}\left(\boldsymbol{\Lambda}_0^{-1}, \mathbf{D}_{22}^{-1}\right)\right). \tag{3.34}$$

Thus, $\left(\boldsymbol{\Lambda}_0^{1/2}, \mathbf{U}_n^*, \mathbf{D}_{22}^{1/2}\mathbf{V}_n^*\right)$ $\left(= \left(\mathbf{U}_n^0, \mathbf{V}_n^0\right), \text{ say}\right)$, has the asymptotic distribution $N_{2p}\left(\mathbf{0}, \sigma_\psi^2 \cdot \mathbf{I}_{2p}\right)$, and we have by (3.32)–(3.34),

$$\begin{aligned} n^{1/2}\left(\hat{\boldsymbol{\theta}}_n^S - \boldsymbol{\theta}\right) &= \gamma^{-1}\boldsymbol{\Lambda}_0^{1/2}\mathbf{U}_n^0 - I(L_n \ge \varepsilon)\gamma^{-1}\mathbf{C}^*\mathbf{D}_{22}^{1/2}\mathbf{V}_n^0 \\ &\quad + (p-3)\gamma^{-1}I(L_n \ge \varepsilon)C^*\mathbf{J}\mathbf{D}_{22}^{1/2}\left(\mathbf{V}_n^0 + \boldsymbol{\tau}^0\right) \\ &\quad \cdot \{(\mathbf{V}_n^0 + \boldsymbol{\tau}^0)'\mathbf{J}^*(\mathbf{V}_n^0 + \boldsymbol{\tau}^0)\}^{-1} \\ &\quad - I(L_n < \varepsilon)\mathbf{C}^*\{n^{1/2}(\hat{\beta}_n - \beta_0)\mathbf{1}_p + \boldsymbol{\tau}\} + \mathbf{o}_p(1), \end{aligned} \tag{3.35}$$

where

$$\mathbf{J}^* = \mathbf{D}_{22}^{1/2}\mathbf{J}'\mathbf{D}_{22}^{-1}\mathbf{J}\mathbf{D}_{22}^{1/2}, \quad \boldsymbol{\tau}^0 = \mathbf{D}_{22}^{-1/2}\boldsymbol{\tau}, \tag{3.36}$$

and $\mathbf{J}^*$ is an idempotent matrix of rank $p-1$. Though this expression is too complicated to put down the asymptotic distribution of $n^{1/2}(\hat{\theta}_n^S - \theta)$ (under $\{K_n\}$ or H_0) in a closed compact form, in the next section, we shall see that it provides the necessary tools for the study of asymptotic (distributional) risk in a convenient form.

4. ASYMPTOTIC DISTRIBUTIONAL RISKS

In the same way as the asymptotic relative efficiency (A.R.E.) in the Pitman sense is solely based on the asymptotic distribution of a statistic under consideration, here also, we would like to employ the asymptotic distributions in Section 3, for suitable definitions of risks, and incorporate them in our main study of the relative dominance pictures of these estimators.

For an estimator $\boldsymbol{\theta}_n^*$ of $\boldsymbol{\theta}$, assuming that under $\{K_n\}$, $n^{1/2}\left(\boldsymbol{\theta}_n^* - \boldsymbol{\theta}\right)$ has a limiting distribution, we denote by

$$G_p^*(\mathbf{x}) = \lim_{n\to\infty} P\left\{n^{1/2}\left(\boldsymbol{\theta}_n^* - \boldsymbol{\theta}\right) \le x \mid K_n\right\}, \quad x \in R^p. \tag{4.1}$$

Then let

$$\mathbf{V}^* = \int\int \mathbf{x}\mathbf{x}' dG_p^*(\mathbf{x}) \tag{4.2}$$

be the dispersion matrix (computed from G_p^*) of this asymptotic distribution, and consider a $p \times p$ positive definite matrix $\mathbf{W}$, so that a (quadratic error loss function based) asymptotical distributional risk (ADR) for $\boldsymbol{\theta}_n^*$ may be defined as

$$R(\boldsymbol{\theta}^*; \mathbf{W}) = \mathrm{Tr}(\mathbf{W}\mathbf{V}^*). \tag{4.3}$$

In the light of this ADR, we say that an estimator $\boldsymbol{\theta}_n^*$ *dominates* over another $\boldsymbol{\theta}_n^{**}$, *asymptotically*, if

$$R(\boldsymbol{\theta}^*; \mathbf{W}) \le R(\boldsymbol{\theta}^{**}; \mathbf{W}), \qquad \forall\ \boldsymbol{\theta}, \boldsymbol{\tau}, \tag{4.4}$$

with the strict inequality sign holding for at least some $(\boldsymbol{\theta}, \boldsymbol{\tau})$. This concept is also related to *asymptotic (distributional) admissibility* (ADA): $\boldsymbol{\theta}_n^*$ is ADA if for no other $\boldsymbol{\theta}^{**}$, the opposite inequality in (4.4) holds for every $(\boldsymbol{\theta}, \boldsymbol{\tau})$. Under more restrictive regularity conditions, ADR results agree with *asymptotic risk* (AR) results, where instead of $\mathbf{V}^*$, in (4.3), the limit of the actual dispersion matrix of $n^{1/2}\left(\boldsymbol{\theta}_n^* - \boldsymbol{\theta}\right)$ is used. This, however, is not needed, and, for our main results, we shall confine ourselves to ADR considerations only.

From (3.16), we obtain by using (4.3) that

$$R\left(\tilde{\boldsymbol{\theta}}; \mathbf{W}\right) = \nu^2\ \mathrm{Tr}(\mathbf{W}\mathbf{D}_{11}) \ \text{ for every } \ \boldsymbol{\tau} \in R^p, \ \ \boldsymbol{\theta} \in R^p. \tag{4.5}$$

Similarly, by (3.24) and (4.3), we have

$$R(\hat{\boldsymbol{\theta}}; \mathbf{W}) = \nu^2 \mathrm{Tr}(\mathbf{W}\mathbf{D}_{11}^*) + \boldsymbol{\tau}'\mathbf{J}'\mathbf{C}^*\mathbf{W}\mathbf{C}^*\mathbf{J}\boldsymbol{\tau}. \tag{4.6}$$

For the PTME, $\hat{\boldsymbol{\theta}}_n^{PT}$, we use (3.28). For this d.f., the computation of V^* follows from Sen and Saleh (1979), and we have (on omitting these steps) that

$$\begin{aligned} R(\hat{\boldsymbol{\theta}}^{PT}; \mathbf{W}) = \nu^2\{&(1 - H_{p+1}(\chi^2_{p-1,\alpha}; \Delta)) \operatorname{Tr}(\mathbf{W}\mathbf{D}_{11}) \\ &+ H_{p+1}(\chi^2_{p-1,\alpha}; \Delta) \operatorname{Tr}(\mathbf{W}\mathbf{D}^*_{11})\} \\ &+ \boldsymbol{\tau}'\mathbf{J}'\mathbf{C}^*\mathbf{W}\mathbf{C}^*\mathbf{J}\boldsymbol{\tau}\{2H_{p+1}(\chi^2_{p-1,\alpha}; \Delta) \\ &- H_{p+3}(\chi^2_{p-1,\alpha}; \Delta)\}. \end{aligned} \tag{4.7}$$

For the computation of $R(\hat{\theta}^S; \mathbf{W})$, in (3.5), we now take ε (> 0) to be arbitrarily small, so that

$$P\{L_n < \varepsilon \mid K_n\} \sim H_{p-1}(\varepsilon; \Delta)(\leq H_{p-1}(\varepsilon; 0) = O(\varepsilon^{(p-1)/2}))$$

can also be made arbitrarily small. Then, the last line in (3.35) can be neglected (in probability), and hence, the first terms on the right hand side of (3.35) may be used to compute the ADR. Since this expression is very much the parallel expression in the case of shrinkage LSE, treated in detail by Saleh and Sen (1985b), we omit the details of manipulation and claim that, for ε sufficiently small,

$$\begin{aligned} R(\hat{\boldsymbol{\theta}}^S; \mathbf{W}) = \nu^2 \{&\operatorname{Tr}(\mathbf{W}\mathbf{D}_{11}) - 2(p-3)[\operatorname{Tr}(\mathbf{A})E(\chi^{-2}_{p+1}(\delta)) \\ &+ (\boldsymbol{\tau}^0\mathbf{A}\boldsymbol{\tau}^0)E(\chi^{-2}_{p+3}(\Delta)) + (\boldsymbol{\tau}^{0\prime}\mathbf{A}\boldsymbol{\tau}^0)E(\chi^{-2}_{p+1}(\Delta))] \\ &+ (p-3)^2[\operatorname{Tr}(\mathbf{A})E(\chi^{-4}_{p+1}(\Delta)) \\ &+ (\boldsymbol{\tau}'_0\mathbf{A}\boldsymbol{\tau}^0)E(\chi^{-4}_{p+3}(\Delta))]\}, \end{aligned} \tag{4.8}$$

where

$$\Delta = \nu^{-2}\left(\boldsymbol{\tau}'\mathbf{D}^{-1}_{22}\boldsymbol{\tau}\right), \qquad \mathbf{A} = \mathbf{A}^{*\prime}\mathbf{J}^*\mathbf{D}^{1/2}_{22}\mathbf{C}^*\mathbf{W}\mathbf{C}^*\mathbf{D}^{1/2}_{22}\mathbf{J}^*\mathbf{A}^* \tag{4.9}$$

and $\mathbf{A}^*$ is defined by $\mathbf{A}^{*\prime}\mathbf{J}^*\mathbf{A}^* - \mathbf{I}_{p-1}$, $\mathbf{1}'_p\mathbf{D}^{-1/2}_{22}\mathbf{A}^* = \mathbf{0}$; note that $\chi_r^{-2}(\delta)$ stands for the reciprocal of $\chi^2_r(\delta)$, and $\chi^2_r(\delta)$ has the d.f. $H_r(x, \delta), r \geq 0, \delta \geq 0$. It follows from Saleh and Sen (1985b) that

$$\boldsymbol{\tau}^{0\prime}\mathbf{A}\boldsymbol{\tau}^0/\Delta \leq \operatorname{Tr}(\mathbf{D}_{11}\mathbf{W}), \tag{4.10}$$

$$\operatorname{Tr}(\mathbf{A}) = \operatorname{Tr}((\mathbf{D}_{11} - \mathbf{D}^*_{11})\mathbf{W}) < \operatorname{Tr}(\mathbf{D}_{11}\mathbf{W}). \tag{4.11}$$

These expressions will be useful in the sequel.

5. ASYMPTOTIC DOMINANCE PICTURES

First, comparing (4.5) and (4.8), and using the Stein identities on the reciprocal moments of $\chi_r^2(\delta)$, we have

$$R(\hat{\boldsymbol{\theta}}^S; \mathbf{W})/R(\tilde{\boldsymbol{\theta}}; \mathbf{W}) \leq 1, \forall \boldsymbol{\tau}, \boldsymbol{\theta}, \qquad p \geq 4, \tag{5.1}$$

where the upper limit 1 is attained when $\Delta \to \infty$. This shows the asymptotic dominance of $\hat{\boldsymbol{\theta}}_n^S$ over $\tilde{\boldsymbol{\theta}}_n$ (in the light of the ADR). Next, we rewrite (4.6) as

$$R(\hat{\boldsymbol{\theta}}; \mathbf{W}) = \nu^2 \text{Tr}(\mathbf{W}\mathbf{D}_{11}) - \nu^2 \text{Tr}(\mathbf{A}) + \boldsymbol{\tau}' \mathbf{C}^* \mathbf{W} \mathbf{C}^* \boldsymbol{\tau}. \tag{5.2}$$

We may then compare (4.8) and (5.2) to study the dominance picture of $\hat{\boldsymbol{\theta}}_n^S$ relative to $\hat{\boldsymbol{\theta}}_n$. Note that under H_0, $\Delta = 0$, so that on using the fact that $E(\chi_q^2(0)) = (q-2)^{-1}$, $\forall q > 2$, along with similar results for higher order negative moments, we obtain that under $H_0 : \boldsymbol{\tau} = \mathbf{0}$ (i.e., $\Delta = 0$),

$$\begin{aligned} R(\hat{\boldsymbol{\theta}}^S; \mathbf{W}) &= R(\hat{\theta}; \mathbf{W}) - 2(p-3)\text{Tr}(\mathbf{A})(p-1)^{-1}\nu^2 \\ &\quad + \nu^2 (p-3)^2/(p-1)(p-3) \cdot \text{Tr}(\mathbf{A}) + \nu^2 \text{Tr}(\mathbf{A}) \\ &= R(\hat{\theta}; \mathbf{W}) + \nu^2 \text{Tr}(\mathbf{A}) 2(p-1)^{-1} > R(\hat{\theta}; \mathbf{W}). \end{aligned} \tag{5.3}$$

Thus, $\hat{\boldsymbol{\theta}}_n^S$ fails to dominate over $\hat{\boldsymbol{\theta}}_n$ when H_0 holds. On the other hand, as $\boldsymbol{\tau}$ moves away from $\mathbf{0}$ (i.e., $\Delta \to \infty$), the picture changes; an opposite inequality holds for $\Delta \geq$ some Δ_0. In the above setup, we assume that $\mathbf{C}^* \neq \mathbf{0}$. If $\mathbf{C}^* = \mathbf{0}$, then all four estimators (i.e., $\tilde{\theta}_n, \hat{\theta}_n, \hat{\theta}^{PT}$ and $\hat{\theta}_n^S$) will have the same ADR, so that no comparison is really needed. Generally Δ_0 is small and the excess is (5.3), i.e., $2(p-1)^{-1}\text{Tr}(A)/\text{Tr}(\mathbf{W}\mathbf{D}_{11}^*)$ is also not very large, and hence, $\hat{\boldsymbol{\theta}}_n^S$ may be advocated over $\hat{\boldsymbol{\theta}}_n$, though either fails to dominate over the other. Likewise, $\hat{\boldsymbol{\theta}}_n$ dominates over $\tilde{\boldsymbol{\theta}}_n$ near $\Delta = 0$, but as Δ increases, the opposite picture holds. This is mainly due to the fact that in (5.2), the last term on the right hand side blows up as $\boldsymbol{\tau}$ moves away from $\mathbf{0}$, while (4.5) is not affected by the fluctuation of $\boldsymbol{\tau}$.

Finally, we may compare (4.7) and (4.8) to study the relative dominance picture of PTME and SME. Under H_0 (i.e., $\boldsymbol{\tau} = \mathbf{0}$, $\Delta = 0$), we obtain that

$$R(\hat{\boldsymbol{\theta}}^S; \mathbf{W}) = R(\hat{\boldsymbol{\theta}}^{PT}; \mathbf{W}) + \nu^2 \text{Tr}(\mathbf{A}) \left\{ H_{p+1}\left(\chi^2_{p-1,\alpha}; 0\right) - \frac{p-3}{p+1} \right\}, \tag{5.4}$$

so that

$$R(\hat{\boldsymbol{\theta}}^S; \mathbf{W}) \geq R(\hat{\boldsymbol{\theta}}^{PT}; \mathbf{W}) \text{ if } H_{p+1}\left(\chi^2_{p-1,\alpha}; 0\right) > \frac{p-3}{p-1}. \tag{5.5}$$

Thus, for any given p (≥ 4), there exists a range ofα ($0 < \alpha \leq \alpha_p$) such that $R(\hat{\boldsymbol{\theta}}^S; \mathbf{W}) \geq R(\hat{\boldsymbol{\theta}}^{PT}; \mathbf{W})$ under H_0. Of course, as α increases, the opposite inequality holds. For $\boldsymbol{r}$ close to $\mathbf{0}$, again, for a small range of α, $\hat{\boldsymbol{\theta}}_n^{PT}$ dominates over of $\hat{\boldsymbol{\theta}}_n^S$. On the other hand, there exists a Δ^0, such that for all $\boldsymbol{r}$ for which $\Delta \geq \Delta^0$, $R(\hat{\boldsymbol{\theta}}^S; \mathbf{W}) \leq R(\hat{\theta}^{PT}; W)$. Thus, none of $\hat{\boldsymbol{\theta}}_n^{PT}$ and $\hat{\boldsymbol{\theta}}_n^S$ dominates over the other (in light of the ADR). For some numerical studies of this relative dominance picture, we may again refer to Saleh and Sen (1985b). Note that for the PTE, we may not need $p \geq 4$, which we do for the SME.

ACKNOWLEDGMENTS

This work was partially supported by NSERC, Canada, Grant A3088, Faculty of Graduate Studies and Research, Carleton University, Grant GR-5, and partially by the U.S. Office of Naval Research, Contract: N00014-83-K0387.

REFERENCES

Akritas, M., A. K. Md. E. Saleh, and P. K. Sen (1985), "Nonparametric estimation of intercepts after a preliminary test on parallelism of several regression lines". In *Biostatistics: B. G. Greenberg Volume*, ed. P. K. Sen, pp. 221–235. Amsterdam: North-Holland.

James, W., and C. Stein (1961), "Estimation with quadratic loss". *Proceedings of the Fourth Berkeley Symposium on Mathematical Statistics and Probability* **1**, 381–379.

Judge, G. G., and M. E. Bock (1978), *The Statistical Implications of Pre-Test and Stein-Rule Estimators in Econometrics*. Amsterdam: North-Holland.

Jurečková, J. (1977), "Asymptotic relations of M-estimates and R-estimates in linear regression model". *Annals of Statistics* **5**, 664–672.

Lambert, A., A. K. Md. E. Saleh, and P. K. Sen (1985), "On least squares estimation of intercepts after a preliminary test on parallelism of regression lines". *Communication in Statistics A, Theory and Methods* **14**, 793–807.

Lindley, D. V. (1962), "Discussion on Professor Stein's paper". *Journal of the Royal Statistical Society*, Series B **24**, 285–287.

Saleh, A. K. Md. E., and P. K. Sen (1985a), "Nonparametric shrinkage estimation in a parallelism problem". *Sankhya*, Series A **47**, 156–165.

Saleh, A. K. Md. E., and P. K. Sen (1985b), "On shrinkage least squares estimation in a parallelism problem". *Carleton University Math. Tech. Ser. No. 215.*

Sen, P. K. (1969), "On a class of rank order tests for the parallelism of several regression lines". *Annals of Statistics* **40**, 1668–1683.

Sen, P. K. (1971), "Robust statistical procedures in problems of linear regression with special reference to quantitative bio-assays, I". *Review of the International Statistical Institute* **39**, 21–38.

Sen, P. K. (1981), *Sequential Nonparametrics: Invariance Principles and Statistical Inference*. New York: Wiley and Sons.

Sen, P. K. (1982), "On *M* tests in linear models". *Biometrika* **69**, 245–248.

Sen, P. K. (1984), "A James-Stein detour of *U*-statistics". *Communications in Statistics A, Theory and Methods* **13**, 2725–2747.

Sen, P. K., and A. K. Md. E. Saleh (1979), "Nonparametric estimation of location parameter after a preliminary test on regression in the multivariate case". *Journal of Multivariate Analysis* **9**, 322–331.

Sen, P. K., and A. K. Md. E. Saleh (1984), "Nonparametric shrinkage estimators of location in a multivariate simple regression model". *Proc. 4th Pannonian Conf. Math. Statist.* pp. 311–324.

Sen, P. K., and A. K. Md. E. Saleh (1985), "On some shrinkage estimators of multivariate location". *Annals of Statistics* **13**, 272–281.

Singer, J. M., and P. K. Sen, (1985), "*M*-methods in multivariate linear models". *Journal of Multivariate Analysis* **17**, 168–184.

Stein, C. (1981), "Estimation of the mean of a multivariate normal distribution". *Annals of Statistics* **9**, 1135–1151.

D. A. S. Fraser [1] and H. Massam [2]

AN ALGORITHM FOR CONCAVE REGRESSION

ABSTRACT

Concave, isotonic, and Lipchitz regression analyses are examples of linear regression in which the coefficients are subject to linear restrictions. Computation of the maximum likelihood (normality) or least squares fit is available for the isotonic case in a bounded number of steps. For the concave case most of the algorithms have been difficult to implement or had no assured convergence; Dykstra (1983) gave an algorithm easy to implement and converging to the least squares solution as $n \to \infty$. Concave regression analysis is examined here from a geometrical viewpoint as a special case of cone-cylinder regression, in which the set of possible regression functions is the Cartesian product of a convex cone and a cylinder. The maximum likelihood/least squares solution is characterized in terms of geometrical properties of the cone-cylinder and the resulting computational procedure leads to the exact solution in a finite number of one dimensional projections.

1. INTRODUCTION

A statistical model arising frequently in economics involves a dependent variable y with a mean $\eta(x)$ that is not taken explicitly to be linear, quadratic, or more general but simply to be *concave* in x. Variations on this can require that $\eta(x)$ be *isotonic* (monotone non-decreasing) or satisfy a *Lipschitz condition* on the first derivative. For some background see Barlow *et al.* (1972) and Wu (1982).

For notation, let $x_1, \ldots, x_k$ be values for the independent variable and y_{ji}, $j = 1, \ldots, m_i$ be the y-values corresponding to the value x_i; then let

[1] Department of Statistics, University of Toronto, Toronto, Ontario M5S 1A1; also University of Waterloo and York University

[2] Department of Mathematics, York University, Downsview, Ontario M3J 1P3

I. B. MacNeill and G. J. Umphrey (eds.), Foundations of Statistical Inference, 121–132.

$y_i = \sum_{j=1}^{m_i} y_{ji}/m_i$ and $\eta_i = \eta(x_i)$ and assume that the x's have been ordered so that $x_1 < \cdots < x_k$.

For the properties it is convenient to use difference notation:

$$\eta[x_1, x_2] = (\eta(x_2) - \eta(x_1))/(x_2 - x_1)$$
$$\eta[x_1, x_2, x_3] = (\eta[x_2, x_3] - \eta[x_1, x_2])/(x_3 - x_1).$$

Then a regression function η is isotonic (monotone nondecreasing) if

$$\eta[x_i, x_{i+1}] \geq 0, \qquad i = 1, \ldots, k-1, \tag{1.1}$$

concave if

$$\eta[x_i, x_{i+1}, x_{i+2}] \leq 0, \qquad i = 1, \ldots, k-2, \tag{1.2}$$

and, trivially, *nonnegative* if

$$\eta(x_i) \geq 0, \qquad i = 1, \ldots, k. \tag{1.3}$$

In each case the regression set $\mathcal{N} = \{\eta = (\eta_1, \ldots, \eta_k)\}$, often called the *feasible set*, is a closed convex cone in R^k, and is not confined to a proper subspace.

The maximum likelihood (normal $(0, \sigma^2)$ error) or least squares fit $\hat{\eta}$ is obtained by minimizing $\sum_{i=1}^{k} \sum_{j=1}^{m_i} (y_{ji} - \eta_i)^2$ or equivalently minimizing

$$\sum_{i=1}^{k} m_i (y_i - \eta_i)^2, \tag{1.4}$$

where $y_i = \sum_{j=1}^{m_i} y_{ji}/m_i$, subject to η lying in a region $\mathcal{N}$ given by linear restrictions such as (1.1) or (1.2) or (1.3).

For the isotonic-regression case, graphical or tabular procedures are discussed by Barlow *et al.* (1972); one involves the *greatest common minorant* (GCM) for a *cumulative sum diagram* (CSM), the other involves a *pool-adjacent-violators* iteration, a variation of Kruskal's *up-and-down-blocks* iteration; Wu (1982) discussed a variation on the pool-adjacent-violators method. These procedures converge in a bounded number of steps.

For the concave-regression case, Hildreth (1954) reformulated the problem as a quadratic programming problem and gave an algorithm for solving it. To avoid the special programming problem, Dent *et al.* (1977) proposed an algorithm that gives an approximate solution; also see Dent (1973), Hanson and Pledger (1976), and Holloway (1979). Wu (1982) proposed two algorithms, one involving some quadratic programming, the other being a modification of Dent *et al.* (1977); one is easier to implement than the other but neither has assured convergence. More recently, Dykstra (1983) gave a convergent algorithm involving an infinite sequence of projections on subspaces. The algorithm presented in this paper consists of a finite number of linear projections along a fixed line. The computations are simple, do not involve the minimization of a function, and lead to the exact solution in a finite number of steps.

In Section 2, we describe the geometrical structure of the concave regression problem. The feasible set of vectors $\mathcal{N}$ is determined as a cone-cylinder. The cross-section cone K of the cylinder is the positive orthant in a basis made up of piecewise-linear concave vectors. This structure and the structure of a cone-cylinder in general allows the space to be partitioned into subsets in correspondence with the faces of the cone K; this is examined in Section 3. The computational solution to the cone-cylinder regression problem follows immediately from this geometrical analysis; this is developed in Section 4. Finally in Section 5, we illustrate the method with the example examined previously by Hildreth (1954) and Wu (1982).

2. CONCAVE REGRESSION

2.1 The Concave Regression Problem

Consider the concave regression problem introduced in Section 1. Let $x_1, \ldots, x_k$ be values of the independent variable ordered so that $x_1 < \cdots < x_k$, and let y_{ji} be the y-values corresponding to the value x_i and $y_i = \sum_{j=1}^{m_i} y_{ji}/m_i$ be the corresponding average. The estimation problem is to find the concave function $\hat{\eta}(x)$ with values $\hat{\eta}_i = \hat{\eta}(x_i)$ such that

$$\sum_{i=1}^{k}\sum_{j=1}^{m_i}(y_{ji} - \eta_i)^2 \qquad \text{or} \qquad \sum_{i=1}^{k} m_i(y_i - \eta_i)^2$$

is a minimum. Consider $y = (y_1, \ldots, y_k)'$ and $\eta = (\eta_1, \ldots, \eta_k)'$ as points in R^k, and let $\mathcal{N} = \{\eta\}$ be the set of all concave regression vectors. The regression problem then is to find $\hat{\eta} \in \mathcal{N}$ to minimize

$$\| y - \eta \|^2 = \langle y - \eta,\ y - \eta \rangle \ = (y - \eta)' M (y - \eta), \tag{2.1}$$

where M is a positive definite matrix giving the needed inner product $\langle \cdot, \cdot \rangle$ on R^k. For the problem as specified by (1.5) the matrix $M =$ diag (m_i) is a diagonal matrix with values m_i on the diagonal, but for the discussion that follows it can be a general inner product matrix.

We now characterize the region $\mathcal{N}$ of "concave" vectors η. The relations (1.2) for a concave vector $\eta = (\eta_1, \ldots, \eta_k)' = (\eta(x_1), \ldots, \eta(x_k))'$ can be rewritten as

$$\frac{\eta_2 - \eta_1}{x_2 - x_1} \geq \frac{\eta_3 - \eta_2}{x_3 - x_2} \geq \cdots \geq \frac{\eta_k - \eta_{k-1}}{x_k - x_{k-1}} \tag{2.2}$$

or equivalently as

$$\Delta x_2 \eta_1 - (\Delta x_1 + \Delta x_2)\, \eta_2 + \Delta x_1 \eta_3 \leq 0$$

$$\vdots$$

$$\Delta x_{k-1} \eta_{k-2} - (\Delta x_{k-2} + \Delta x_{k-1})\, \eta_{k-1} + \Delta x_{k-2} \eta_k \leq 0,$$

which can be rewritten as

$$\alpha_i' \eta \leq 0 \qquad (i = 1, \ldots, k-2), \tag{2.3}$$

where

$$\alpha_i = \left(0, \cdots 0, \Delta x_{i+1}, -\left(\Delta x_i + \Delta x_{i+1}\right), \Delta x_i, 0, \ldots, 0\right)' \tag{2.4}$$

(with $i-1$ zeros on the left and $k-i-2$ on the right) are vectors in the dual space of linear functions on R^k; the α_i are easily seen to be linearly independent.

The feasibility set $\mathcal{N}$ is thus defined by $k-2$ linear inequalities,

$$\mathcal{N} = \{\eta : \alpha_i' \eta \leq 0,\ i = 1, \ldots, k-2\}. \tag{2.5}$$

As part of characterizing $\mathcal{N}$ we will find that $\mathcal{N}$ can be viewed as the product of a two-dimensional subspace (spanned by a 1-vector and an x-vector as defined below) and a convex, closed $k-2$ dimensional cone.

Wu (1982) showed that for the special case where y satisfies $\alpha_i' y_i \geq 0$, $i = 1, \ldots, k-2$ (none of the restrictions (2.3) holding strictly), the least squares solution is just the straight-line solution $\eta_i = \bar{y} + b(x_i - \bar{x})$ where $\bar{y} = \sum m_i y_i / \sum m_i$, $\bar{x} = \sum m_i x_i / \sum m_i$ and $b = \sum m_i (x_i - \bar{x}) y_i / \sum m_i (x_i - \bar{x})^2$, for the case of a diagonal matrix M. In this special case the solution belongs to the 2-dimensional subspace spanned by affine functions; it has no concave components.

2.2 A Set of Generators for the Cone-Cylinder

Towards determining generators for the region $\mathcal{N}$ we reexpress (2.3) in terms of the M inner product:

$$\gamma_i' M \eta \leq 0, \qquad i = 1, \ldots, k-2, \tag{2.6}$$

where $\gamma_i = M^{-1} \alpha_i$. We then add two further vectors having zero concavity– thus orthogonal to the preceding vectors γ_i; in particular we choose the unit vectors corresponding to the constant and (orthogonal) linear function:

$$\gamma_{k-1} = 1 / \ \| 1 \|, \qquad \gamma_k = (x - \bar{x}1) / \ \| x - \bar{x}1 \|, \tag{2.7}$$

where $1 = (1, \ldots, 1)'$, $\| 1 \| = (\sum m_i)^{\frac{1}{2}}$, $x = (x_1, \ldots, x_k)'$, $\bar{x} = \sum m_i x_j / \sum m_i$. Let $C = (\gamma_1 \cdots \gamma_k)$; then (2.6) states that

$$C' M \eta \tag{2.8}$$

has its first $k-2$ coordinates ≤ 0.

Now let $B = (\beta_1 \cdots \beta_k)$ be a basis

$$C' M B = -I \tag{2.9}$$

or $B = -M^{-1}(C')^{-1}$, $C = -M^{-1}(B')^{-1}$; in particular $\beta_{k-1} = -\gamma_{k-1}$, $\beta_k = -\gamma_k$. For a point $\eta = \sum b_i \beta_i = Bb$ we have that $b = B^{-1}\eta = -C' M \eta$ and thus (2.6) can be reexpressed (using (2.8)) as

$$b_i \geq 0, \qquad i = 1, \ldots, k-2. \tag{2.10}$$

It follows that the feasible set $\mathcal{N}$ is

$$\begin{aligned} \mathcal{N} &= \left\{ \sum_{i=1}^{k} b_i \beta_i : b_1 \geq 0, \ldots, b_{k-2} \geq 0 \right\} \\ &= K \otimes L^{\perp}(K), \end{aligned} \tag{2.11}$$

where $K = \left\{ \sum_{i=1}^{k-2} b_k \beta_i : b_i \geq 0 \right\}$ is an orthant cone in $L(\beta_1, \ldots, \beta_{k-2})$ and $L^{\perp}(K) = L(\beta_{k-1}, \beta_k)$. The term orthant indicates that K is the positive orthant with respect to the basis $\beta_1, \ldots, \beta_{k-2}$ and thus the cone has the minimum number of generators for the given dimension.

2.3 Remarks

The preceding gives an algebraic and computational procedure for obtaining the generators for the feasible set $\mathcal{N}$. We can, however, provide a direct argument leading to the generators. The following functions

$$\begin{aligned} w_1(x) = -\mid x - x_2 \mid, \quad w_2(x) &= -\mid x - x_3 \mid, \ldots, \\ w_{k-2}(x) &= -\mid x - x_{k-1} \mid, \end{aligned} \tag{2.12}$$

are $k-2$ linearly independent concave functions over the input values $x_1, \ldots, x_k$. A particular function with a positive coefficient provides a second difference contribution at the corresponding point and none at other points. These functions together with the constant and linear functions 1, x generate the full set of piecewise linear concave functions relative to the values $x_1, x_2, \ldots, x_k$. The functions (2.12) can then be orthogonalized to the 1 and

x vectors giving $\beta_1, \ldots, \beta_{k-2}$ except for scale factors. This can be checked by noting that

$$(\alpha_1 \cdots \alpha_{k-2})' (w_1 \cdots w_{k-2})$$

is diagonal where $w_1, \ldots, w_{k-2}$ are the vectors recording values for the functions (2.12). The general concave function vector can thus be written

$$\sum_{i=1}^{k} b_i \beta_i = \sum_{i=1}^{k-2} d_i w_i + a1 + bx$$

where $b_i \geq 0$ $(i = 1, \ldots, k-2)$ and $d_i \geq 0$.

In the next section on general cone-cylinder regression, we determine the relationship between a point y in R^k and the nearest point $\hat{\eta}$ on an orthant cone-cylinder $\mathcal{N}$.

3. CONE-CYLINDER PROJECTIONS

The feasible set in the concave-regression problem is the Cartesian product of a $k-2$ dimensional cone and a 2 dimensional space. In this section, we assume the feasible set to be a general cone-cylinder, the Cartesian product of an l dimensional cone and a $k-l$ dimensional subspace.

Consider a point y in R^k and a feasible set

$$\mathcal{N} = \left\{ \sum_{i=1}^{k} b_i \beta_i : b_i \geq 0, \quad i = 1, \ldots, l \right\}.$$

For this we assume that $\beta_{l+1}, \ldots, \beta_k$ are orthonormal vectors orthogonal to $L(\beta_1, \ldots, \beta_l)$. The set $\mathcal{N}$ then has the structure

$$\mathcal{N} = K \otimes L^{\perp}(K),$$

the Cartesian product on an orthant cone K and a $k-l$ dimensional Euclidean space. In this section, we partition R^k into 2^l regions and correspondingly $\mathcal{N}$ into 2^l regions such that there is a one-one correspondence between regions, and on corresponding regions $y \rightarrow \hat{\eta}$ is an orthogonal linear projection.

As a preliminary we note that the problem can be separated according to K and $L^{\perp}(K)$. The point η in $\mathcal{N}$ which minimizes $\| y - \eta \|$ is the projection of y on $\mathcal{N}$. Since $L^{\perp}(K)$ is orthogonal to the smallest subspace $L(K)$ containing K,

$$P_{\mathcal{N}} y = P_{L^{\perp}(K)} y + P_K y$$

where $P_B y$ denotes the projection of y on the set B. The point $\eta_K^\perp = P_{L^\perp(K)} y$ is the projection of y on the subspace $L^\perp(K)$ and can be computed by the standard methods. The point $\eta_K = P_K y$ is the projection of y on the cone K and we shall now indicate how to obtain it.

As a first case consider a point $\eta = \sum_1^l b_i\beta_i + \sum_{l+1}^k b_i\beta_i$ with $b_1 > 0, \ldots, b_l > 0$, in the interior

$$F_0 = \left\{ \sum_1^k b_i\beta_i : b_i > 0, \quad i = 1, \ldots, l \right\} \tag{3.2}$$

of the cone-cylinder. As η has an open neighbourhood interior to $\mathcal{N}$, it follows that only $y = \eta$ will have η as projection point. Thus the interior F_0 of $\mathcal{N}$ corresponds identically to $S_0 = F_0$ in R^k as the preimage under the projection map $y \to \hat{\eta}$.

As a second case consider a face F_i, say F_1, of the region $\mathcal{N}$:

$$F_1 = \left\{ \eta = \sum_{i=2}^k b_i\beta_i : b_i > 0, \quad i = 2, \ldots, l \right\}. \tag{3.3}$$

From (2.9) we have that γ_1 is orthogonal to F_1 and has direction exterior to F_0. A point $\eta = \sum_{i=2}^k b_i\beta_i$ with $b_i > 0, i = 2, \ldots, l$, corresponds to points

$$y = c_1\gamma_1 + \sum_{i=2}^k b_i\beta_i, \qquad c_1 \geq 0. \tag{3.4}$$

Similarly a neighbourhood of η on the face F_1 corresponds to the cylinder $(c_1\gamma_1, c_1 \geq 0)$ –set over that neighbourhood, and the projection $y \to \hat{\eta}$ is entirely a linear projection to $L(\beta_2 \cdots \beta_k)$. Thus F_1 corresponds to

$$S_1 = \{c_1\gamma_1 + F_1 : c_1 \geq 0\} \tag{3.5}$$

as the preimage under the projection map $y \to \hat{\eta}$.

As a third and general case consider a face F_J where $J = \{i_1, \ldots, i_r\}$ is a subset of $L = \{1, \ldots, l\}$ and for simplicity suppose $J = \{1, \ldots, r\}$:

$$F_J = \left\{ \eta = \sum_{i=r+1}^k b_i\beta_i : b_i > 0, \quad i = r+1, \ldots, l \right\}. \tag{3.6}$$

From (2.9) we have that $\gamma_1, \ldots, \gamma_r$ are orthogonal to F_J and point exterior to F_0. A point $\eta = \sum_{i=r+1}^k b_i\beta_i$ with $b_i > 0, \quad i = r+1, \ldots, l$, corresponds to points

$$y = \sum_{i=1}^r c_i\gamma_i + \sum_{i=r+1}^k b_i\beta_i \qquad c_i \geq 0, \quad i = 1, \ldots, r. \tag{3.7}$$

Similarly a neighbourhood of η on the face F_J corresponds to the cylinder $(c_1\gamma_1 + \cdots + c_r\gamma_r, c_i \geq 0)$-set over that neighbourhood, and the projection $y \to \hat{\eta}$ is entirely an orthogonal linear projection within the cylinder to the base on $L(\beta_{r+1} \cdots \beta_k)$. Thus F_J corresponds to

$$S_J = \{c_1\gamma_1 + \cdots + c_r\gamma_r + F_J : c_i \geq 0\} \tag{3.8}$$

as the preimage under the projection map $y \to \hat{\eta}$.

We summarize the results as follows. Any point y in R^k is in one of the 2^l regions each of which consists of a face of $\mathcal{N}$ and all the points projecting on it. Any point y can be written as

$$y = \sum_{i \in J} c_i\gamma_i + \sum_{i \in L-J} b_i\beta_i + \sum_{i=l+1}^{k} b_i\beta_i$$

with $c_i \geq 0$, $i \in J$ and $b_i > 0$, $i \in L - J$ where $L = \{1, \ldots, l\}$ and J is some subset of L. The point in K that is closest to y is $\eta_K = \sum_{i \in L-J} b_i\beta_i$ and the point η in the cone-cylinder that is closest to y is then

$$\hat{\eta} = \sum_{i \in L-J} b_i\beta_i + \sum_{i=l+1}^{k} b_i\beta_i,$$

obtained by omitting the γ terms.

A formal proof of the preceding can be given by expanding the development in this section, or by using the definitions of the tangent cone $S_K(\hat{\eta})$ and the normal cone $N_K(\hat{\eta})$ of a point $\hat{\eta}$ of a convex set K, and the optimization theory result that the minimum distance point is $\hat{\eta}$ if and only if $y - \hat{\eta}$ belongs to $N_K(\hat{\eta})$. The definitions are given in the Appendix, Section 6.

4. CALCULATING THE PROJECTION

The preceding section shows that R^k can be partitioned into 2^l regions each of which is the Cartesian product of the positive orthant in a basis $B_J^0 = \{\gamma_i, i \in J; \beta_i, i \in L - J\}$ and the subspace spanned by $\beta_i, i = l + 1, \ldots, k$. Let the full basis be $B_J = \{\gamma_i, i \in J; \beta_i, i \in L - J; \beta_i, i = l + 1, \ldots, k\}$. If y can be expressed in B_J as

$$y = \sum_{i \in J} c_i\gamma_i + \sum_{i \in L-J} b_i\beta_i + \sum_{i=l+1}^{k} b_i\beta_i \tag{4.1}$$

where $c_i \geq 0$, $i \in J$ and $b_i > 0$ for $i \in L - J$, then the nearest point $\hat{\eta}$ in the cone-cylinder $\mathcal{N}$ is

$$\hat{\eta} = \sum_{i \in L-J} b_i \beta_i + \sum_{i=l+1}^{k} b_i \beta_i. \tag{4.2}$$

Thus to find the projection $\hat{\eta}$ of y it suffices to determine which of the 2^l bases B_J gives the representation (4.1), and then to omit the γ components thus obtaining the projection (4.2).

For this, consider some initial point y_0 with the special restricted coordinates (4.1) for *some* basis $B_{(0)}$; for an initial point we could choose $y_0 = \sum_1^l \beta_i$ which has the special restricted coordinates with respect to $B_{(0)} = B$. If the data point y also has nonnegative coordinates with respect to $B_{(0)}$, then $B_{(0)}$ is the required basis and the point is its own projection.

If y does not have nonnegative coordinates with respect to $B_{(0)}$ then project y_0 towards y (thus parallel to $y - y_0$) onto the boundary of the positive orthant of $B_{(0)}$; let y_1 be the resulting projection. The point y_1 on the boundary thus adjoins a different positive orthant with basis say $B_{(1)}$. If y has the special restricted coordinates with respect to $B_{(1)}$, then $B_{(1)}$ is the required basis and the projection is obtained by omitting the γ components. If not, then y_1 is treated as the preceding y_0 and the projection procedure is repeated.

In overview, an initial point y_0 is projected to y along the connecting line segment and at each boundary face the orthant basis for a preceding region is replaced by that for the succeeding region.

The projection procedure is one-dimensional and produces a finite sequence $y_1, y_2, \ldots, y_m, y$ of points along the line segment joining y_0 to y; a new point $y_{s+1} = P_{s+1} y_s$ is a projection onto the boundary of the $B_{(s)}$ orthant parallel to the vector $y - y_0$.

The computation for y_{s+1} and for the new orthant proceeds as follows. Let d_i^s and d_i be the coordinates of y_s and y respectively in the basis $B_{(s)}$. The point $y_s + t(y - y_s)$ is at a proportion t along the segment joining y_s to y_1, and has coordinate $d_i^s + t(d_i - d_i^s)$ with respect to the ith basis vector. The value of t must be the first value as t increases from zero such that one of the coordinates $d_i^s + t(d_i - d_i^s)$, $i = 1, \ldots, l$, is zero and all the others are nonnegative. Since the value of t is between 0 and 1, the only coordinates $d_i^s + t(d_i - d_i^s)$ that can decrease from a nonnegative value to a negative value are those for which $d_i < 0$. Thus the required value of t is

$$t = \min\left\{d_i^s / (d_i^s - d_i), 1; d_i < 0\right\}.$$

Let j be the index having $t = d_j^s/(d_j^s - d_j)$; then the basis $B_{(s+1)}$ is obtained from $B_{(s)}$ by replacing the corresponding β_j or γ_j by its opposite γ_j or β_j. The point y is reached when $t = 1$ and there is no corresponding j.

A measure theoretic point can arise if two values of j produce the minimum t value; this corresponds to the line segment intersecting a boundary "edge" and it is then not immediately clear which adjacent orthant is next along the segment. The difficulty can be avoided by adding a random deviation to the initial point y_0, an error small enough to retain the point in the initial orthant.

5. EXAMPLE

Consider the concave regression example by Hildreth (1954) and Wu (1982). In this example y_i is the average corn yield, x_i is the amount of nitrogen fertilizer and $k = 8$, $l = 6$. The data are:

x_i	0	20	40	60	80	120	160	180
m_i	27	9	8	10	9	19	10	8
y_i	22.94	41.58	65.64	58.81	81.74	82.15	96.59	94.01

Here $k = 8$ and $l = 2$ and the matrix M is diag(m_i).
The matrix $B = (B_1 \cdots B_8)$ is shown in Table 1.

Table 1. *Elements of the Matrix B*

1	2	3	4	5	6	7	8
−.293	−.487	−.520	−.366	−.118	−.479	−.1	.114
.409	.169	−.008	−.064	−.045	−.021	−.1	.082
.331	.828	.504	.238	.028	.006	−.1	.050
.254	.638	1.016	.539	.101	.033	−.1	.018
.177	.451	.731	.841	.174	.060	−.1	−.014
.023	.763	.160	.239	.320	.114	−.1	−.078
−.132	−.298	−.411	−.363	−.183	.168	−.1	−.142
−.209	−.485	−.696	−.664	−.434	−.408	−.1	−.174

The components of y in this basis B are:

−6.72857, 36.2041, −37.0879, 37.71, −21.6067, 16.2622, −611.987, −261.074.

The index j for the first projection is 3.

The components of y in the resulting basis $B_{\{3\}} = (\beta_1, \beta_2, \gamma_3, \beta_4, \beta_5, \beta_6, \beta_7, \beta_8)$ are:

0.735345, 11.3894, 37.0879, 9.41632, −13.6496, 16.2622, −611.987, −261.074.

The index j for the second projection is 5.

The components of y in the resulting basis $B_{\{3,5\}} = (\beta_1, \beta_2, \gamma_3, \beta_4, \gamma_5, \beta_6, \beta_7, \beta_8)$ are:

0.117553, 13.4434, 34.0181, 3.73951, 14.3082, 8.85289, −611.987, −261.074.

The first 6 components are positive and thus y is expressed as in (4.1). The projection $\hat{\eta}$ is obtained from y by setting the 3rd and 5th coordinates to 0. The components of $\hat{\eta}$ in the original standard basis are:

22.98, 41.4984, 60.1084, 67.3106, 74.5317 84.4668, 94.402, 94.0346.

The minimum $\sum_{i=1}^{k} m_i(y_i - \hat{\eta}_i)^2$ is 1569.33.

The value of $\sum_{i=1}^{k} m_i(y_i - u_i)^2$ for the optimum point obtained by Wu (1982) is 1570.81.

6. APPENDIX

The tangent or support cone of a closed convex set K at a point η in K is the smallest closed convex cone with vertex at the origin containing $K - \eta$. The tangent cone is designated $S_K(\eta)$ and can be verified as the intersection of the supporting half-spaces tangent to K at η; such a half-space contains $K - \eta$ and has boundary face containing the origin.

In the case where K is the positive orthant with respect to the first l of the basis vectors $\beta_1, \ldots, \beta_k$ and

$$\eta = \sum_{i \in L - J} d_i \beta_i + \sum_{i=l+1}^{k} b_i \beta_i$$

with $d_i > 0$, the supporting half-spaces to K at η are $H_i = \{u : \langle \gamma_i, u \rangle \leq 0\}$ for $i \in J$ and the tangent cone

$$S_k(\eta) = \cap H_i = \left\{u : \sum c_i \langle \gamma_i, u \rangle \leq 0, \; c_i \geq 0 \right\}.$$

The normal cone $N_k(\eta)$ to a closed convex set K at the point η is the dual $S_k^{\perp}(\eta)$ of $S_k(\eta)$:

$$N_K(\eta) = S_K^{\perp}(n) = \left\{u^{\perp} : \langle u^{\perp}, u \rangle \leq 0, u \in S_K(\eta)\right\}.$$

For the present problem the normal cone is the set of all convex combinations of the γ_i vectors:

$$N_K(\eta) = \left\{ \sum_{i \in J} c_i \gamma_i : c_i \geq 0 \right\}.$$

A point $\hat{\eta}$ minimizes the distance from y to a closed convex cone if and only if $y - \hat{\eta}$ belongs to $N_K(\hat{\eta})$ or equivalently y belongs to $\hat{\eta} + N_k(\hat{\eta})$. For the proof of this optimization result see Rockafellar (1970) or Zarantonello (1971). For the present problem

$$\hat{\eta} + N_K(\hat{\eta}) = \hat{\eta} + \left\{ \sum_{i \in J} c_i \gamma_i : c_i \geq 0 \right\},$$

giving the needed result in Section 3.

ACKNOWLEDGMENT

This research has been partially funded by NSERC grants A3011 and A8592.

REFERENCES

Barlow, R. E., D. J. Bartholomew, J. M. Brenner, and H. D. Brunk (1972), *Statistical Inference Under Order Restriction.* New York: Wiley and Sons.

Dent, W. (1973), "A note on least squares fitting of function constrained to be either nonnegative, nondecreasing or convex." *Management Science* **20,** 130–132.

Dent, W., T. Robertson, and M. Johnson (1977), "Concave regression." Unpublished manuscript, University of Iowa.

Dykstra, R. L. (1983), "An algorithm for restricted least squares regression." *Journal of the American Statistical Association* **78,** 837–842.

Hanson, D. L., and G. Pledger (1976), "Consistency in concave regression." *Annals of Statistics* **4,** 1038–1050.

Hildreth, C. (1954), "Point estimates of ordinates of concave functions." *Journal of the American Statistical Association* **49,** 598–619.

Holloway, C. A. (1979), "On the estimation of convex functions." *Operations Research* **27,** 401–407.

Rockafellar, R. T. (1970), *Convex Analysis.* Princeton, New Jersey: Princeton University Press.

Wu, C. F. (1982), "Some algorithms for concave and isotonic regression." *Studies in the Management Sciences* **19,** 105–116.

Zarantonello, E. H. (1971), "Projections on convex sets in Hilbert spaces and spectral theory, Part I: Projections on convex sets." In *Contributions to Nonlinear Functional Analysis,* pp. 237–241. New York: Academic Press.

M. Safiul Haq [1]

ON THE PREDICTION OF THE DIFFERENCE BETWEEN RESPONSES FROM TWO LINEAR MODELS

ABSTRACT

Two linear models are considered, and their structures are utilised to generate prediction distributions of future responses. Then, by a simple linear transformation, the prediction distributions of the differences between responses are obtained. The usual method of derivation involves a cumbersome integration procedure which is avoided here by the use of structural relations.

1. INTRODUCTION

Consider the responses from the following two regression models:

$$\mathbf{y}_1 = X_1\boldsymbol{\beta}_1 + \sigma\mathbf{e}_1$$

and

$$\mathbf{y}_2 = X_2\boldsymbol{\beta}_2 + \sigma\mathbf{e}_2, \tag{1.1}$$

where: $\mathbf{y}'_j = (y_{j1}, y_{j2}, \ldots, y_{jn})$, $j = 1, 2$, are the responses; $\mathbf{e}'_j = (e_{j1}, e_{j2}, \ldots, y_{jn})$, $j = 1, 2$, are the error variables; and $\boldsymbol{\beta}_j = (\beta_{j1}, \beta_{j2}, \ldots, \beta_{jp})$, $j = 1, 2$, and σ (> 0) are the parameters which generate the responses $\mathbf{y}_j$ through X_j, $j = 1, 2$, the design matrices of order $n \times p$, and the error variables $\mathbf{e}_j$, $j = 1, 2$. The responses may be written in a single equation as follows:

$$\begin{bmatrix} \mathbf{y}_1 \\ \mathbf{y}_2 \end{bmatrix} = \begin{bmatrix} X_1 & 0 \\ 0 & X_2 \end{bmatrix} \begin{bmatrix} \boldsymbol{\beta}_1 \\ \boldsymbol{\beta}_2 \end{bmatrix} + \sigma \begin{bmatrix} \mathbf{e}_1 \\ \mathbf{e}_2 \end{bmatrix}$$

or

[1] Department of Statistical and Actuarial Sciences, The University of Western Ontario, London, Ontario N6A 5B9

I. B. MacNeill and G. J. Umphrey (eds.), Foundations of Statistical Inference, 133–138.

$$\mathbf{y} = X\boldsymbol{\beta} + \sigma\mathbf{e}. \tag{1.2}$$

In this paper the prediction distribution of the difference between two future responses $\mathbf{Y}_{1f}$, and $\mathbf{Y}_{2f}$ corresponding to the design matrices X_{1f} and X_{2f} given the responses $\mathbf{y}_1$ and $\mathbf{y}_2$ are derived. Here, the future response vectors $\mathbf{Y}_{jf}$ $(j = 1,2)$ are of order $m \times 1$ and the design matrices X_{jf}, $(j = 1,2)$ are of order $(m \times p)$. The derivation is based on the use of the linear relation (1.2), which is described as a structural relation in the next section. Haq (1978, 1982) has demonstrated the use of structural relations for the derivation of prediction distributions for location/scale parameters in multivariate models.

2. THE LINEAR STRUCTURAL MODEL AND ITS USE IN PREDICTION

The linear model described in (1.2) may be represented as

$$\begin{aligned}\mathbf{y} &= [\boldsymbol{\beta}_1', \boldsymbol{\beta}_2', \sigma]_X \mathbf{e} \\ &= [\boldsymbol{\beta}', \sigma]_X \mathbf{e},\end{aligned} \tag{2.1}$$

where $[\boldsymbol{\beta}', \sigma]$ is an element of the group of transformations, $G = \{[\mathbf{a}, c]\} = \{[\mathbf{a}_1, \mathbf{a}_2, c] \mid -\infty < \mathbf{a}_1 < \infty,\ -\infty < \mathbf{a}_2 < \infty; \sigma > 0\}$. The group G operates on the sample space of $\mathbf{e}$ or $\mathbf{y}$ through a design matrix Z in the following way:

$$\begin{aligned}\mathbf{e} &= [\mathbf{a}', c]_Z \mathbf{e}_0 \\ &= Z\mathbf{a} + c\mathbf{e}_0;\end{aligned} \tag{2.2}$$

that is,

$$\begin{bmatrix} \mathbf{e}_1 \\ \mathbf{e}_2 \end{bmatrix} = \begin{bmatrix} Z_1 & 0 \\ 0 & Z_2 \end{bmatrix} \begin{bmatrix} \mathbf{a}_1 \\ \mathbf{a}_2 \end{bmatrix} + c \begin{bmatrix} \mathbf{e}_{10} \\ \mathbf{e}_{20} \end{bmatrix},$$

where $\mathbf{e}$ and $\mathbf{e}_0$ belong to the sample space, and the partitioning of matrices and vectors corresponds to (1.2). It is to be noted that $[\mathbf{a}, c]^{-1} = [-\mathbf{a}'c^{-1};\ c^{-1}] = [-\mathbf{a}_1'c^{-1}, -\mathbf{a}_2c^{-1};\ c^{-1}]$ and the identity element is $[\mathbf{0}', 1]$.

Let the probability element of $\mathbf{e}$ be denoted by

$$p(\mathbf{e}) = f(\mathbf{e}_1, \mathbf{e}_2)d\mathbf{e}_1\ d\mathbf{e}_2. \tag{2.3}$$

The model described by (2.1) and (2.3) is a structural model. Then straightforward structural analysis yields the following reduced model:

$$[\mathbf{b}'(y), s(y)] = [\boldsymbol{\beta}, \sigma][\mathbf{b}'(e), s(e)], \tag{2.4}$$

and the conditional probability element for $\mathbf{b}(e)$ and $s(e)$ given $\mathbf{d}$ is

$$p(\mathbf{b}(e), s(e)|\mathbf{d}) = k(\mathbf{d}) f([\mathbf{b}_1(e), s(e)]_{X_1}\mathbf{d}_1, [\mathbf{b}_2(e), s(e)]_{X_2}\mathbf{d}_2) \times [s(e)]^{n-p} d\mathbf{b}_1(e)\, d\mathbf{b}_2(e)\, ds(e),$$

where

$$\mathbf{b}(e) = \begin{bmatrix} \mathbf{b}_1(e) \\ \mathbf{b}_2(e) \end{bmatrix} = \begin{bmatrix} (X_1'X_1^{-1})X_1'\mathbf{e}_1 \\ (X_2'X_2^{-1})X_2'\mathbf{e}_2 \end{bmatrix};$$

$$\begin{aligned} [s(e)]^2 &= (\mathbf{e} - X\mathbf{b}(e))'(\mathbf{e} - X\mathbf{b}(e)) \\ &= (\mathbf{e}_1 - X_1\mathbf{b}_1(e))'(\mathbf{e}_1 - X_1\mathbf{b}_1(e)) \\ &\quad + (\mathbf{e}_1 - X_2\mathbf{b}_2(e))'(\mathbf{e}_2 - X_2\mathbf{b}_2(e)) \end{aligned} \tag{2.5}$$

and

$$\begin{aligned} \mathbf{d} &= (\mathbf{e} - X\mathbf{b}(e))/s(e) \\ &= \frac{1}{s(e)} \begin{bmatrix} \mathbf{e}_1 - X_1\mathbf{b}_1(e) \\ \mathbf{e}_2 - X_2\mathbf{b}_2(e) \end{bmatrix}, \\ &= \frac{1}{s(y)} \begin{bmatrix} \mathbf{y}_1 - X_1\mathbf{b}_1(y) \\ \mathbf{y}_2 - X_2\mathbf{b}_2(y) \end{bmatrix} \\ &= \begin{bmatrix} \mathbf{d}_1 \\ \mathbf{d}_2 \end{bmatrix}. \end{aligned} \tag{2.6}$$

The reduced structural equation (2.4) yields:

$$\begin{aligned} [\boldsymbol{\beta}', \sigma] &= [\mathbf{b}'(y), s(y)][\mathbf{b}'(e), s(e)]^{-1} \\ &= \left[\mathbf{b}'(y) - \mathbf{b}'(e)\frac{s(y)}{s(e)}, \frac{s(y)}{s(e)}\right]. \end{aligned} \tag{2.7}$$

Let

$$\mathbf{Y}_{1f} = X_{1f}\boldsymbol{\beta}_1 + \sigma\mathbf{e}_{1f}$$

and

$$\mathbf{Y}_{2f} = Y_{2f}\boldsymbol{\beta}_2 + \sigma\mathbf{e}_{2f} \tag{2.9}$$

be the set of future responses from the model (1.2) corresponding to the design matrices X_{1f} and X_{2f} respectively, where $\mathbf{Y}_{jf}$, $j = 1, 2$, is of order $m \times 1$ and $\mathbf{X}_{jf}$, $j = 1, 2$, is of order $m \times p$. As before, in group notation, the future responses may be expressed as follows:

$$\begin{aligned} \mathbf{Y}_f &= [\boldsymbol{\beta}, \sigma]_{X_f}\mathbf{e}_f \\ &= \left[\mathbf{b}'(y) - \mathbf{b}'(e)\frac{s(y)}{s(e)}, \frac{s(y)}{s(e)}\right]_{X_f}\mathbf{e}_f \\ &= X_f\mathbf{b}(y) + \frac{s(y)}{s(e)}(e_f - X_f\mathbf{b}(e)). \end{aligned} \tag{2.10}$$

Componentwise, the future response may be expressed as

$$\mathbf{Y}_{1f} = X_{1f}\mathbf{b}_1(y) + \frac{s(y)}{s(e)}(\mathbf{e}_{1f} - X_{1f}\mathbf{b}_1(e))$$

and

$$\mathbf{Y}_{2f} = X_{2f}\mathbf{b}_2(y) + \frac{s(y)}{s(e)}(\mathbf{e}_{2f} - X_2\mathbf{b}_2(e)).$$

The prediction distribution of $\mathbf{Y}_f$ may be obtained from the distribution of $(\mathbf{e}_f - X_f\mathbf{b}(e))/s(e)$, and, from the distribution of $\mathbf{Y}_f$, the distribution of $\mathbf{Y}_{1f} - \mathbf{Y}_{2f}$ can easily be obtained. If the design matrices were the same for $\mathbf{y}_1$ and $\mathbf{y}_2$, one could form a transformed structural model for $\mathbf{y}_1 - \mathbf{y}_2$ and the prediction distribution could be obtained directly from $\mathbf{e}_{1f} - \mathbf{e}_{2f}$. The result obtained by the transformed responses would be the same as indicated by Dawid and Stone (1982).

3. THE NORMAL ERROR MODEL

Let the probability distribution of the error variable of the structural model described by (2.1) and (2.3) be

$$p(\mathbf{e}_1, \mathbf{e}_2) = (2\pi)^{-n} \exp\left[-\frac{1}{2}\mathbf{e}_1'\mathbf{e}_1 - \frac{1}{2}\mathbf{e}_2'\mathbf{e}_2\right] d\mathbf{e}_1\, d\mathbf{e}_2. \tag{3.1}$$

Then, it is readily seen that $\mathbf{e}_f - X_f\mathbf{b}(e) = ((\mathbf{e}_{1f} - X_{1f}\mathbf{b}_e(e))', (e_{2f} - Y_{2f}\mathbf{b}_2(e))')'$ has a multivariate normal distribution with mean $\mathbf{0}$ and covariance matrix

$$\begin{aligned} R &= \begin{bmatrix} R_1 & 0 \\ 0 & R_2 \end{bmatrix} \\ &= \begin{bmatrix} I_m + X_{1f}(X_1'X_1)^{-1}X_{1f} & 0 \\ 0 & I_m + X_{2f}(X_2'X_2)^{-1}X_{2f} \end{bmatrix}, \end{aligned} \tag{3.2}$$

where I_m is $m \times m$ identity matrix. Then following Cornish (1954) and Fraser (1964), it is readily seen that

$$\begin{aligned} \mathbf{t} &= \frac{2(n-p)}{s(e)}(\mathbf{e}_f - X\mathbf{b}(e)) \\ &= \frac{2(n-p)}{s(e)} \begin{bmatrix} \mathbf{e}_{1f} - X_{1f}\mathbf{b}_1(e) \\ \mathbf{e}_{2f} - X_{2f}\mathbf{b}_2(e) \end{bmatrix} \\ &= \begin{bmatrix} \mathbf{t}_1 \\ \mathbf{t}_2 \end{bmatrix} \end{aligned}$$

has a $2m$ dimensional multivariate t-distribution with $2(n-p)$ degrees of freedom. Its probability density function is given by

$$p(\mathbf{t}) = \frac{\Gamma(m+n-p)|R|^{-1/2}}{(2\pi(n-p))^m\Gamma(n-p)}\left\{1+\frac{1}{2(n-p)}\mathbf{t}'R^{-1}\mathbf{t}\right\}^{-(m+n-p)} . \tag{3.3}$$

Then, following Press (1972, p. 128), the probability density function of $\mathbf{u} = \mathbf{t}_1 - \mathbf{t}_2 = s^{-1}(y)(Y_{1f} - Y_{2f} - X_{1f}\mathbf{b}_1(y) + X_{2f}\mathbf{b}_2(y))\sqrt{2(n-p)}$ is obtained as

$$p(\mathbf{u}) = \frac{\Gamma\left(m/2+(n-p)\right)}{(2\pi(n-p))^{m/2}\Gamma(n-p)}\left\{1+\frac{1}{2(n-p)}\mathbf{u}Q^{-1}\mathbf{u}\right\}^{-m/2-(n-p)} , \tag{3.4}$$

where

$$Q = CRC'$$

and

$$C = \begin{bmatrix} 1 & 0 & \cdots & 0 & -1 & 0 & \cdots & 0 \\ 0 & 1 & & 0 & 0 & -1 & & 0 \\ \vdots & & \ddots & & \vdots & & \ddots & \\ 0 & 0 & & 1 & 0 & 0 & & -1 \end{bmatrix} .$$

Thus, the prediction distribution of $Y_{1f} - Y_{2f}$ is an m-dimensional multivariate t with $2(n-p)$ degrees of freedom. The expected value of $Y_{1f} - Y_{2f}$ is $X_{1f}\mathbf{b}_1(y) - X_{2f}\mathbf{b}_2(y)$, and the covariance matrix is $\frac{n-p}{n-2-p}Q$. The density (3.4) is a monotonically increasing function of the quadratic form $\mathbf{u}'Q^{-1}\mathbf{u}$ so that the contours of $p(\mathbf{y}_{1f} - \mathbf{y}_{2f} \mid \mathbf{y})$ are ellipsoid in the sample space. Furthermore, the quantity

$$\begin{aligned}\frac{1}{m}\mathbf{u}'Q^{-1}\mathbf{u} = \frac{2(n-p)}{ms^2(y)}&(Y_{1f} - Y_{2f} - X_{1f}\mathbf{b}_1(y) + X_{2f}\mathbf{b}_2(y))' \\ &\times Q^{-1}(Y_{1f} - Y_{2f} - X_{1f}\mathbf{b}_1(y) + X_{2f}\mathbf{b}_2(y))\end{aligned}$$

is distributed as an F-variate with m and $2(n-p)$ degrees of freedom. Therefore, the region defined by

$$\begin{aligned}s^{-2}(y)&(Y_{1f} - Y_{2f} - X_{1f}\mathbf{b}_1(y) + Y_{2f}b_2(y))' \\ &\times Q^{-1}(Y_{1f} - Y_{2f} - X_{1f}\mathbf{b}(y) + Y_{2f}\mathbf{b}_2(y)) \\ &\leq \frac{m}{2(n-p)}F_\alpha(m, 2(n-p))\end{aligned}$$

will give a $(1-\alpha)100\%$ prediction region for the difference between the two future responses, where $F_\alpha(m, 2(n-p))$ is the $100(1-\alpha)$ percentile of the F-distribution.

REFERENCES

Cornish, E. A. (1954), "The multivariate t-distribution associated with a set of normal sample deviates". *Australian Journal of Physics* **7**, 531–542.

Dawid, A. P., and M. Stone (1982), "The functional model basis for fiducial inference". *Annals of Statistics* **10**, 1054–1067.

Fraser, D. A. S. (1964), "Fiducial inference for location and scale parameters". *Biometrika* **51**, 17–24.

Haq, M. Safiul (1978), "On prediction from the location-scale model with equicorrelated response". *Communications in Statistics A, Theory and Methods* **7**, 857–866.

Haq, M. Safiul (1982), "Structural relations and prediction for multivariate models". *Statistische Hefte* **23**, 218–227.

Press, S. J. (1972), *Applied Multivariate Analysis*. New York: Holt, Rinehart and Winston.

D. S. Tracy and K. G. Jinadasa [1]

ON ULTRASTRUCTURAL RELATIONSHIPS MODELS

ABSTRACT

In the long history of the errors-in-variables model, two types of models were considered, namely structural relationships and functional relationships models. Dolby (1976) combined these two models in the univariate case into a single model called the ultrastructural relationships model and found the maximum likelihood estimators of the parameters. We extend Dolby's ultrastructural relationships model to the multivariate case, assuming the existence of an estimator for the error covariance matrix, independent of the observations. We find that the maximum likelihood method is unable to distinguish between the multivariate ultrastructural relationships model and the multivariate functional relationships model considered by Amemiya and Fuller (1984).

1. INTRODUCTION

When variables are measured with possible errors of measurement, as is the usual case in practice, models are constructed to find the relationships among such variables. In the literature two types of models are considered, namely structural relationships and functional relationships. Anderson (1951) gave estimates of parameters and their properties for these two models separately in the multivariate case, and the results are summarized by Anderson (1984). Dolby (1976) combined these two models for the univariate case, calling it the ultrastructural relationships model, and found the maximum likelihood estimators of the parameters. Villegas (1982) has found the maximum likelihood estimators for the multivariate functional relationships model assuming the existence of an estimator for the error covariance matrix. Amemiya and Fuller (1984) and Amemiya (1982) have found the maximum

[1] Department of Mathematics and Statistics, University of Windsor, Windsor, Ontario N9B 3P4 (both authors)

I. B. MacNeill and G. J. Umphrey (eds.), Foundations of Statistical Inference, 139–147.

likelihood estimators and their asymptotic properties for the multivariate functional and structural relationships models, assuming the same set up as Villegas. Except by Dolby (1976), the two models have been considered separately.

It is natural to attempt to combine the two multivariate models into the multivariate ultrastructural relationships model. We find that the maximum likelihood method fails to distinguish between the ultrastructural model and the functional relationships model under the assumption of the existence of an estimator for the error covariance matrix.

2. MATRIX RESULTS

The following propositions are extensions of Theorem 2.1 of Amemiya (1982, p. 26) and are useful in our calculations.

Proposition 1. Let $\Sigma_{\epsilon\epsilon}$ be a symmetric positive definite matrix and Σ_{xx} be a symmetric non-negative definite matrix of lower order than $\Sigma_{\epsilon\epsilon}$. Let

$$\Sigma_{\epsilon\epsilon} = \begin{pmatrix} \Sigma_{ee} & \Sigma_{eu} \\ \Sigma_{ue} & \Sigma_{uu} \end{pmatrix},$$

where the order of Σ_{uu} is the same as of Σ_{xx}. Let

$$\Sigma_{ZZ} = \Sigma_{\epsilon\epsilon} + (\beta', I)'\Sigma_{xx}(\beta', I),$$

where I is the unit matrix and β is a matrix such that the above operations are conformable. Then

$$\left[(\beta', I)\Sigma_{ZZ}^{-1}(\beta', I)'\right]^{-1} = \Sigma_{uu} + \Sigma_{xx} - \Sigma_{uv}\Sigma_{vv}^{-1}\Sigma_{vu},$$

where

$$\Sigma_{uv} = \Sigma'_{vu} = \Sigma_{ue} - \Sigma_{uu}\beta' \text{ and } \Sigma_{vv} = (I, -\beta)\Sigma_{\epsilon\epsilon}(I, \beta)'.$$

Proof. $\Sigma_{ZZ} = \Sigma_{\epsilon\epsilon} + (\beta', I)'\Sigma_{xx}(\beta', I) = \begin{pmatrix} \Sigma_{ee} + \beta\Sigma_{xx}\beta' & \Sigma_{eu} + \beta\Sigma_{xx} \\ \Sigma_{ue} + \Sigma_{uu}\beta' & \Sigma_{uu} + \Sigma_{xx} \end{pmatrix}.$

Since Σ_{ZZ} is positive definite, using the formula for the inverse of a partitioned positive definite matrix we have

$$\Sigma_{ZZ}^{-1} = \begin{pmatrix} P^{-1} & A_1 \\ A_2 & A_3 \end{pmatrix},$$

where

$$\begin{aligned} P^{-1} &= \Sigma_{ee} + \beta\Sigma_{xx}\beta' - (\Sigma_{eu} + \beta\Sigma_{xx})(\Sigma_{uu} + \Sigma_{xx})^{-1}(\Sigma_{ue} + \Sigma_{xx}\beta'), \\ A_1 &= -P^{-1}(\Sigma_{eu} + \beta\Sigma_{xx})(\Sigma_{uu}\Sigma_{xx})^{-1}, \\ A_2 &= -(\Sigma_{uu} + \Sigma_{xx})^{-1}(\Sigma_{ue} + \Sigma_{xx}\beta')P^{-1}, \end{aligned}$$

and

$$\begin{aligned} A_3 = {} & (\Sigma_{uu} + \Sigma_{xx})^{-1}(\Sigma_{ue} + \Sigma_{xx}\beta)P^{-1}(\Sigma_{eu} + \beta\Sigma_{xx}) \\ & \times (\Sigma_{uu} + \Sigma_{xx})^{-1} + (\Sigma_{uu} + \Sigma_{xx})^{-1}. \end{aligned}$$

Thus

$$\begin{aligned} (\beta', I)\Sigma_{ZZ}^{-1}(\beta', I)' = {} & (\Sigma_{uu} + \Sigma_{xx})^{-1} \\ & + \left[\beta' - (\Sigma_{uu} + \Sigma_{xx})^{-1}(\Sigma_{ue} + \Sigma_{xx}\beta)\right] P^{-1} \\ & \times \left[\beta - (\Sigma_{eu} + \beta\Sigma_{xx})(\Sigma_{uu} + \Sigma_{xx})^{-1}\right]. \end{aligned}$$

Now using the formula for the inverse of the sum of two matrices, we can write

$$\begin{aligned} & [(\beta', I)\Sigma_{ZZ}^{-1}(\beta', I)']^{-1} \\ & \quad = \Sigma_{uu} + \Sigma_{xx} - (\Sigma_{uu} + \Sigma_{xx}) \\ & \quad \times \left[\beta' - (\Sigma_{uu} + \Sigma_{xx})^{-1}(\Sigma_{ue} + \Sigma_{xx}\beta')\right] \\ & \quad \times \left\{P + \left[\beta - (\Sigma_{eu} + \beta\Sigma_{xx})(\Sigma_{uu} + \Sigma_{xx})^{-1}\right]\right. \\ & \quad \times \left.(\Sigma_{uu} + \Sigma_{xx})\left[\beta' - (\Sigma_{uu} + \Sigma_{xx})^{-1}(\Sigma_{ue} + \Sigma_{xx}\beta')\right]\right\}^{-1} \\ & \quad \times \left[\beta - (\Sigma_{eu} + \beta\Sigma_{xx})(\Sigma_{uu} + \Sigma_{xx})^{-1}\right](\Sigma_{uu} + \Sigma_{xx}). \end{aligned} \tag{1}$$

Direct multiplication gives us

$$(\Sigma_{uu} + \Sigma_{xx})\left[\beta' - (\Sigma_{uu} + \Sigma_{xx})^{-1}(\Sigma_{eu} + \Sigma_{xx}\beta')\right] = -\Sigma_{uv}. \tag{2}$$

Finally substituting (2) and the expression for P into (1) gives the result.

Proposition 2. Let $\Sigma_{\epsilon\epsilon}$ and β be as in Proposition 1. Then

$$\begin{aligned} \Sigma_{\epsilon\epsilon}^{-1} - \Sigma_{\epsilon\epsilon}^{-1}(\beta', I)'\left[(\beta', I)\Sigma_{\epsilon\epsilon}^{-1}(\beta', I)'\right]^{-1}(\beta', I)\Sigma_{\epsilon\epsilon}^{-1} \\ = (I, -\beta)'\Sigma_{vv}^{-1}(I, -\beta). \end{aligned}$$

Proof. By Proposition 1, with $\Sigma_{xx} = 0$, we have

$$\left[(\beta', I)\Sigma_{\epsilon\epsilon}^{-1}(\beta', I)'\right]^{-1} = \Sigma_{uu} - \Sigma_{uv}\Sigma_{vv}^{-1}\Sigma_{vu}. \tag{3}$$

Now

$$\Sigma_{ee}^{-1} = \begin{pmatrix} Q^{-1} & -Q^{-1}\Sigma_{eu}\Sigma_{uu}^{-1} \\ -\Sigma_{uu}\Sigma_{ue}Q^{-1} & \Sigma_{uu}^{-1} + \Sigma_{uu}^{-1}\Sigma_{ue}Q^{-1}\Sigma_{eu}\Sigma_{uu}^{-1} \end{pmatrix},$$

where

$$Q = \Sigma_{ee} - \Sigma_{eu}\Sigma_{uu}^{-1}\Sigma_{ue}.$$

Note that

$$\Sigma_{vv} - Q = \Sigma_{vu}\Sigma_{uu}^{-1}\Sigma_{uv}. \tag{4}$$

Now

$$\Sigma_{ee}^{-1}(\beta', I)' = \begin{pmatrix} Q^{-1}\beta - Q^{-1}\Sigma_{eu}\Sigma_{uu}^{-1} \\ -\Sigma_{uu}^{-1}\Sigma_{ue}Q^{-1}\beta + \Sigma_{uu}^{-1} + \Sigma_{uu}^{-1}\Sigma_{ue}Q^{-1}\Sigma_{eu}\Sigma_{uu}^{-1} \end{pmatrix}. \tag{5}$$

By pre- and post-multiplication of (3) by (5) and by the transpose of (5) respectively, and expressing in partitioned form, we get

$$\Sigma_{ee}^{-1}(\beta', I)'\left[(\beta', I)\Sigma_{ee}^{-1}(\beta', I)'\right]^{-1}(\beta', I)\Sigma_{ee}^{-1} = \begin{pmatrix} H_{11} & H_{12} \\ H_{21} & H_{22} \end{pmatrix},$$

where

$$\begin{aligned} H_{11} &= Q^{-1}\left(\beta - \Sigma_{eu}\Sigma_{uu}^{-1}\right)\left(\Sigma_{uu} - \Sigma_{uv}\Sigma_{vv}^{-1}\Sigma_{vu}\right)\left(\beta' - \Sigma_{uu}^{-1}\Sigma_{ue}\right)Q^{-1}, \\ H_{12} &= Q^{-1}\left(\beta - \Sigma_{eu}\Sigma_{uu}^{-1}\right)\left(\Sigma_{uu} - \Sigma_{uv}\Sigma_{vv}^{-1}\Sigma_{vu}\right) \\ &\quad \times \left[\Sigma_{uu}^{-1} - \left(\beta' - \Sigma_{uu}^{-1}\Sigma_{ue}\right)Q^{-1}\Sigma_{eu}\Sigma_{uu}^{-1}\right], \\ H_{21} &= H_{12}', \end{aligned}$$

and

$$\begin{aligned} H_{22} &= \left[\Sigma_{uu}^{-1} - \Sigma_{uu}^{-1}\Sigma_{ue}Q^{-1}\left(\beta - \Sigma_{eu}\Sigma_{uu}^{-1}\right)\right]\left(\Sigma_{uu} - \Sigma_{uv}\Sigma_{vv}^{-1}\Sigma_{vu}\right) \\ &\quad \times \left[\Sigma_{uu}^{-1} - \left(\beta' - \Sigma_{uu}\Sigma_{ue}\right)Q^{-1}\Sigma_{eu}\Sigma_{uu}^{-1}\right]. \end{aligned}$$

By direct multiplication and using the definitions of Σ_{uv} and Σ_{vu}, we can write

$$H_{11} = Q^{-1}\left[\Sigma_{vu}\Sigma_{uu}^{-1}\Sigma_{uv} - \Sigma_{vu}\Sigma_{uu}^{-1}\Sigma_{uv}\Sigma_{vv}^{-1}\Sigma_{vu}\Sigma_{uu}^{-1}\Sigma_{uv}\right]Q^{-1}. \tag{6}$$

Now substituting (4) into (6) we get

$$H_{11} = Q^{-1} - \Sigma_{vv}^{-1}.$$

Comparing H_{11} and H_{12}, using the definitions of Σ_{uv} and Σ_{vu} and (4), we can write

$$H_{12} = -H_{11}\Sigma_{eu}\Sigma_{uu}^{-1} - Q^{-1}\Sigma_{vu}\Sigma_{uu}^{-1} + Q^{-1}(\Sigma_{vv} - Q)\left(\Sigma_{vv}^{-1}\Sigma_{vu}\Sigma_{uu}^{-1}\right).$$

Now substituting the final expression for H_{11} and simplifying we get

$$H_{12} = -Q^{-1}\Sigma_{eu}\Sigma_{uu}^{-1} + \Sigma_{vv}^{-1}\beta.$$

Again comparing H_{22} and H_{12} and simplifying we can write

$$\begin{aligned} H_{22} = &\left[\Sigma_{uu}^{-1} - \Sigma_{uu}^{-1}\Sigma_{uv}\Sigma_{vv}^{-1}\Sigma_{vu}\Sigma_{uu}^{-1}\right]\left[I + \Sigma_{uv}Q^{-1}\Sigma_{eu}\Sigma_{uu}^{-1}\right] \\ &- \Sigma_{uu}^{-1}\Sigma_{ue}H_{12}. \end{aligned} \tag{7}$$

Expanding the terms of (7) and substituting the final expression of H_{12}, and simplifying using (4), we get

$$H_{22} = \Sigma_{uu}^{-1} + \Sigma_{uu}^{-1}\Sigma_{ue}Q^{-1}\Sigma_{eu}\Sigma_{uu}^{-1} - \beta'\Sigma_{vv}^{-1}\beta.$$

Using the final expressions of H_{11}, H_{12} and H_{22} we get the required expression.

Proposition 3. Let Σ_{ee} and Σ_{ZZ} be as in Proposition 1. Then

$$\begin{aligned} &\Sigma_{ZZ}^{-1}\left[\Sigma_{ZZ} - (\beta', I)'\left[(\beta', I)\Sigma_{ZZ}^{-1}(\beta', I)'\right]^{-1}(\beta', I)\right]\Sigma_{ZZ}^{-1} \\ &= \Sigma_{ee}^{-1}\left[\Sigma_{ee}^{-1} - (\beta', I)'\left[(\beta', I)\Sigma_{ee}^{-1}(\beta', I)'\right]^{-1}(\beta', I)\right]\Sigma_{ee}^{-1} \\ &= (I, -\beta)'\Sigma_{vv}^{-1}(I, -\beta). \end{aligned}$$

Proof. Apply the result of Proposition 1 along with the definition of Σ_{ZZ} to get

$$\begin{aligned} &\Sigma_{ZZ} - (\beta', I)'\left[(\beta', I)\Sigma_{ZZ}^{-1}(\beta', I)'\right]^{-1}(\beta', I) \\ &= \Sigma_{ee} - (\beta', I)'\left[(\beta', I)\Sigma_{ee}^{-1}(\beta', I)'\right]^{-1}(\beta', I). \end{aligned} \tag{8}$$

Substituting (3) into (8) we get

$$\begin{aligned} &\Sigma_{ZZ} - (\beta', I)'\left[(\beta', I)\Sigma_{ZZ}^{-1}(\beta', I)'\right]^{-1}(\beta', I) \\ &= \Sigma_{ee}^{-1} - (\beta', I)'\left[(\beta', I)\Sigma_{ee}^{-1}(\beta', I)'\right]^{-1}(\beta', I). \end{aligned} \tag{9}$$

Note that

$$\Sigma_{ZZ}^{-1} = \Sigma_{ee}^{-1} - \Sigma_{ee}^{-1}(\beta', I)'\left[\Sigma_{xx}^{-1} + (\beta', I)\Sigma_{ee}^{-1}(\beta', I)'\right]^{-1}(\beta', I)\Sigma_{ee}^{-1}. \tag{10}$$

Now premultiplying (9) by (10) and simplifying we get

$$\begin{aligned} &\Sigma_{ZZ}^{-1}\left\{\Sigma_{ZZ} - (\beta', I)'\left[(\beta', I)\Sigma_{ZZ}^{-1}(\beta', I)'\right]^{-1}(\beta', I)\right\} \\ &= I - \Sigma_{ee}^{-1}(\beta', I)'\left[(\beta', I)\Sigma_{ee}^{-1}(\beta', I)'\right]^{-1}(\beta', I). \end{aligned} \tag{11}$$

Finally post-multiplying (11) by (10) and simplifying we get the first equality. The second equality is Proposition 2.

3. ULTRASTRUCTURAL RELATIONSHIPS MODEL

Let $\mathbf{X}_t$ and $\mathbf{Y}_t$ be observed values of the random vectors $\mathbf{x}_t$ and $\mathbf{y}_t$ with errors of measurement $\mathbf{u}_t$ and $\mathbf{e}_t$ respectively, $t = 1, 2, \ldots, n$. Let the relationship between $\mathbf{x}_t$ and $\mathbf{y}_t$ be $\mathbf{y}_t = \boldsymbol{\beta}_o + \beta \mathbf{x}_t$ and $E(\mathbf{x}_t) = \boldsymbol{\mu}_t$, $t = 1, 2, \ldots, n$. Further let $\mathbf{x}_t = \boldsymbol{\mu}_t + \mathbf{v}_t$ with $E(\mathbf{v}_t) = \mathbf{0}$. Then

$$\mathbf{Z}_t = \begin{pmatrix} \mathbf{y}_t \\ \mathbf{x}_t \end{pmatrix} + \begin{pmatrix} \boldsymbol{\beta}_o \\ \mathbf{0} \end{pmatrix} + \begin{pmatrix} \beta \\ I \end{pmatrix} \boldsymbol{\mu}_t + \begin{pmatrix} \beta \\ I \end{pmatrix} \mathbf{v}_t + \boldsymbol{\varepsilon}_t, \quad \text{where } \boldsymbol{\varepsilon}_t = \begin{pmatrix} \mathbf{e}_t \\ \mathbf{u}_t \end{pmatrix}.$$

Assume that $\mathbf{v}_t \sim N(\mathbf{0}, \Sigma_{xx})$, $\boldsymbol{\varepsilon}_t \sim N(\mathbf{0}, \Sigma_{\varepsilon\varepsilon})$, $\boldsymbol{\varepsilon}_t$ and $\mathbf{v}_t$ are uncorrelated and

$$\Sigma_{\varepsilon\varepsilon} = \begin{pmatrix} \Sigma_{ee} & \Sigma_{eu} \\ \Sigma_{ue} & \Sigma_{uu} \end{pmatrix}.$$

When $\Sigma_{xx} = 0$, we get the functional relationships model, and when $\boldsymbol{\mu}_t = \boldsymbol{\mu}$, $t = 1, 2, \ldots, n$, we get the structural relationships model.

Now

$$\operatorname{Cov}(\mathbf{Z}_t, \mathbf{Z}_t') = \Sigma_{\varepsilon\varepsilon} + \begin{pmatrix} \beta \\ I \end{pmatrix} \Sigma_{xx}(\beta', I).$$

We assume that there exists an estimator $S_{\varepsilon\varepsilon}$ for $\Sigma_{\varepsilon\varepsilon}$, which is independent of $\mathbf{Z}_t$, $t = 1, \ldots, n$, and such that $dS_{\varepsilon\varepsilon}$ has a Wishart distribution with d degrees of freedom and covariance matrix $\Sigma_{\varepsilon\varepsilon}$.

Theorem. If the maximum likelihood estimator of Σ_{xx} in the ultrastructural model is $\hat{\Sigma}_{xx}$, then $\hat{\Sigma}_{xx} = 0$.

Proof. The log-likelihood function of this model is

$$\begin{aligned} \log\ L = C_o &- \frac{n}{2} \log \mid \Sigma_{ZZ} \mid \\ &- \frac{1}{2} \sum_{t=1}^{n} (\mathbf{Z}_t - \boldsymbol{\alpha} - \gamma \boldsymbol{\mu}_t)' \, \Sigma_{ZZ}^{-1} \, (\mathbf{Z}_t - \boldsymbol{\alpha} - \gamma \boldsymbol{\mu}_t) \\ &- \frac{d}{2} \log \mid \Sigma_{\varepsilon\varepsilon} \mid - \frac{d}{2} \operatorname{tr} \left(S_{\varepsilon\varepsilon} \Sigma_{\varepsilon\varepsilon}^{-1} \right), \end{aligned} \tag{12}$$

where

$$\boldsymbol{\alpha} = \begin{pmatrix} \boldsymbol{\beta}_o \\ \mathbf{0} \end{pmatrix}, \qquad \gamma = \begin{pmatrix} \beta \\ I \end{pmatrix}$$

and C_o is a constant.

In order to find the maximum likelihood estimators, we have to maximize the above function with respect to the parameters. First fix β, $\Sigma_{\epsilon\epsilon}$ and Σ_{xx} and maximize with respect to $\boldsymbol{\beta}_o$ and $\boldsymbol{\mu}_t$, $t = 1, 2, \ldots, n$. Equating the derivative of (12) with respect to $\boldsymbol{\beta}_o$ to zero, we get

$$\sum_{t=1}^{n} (\mathbf{Y}_t - \boldsymbol{\beta}_o - \beta\boldsymbol{\mu}_t) = \mathbf{0} \quad \text{and} \quad \sum_{t=1}^{n} (\mathbf{X}_t - \boldsymbol{\mu}_t) = \mathbf{0},$$

which implies

$$\overline{\mathbf{Y}} - \beta\overline{X} = \hat{\boldsymbol{\beta}}_o \quad \text{and} \quad \hat{\boldsymbol{\mu}} = \overline{\mathbf{X}},$$

where $\overline{\mathbf{X}} = \frac{1}{n}\sum_{t=1}^{n} \mathbf{X}_t$ and $\overline{\mathbf{Y}} = \frac{1}{n}\sum_{t=1}^{n} \mathbf{Y}_t$.
Therefore,

$$\hat{\boldsymbol{\beta}}_o = \overline{\mathbf{Y}} - \beta\overline{\mathbf{X}}.$$

Now consider the term of the log-likelihood involving $\boldsymbol{\mu}_t$, namely

$$-(\mathbf{Z}_t - \boldsymbol{\alpha} - \gamma\boldsymbol{\mu}_t)' \Sigma_{ZZ}^{-1} (\mathbf{Z}_t - \boldsymbol{\alpha} - \gamma\boldsymbol{\mu}_t). \tag{13}$$

The derivative of (13) with respect to $\boldsymbol{\mu}_t$ is

$$-2\gamma'\Sigma_{ZZ}^{-1}\gamma\boldsymbol{\mu}_t + 2\gamma'\Sigma_{ZZ}^{-1}(\mathbf{Z}_t - \boldsymbol{\alpha}).$$

Thus

$$\hat{\boldsymbol{\mu}}_t = \left(\gamma'\Sigma_{ZZ}^{-1}\gamma\right)^{-1} \gamma'\Sigma_{ZZ}^{-1}(\mathbf{Z}_t - \boldsymbol{\alpha}), \qquad t = 1, 2, \ldots, n.$$

Now

$$\begin{aligned}
\mathbf{Z}_t - \begin{pmatrix} \hat{\boldsymbol{\beta}}_o \\ \mathbf{0} \end{pmatrix} - \begin{pmatrix} \beta \\ I \end{pmatrix} \hat{\boldsymbol{\mu}}_t &= \mathbf{Z}_t - \begin{pmatrix} \hat{\boldsymbol{\beta}}_o \\ \mathbf{0} \end{pmatrix} - \begin{pmatrix} \beta \\ I \end{pmatrix} \left[(\beta', I)\Sigma_{ZZ}^{-1}(\beta', I)'\right]^{-1} \\
&\quad \times (\beta', I)\Sigma_{ZZ}^{-1} \left(\mathbf{Z}_t - \begin{pmatrix} \hat{\boldsymbol{\beta}}_o \\ \mathbf{0} \end{pmatrix}\right) \\
&= \left\{I - (\beta', I)' \left[(\beta', I)\Sigma_{ZZ}^{-1}(\beta', I)'\right]^{-1} (\beta', I)\Sigma_{ZZ}^{-1}\right\} \\
&\quad \times \left(\mathbf{Z}_t - \begin{pmatrix} \hat{\boldsymbol{\beta}}_o \\ \mathbf{0} \end{pmatrix}\right).
\end{aligned}$$

Hence

$$\begin{aligned}
&\left(\mathbf{Z}_t - \begin{pmatrix} \hat{\boldsymbol{\beta}}_o \\ \mathbf{0} \end{pmatrix} - \begin{pmatrix} \beta \\ I \end{pmatrix} \hat{\boldsymbol{\mu}}_t\right)' \Sigma_{ZZ}^{-1} \left(\mathbf{Z}_t - \begin{pmatrix} \hat{\boldsymbol{\beta}}_o \\ \mathbf{0} \end{pmatrix} - \begin{pmatrix} \beta \\ I \end{pmatrix} \hat{\boldsymbol{\mu}}_t\right) \\
&= \left(\mathbf{Z}_t - \begin{pmatrix} \hat{\boldsymbol{\beta}}_o \\ \mathbf{0} \end{pmatrix}\right)' \left\{I - (\beta', I)' \left[(\beta', I)\Sigma_{ZZ}^{-1}(\beta', I)'\right]^{-1}\right. \\
&\left.\times(\beta', I)\Sigma_{ZZ}^{-1}\right\}' \Sigma_{ZZ}^{-1} \\
&\times \left\{I - (\beta', I)' \left[(\beta', I)\Sigma_{ZZ}^{-1}(\beta', I)'\right]^{-1} (\beta', I)\Sigma_{ZZ}^{-1}\right\} \left(\mathbf{Z}_t - \begin{pmatrix} \hat{\boldsymbol{\beta}}_o \\ \mathbf{0} \end{pmatrix}\right).
\end{aligned} \tag{14}$$

Simplifying (14), using Proposition 3, we get

$$\left(\mathbf{Z}_t - \begin{pmatrix}\hat{\boldsymbol{\beta}}_o\\ \mathbf{0}\end{pmatrix} - \begin{pmatrix}\beta\\ I\end{pmatrix}\hat{\boldsymbol{\mu}}_t\right)' \Sigma_{ZZ}^{-1} \left(\mathbf{Z}_t - \begin{pmatrix}\hat{\boldsymbol{\beta}}_o\\ \mathbf{0}\end{pmatrix} - \begin{pmatrix}\beta\\ I\end{pmatrix}\hat{\boldsymbol{\mu}}_t\right)$$
$$= \left(\mathbf{Z}_t - \begin{pmatrix}\hat{\boldsymbol{\beta}}_o\\ \mathbf{0}\end{pmatrix}\right)' \left\{\Sigma_{\epsilon\epsilon}^{-1} - \Sigma_{\epsilon\epsilon}^{-1}(\beta', I)' \left[(\beta', I)\Sigma_{\epsilon\epsilon}^{-1}(\beta', I)'\right]^{-1}(\beta', I)\Sigma_{\epsilon\epsilon}^{-1}\right\}$$
$$\times \left(\mathbf{Z}_t - \begin{pmatrix}\hat{\boldsymbol{\beta}}_o\\ \mathbf{0}\end{pmatrix}\right). \tag{15}$$

Since

$$\begin{pmatrix}\hat{\boldsymbol{\beta}}_o\\ \mathbf{0}\end{pmatrix} = \overline{\mathbf{Z}} - \begin{pmatrix}\beta\\ I\end{pmatrix}\overline{\mathbf{X}},$$

we can further simplify (15) using Proposition 2, to get

$$\left(\mathbf{Z}_t - \begin{pmatrix}\hat{\boldsymbol{\beta}}_o\\ \mathbf{0}\end{pmatrix} - \begin{pmatrix}\beta\\ I\end{pmatrix}\hat{\boldsymbol{\mu}}_t\right)' \Sigma_{ZZ}^{-1} \left(\mathbf{Z}_t - \begin{pmatrix}\hat{\boldsymbol{\beta}}_o\\ \mathbf{0}\end{pmatrix} - \begin{pmatrix}\beta\\ I\end{pmatrix}\boldsymbol{\mu}_t\right)$$
$$= (\mathbf{Z}_t - \overline{\mathbf{Z}})'(I, -\beta)'\Sigma_{vv}^{-1}(I, -\beta)(\mathbf{Z}_t - \overline{\mathbf{Z}}).$$

Thus after substituting $\hat{\boldsymbol{\mu}}_t$, $t = 1, \ldots, n$, and $\hat{\boldsymbol{\beta}}_o$ into the log-likelihood function (12) we get

$$f = C_o - \frac{n}{2}\log|\Sigma_{ZZ}| - \frac{1}{2}\text{tr}\left(\Sigma_{vv}^{-1}m_{vv}\right) - \frac{d}{2}\log|\Sigma_{\epsilon\epsilon}| - \frac{d}{2}\text{tr}(S_{\epsilon\epsilon}\Sigma_{\epsilon\epsilon}^{-1}),$$

where

$$m_{vv} = (I, -\beta)\left(\sum_{t=1}^{n}(\mathbf{Z}_t - \overline{\mathbf{Z}})(\mathbf{Z}_t - \overline{\mathbf{Z}})'\right)(I, -\beta)'.$$

In the above expression Σ_{xx} is involved only in $\log|\Sigma_{ZZ}|$. The derivative of $\log|\Sigma_{ZZ}|$ with respect to Σ_{xx} is

$$\frac{\partial}{\partial \text{vec}\Sigma_{xx}}\log|\Sigma_{ZZ}| = \left\{\text{vec}(\Sigma_{ZZ}^{-1})'\right\}'\frac{\partial \text{vec}\Sigma_{ZZ}}{\partial \text{vec}\Sigma_{xx}}$$
$$= \left[\text{vec}(\Sigma_{ZZ}^{-1})'\right]'\begin{pmatrix}\beta\\ I\end{pmatrix} \otimes \begin{pmatrix}\beta\\ I\end{pmatrix}$$
$$= \left\{\text{vec}\left[(\beta', I)\Sigma_{ZZ}^{-1}(\beta', I)'\right]\right\}'.$$

In the matrix form we get

$$\frac{\partial \log|\Sigma_{ZZ}|}{\partial \Sigma_{xx}} = (\beta', I)\Sigma_{ZZ}^{-1}(\beta', I)'.$$

Note that this is a symmetric non-singular matrix. In the above calculations we have ignored the symmetry of Σ_{xx}. If we impose symmetry we get the diagonal elements and twice the off-diagonal elements. Thus the derivative with respect to Σ_{xx} cannot be the null matrix. Hence the maximum of the above function could exist only on the boundary of the domain of Σ_{xx}.

The minimum of $\log \mid \Sigma_{\epsilon\epsilon} + (\beta', I)'\Sigma_{xx}(\beta', I) \mid$ exists when $\Sigma_{xx} = 0$, since Σ_{xx} is non-negative definite. Thus the maximum likelihood estimator of Σ_{xx} is 0.

This concludes the proof of the theorem.

As soon as $\hat{\Sigma}_{xx} = 0$, the model reduces to the functional relationships model. Hence we conclude that the maximum likelihood method fails to distinguish between the multivariate functional relationships model and the multivariate ultrastructural relationships model under the assumption of the existence of an estimator for the error covariance matrix.

REFERENCES

Amemiya, Y. (1982), "Estimators for the errors-in-variables model." Ph.D. thesis, Iowa State University.

Amemiya, Y., and W. A. Fuller (1984), "Estimation for the multivariate errors-in-variables model with estimated error covariance matrix". *Annals of Statistics* **12,** 497–509.

Anderson, T. W. (1951), "Estimating linear relations on regression coefficients for multivariate normal distributions". *Annals of Mathematical Statistics* **22,** 327–351.

Anderson, T. W. (1984), "Estimating linear statistical relationships". *Annals of Statistics* **12,** 1–45.

Dolby, G. R. (1976), "The ultrastructural relation: A synthesis of the functional and structural relations". *Biometrika* **63,** 39–50.

Villegas, C. (1982), "Maximum likelihood and least-squares estimation in linear affine functional models". *Annals of Statistics* **10,** 256–265.

Serge B. Provost [1]

TESTING FOR THE NULLITY OF THE MULTIPLE CORRELATION COEFFICIENT WITH INCOMPLETE MULTIVARIATE DATA

ABSTRACT

This paper derives the likelihood ratio statistic to test the nullity of the multiple correlation coefficient between X_1 and $(X_2, \ldots, X_k)$ under the assumption that $(X_1, X_2, \ldots, X_k)$ has a multivariate normal distribution and a sample of size n is available, where for N observation vectors all components are available, while for $M = (n - N)$ observation vectors, the data on the last q components, $(X_{k-q+1}, X_{k-q+2}, \ldots, X_k)$, are missing.

1. INTRODUCTION

In certain areas of research such as the social sciences or medicine, statistical models for handling missing data are needed. Statistical problems connected with such models have been studied by Wilks (1932), Lord (1955), Edgett (1956), Rao (1956), Anderson (1957), Trawinski and Bargmann (1964), Afifi and Elashoff (1966, 1967, 1969a,b), Smith and Pfaffenberger (1970), Morrison (1971), Eaton and Kariya (1974, 1975), Bhargava (1975), Little (1976) and Radhakrishnan (1982), among others.

Eaton and Kariya (1974) derived tests for the independence of two normally distributed sub-vectors when an additional random sample is available on one of them. In this article, a different pattern is considered wherein the data on the last q components of a k-dimensional normal random vector are missing for M observation vectors while no components are missing for N other observation vectors. The likelihood ratio statistic to test the nullity of the multiple correlation coefficient between the first coordinate and the last

[1] Department of Statistical and Actuarial Sciences, The University of Western Ontario, London, Ontario N6A 5B9

I. B. MacNeill and G. J. Umphrey (eds.), Foundations of Statistical Inference, 149–161.

$(k-1)$ components of the normally distributed random vector is derived in Section 3.

2. NOTATION

In order to obtain the maximum likelihood ratio statistic for this test under the pattern specified in the introduction, we partition the k-dimensional normal random vector, $\mathbf{W} = (X_1, X_2, \ldots, X_k)'$, as follows:

$$\mathbf{W} = \begin{pmatrix} X \\ \mathbf{Y} \\ \mathbf{Z} \end{pmatrix} \sim N_{1+p+q}(\boldsymbol{\mu}, \boldsymbol{\Sigma}),$$

that is, $\mathbf{W}$ is distributed as a $(1+p+q)$–dimensional normal random vector with mean $\boldsymbol{\mu}$ and covariance matrix $\boldsymbol{\Sigma}$, where

$$\boldsymbol{\mu} = \begin{pmatrix} \mu_1 \\ \boldsymbol{\mu}_2 \\ \boldsymbol{\mu}_3 \end{pmatrix}, \qquad \boldsymbol{\Sigma} = \begin{pmatrix} \sigma_{11} & \boldsymbol{\Sigma}_{12} & \boldsymbol{\Sigma}_{13} \\ \boldsymbol{\Sigma}_{21} & \boldsymbol{\Sigma}_{22} & \boldsymbol{\Sigma}_{23} \\ \boldsymbol{\Sigma}_{31} & \boldsymbol{\Sigma}_{32} & \boldsymbol{\Sigma}_{33} \end{pmatrix},$$

$$\sigma_{11} : 1 \times 1, \boldsymbol{\Sigma}_{22} : p \times p \text{ and } \boldsymbol{\Sigma}_{33} : q \times q.$$

Let R denote the multiple correlation coefficient between X and $(\mathbf{Y}', \mathbf{Z}')$. Treated in this article is the problem of testing

$$H_0 : R^2 = 0 \text{ against } H_A : R^2 > 0 \tag{1}$$

on the basis of a random sample, $\mathbf{w}_1, \ldots, \mathbf{w}_N$, on $\mathbf{W}$, where

$$\mathbf{w}_i = \begin{pmatrix} x_i \\ \mathbf{y}_i \\ \mathbf{z}_i \end{pmatrix}, \quad i = 1, \ldots, N, \text{ and } N \geq 1 + p + q,$$

and of an independent random sample on $(X, \mathbf{Y}')'$ denoted by

$$\begin{pmatrix} u_1 \\ \mathbf{v}_1 \end{pmatrix}, \ldots, \begin{pmatrix} u_M \\ \mathbf{v}_M \end{pmatrix}, \text{ where } M \geq 1 + p.$$

Since

$$R^2 = \frac{1}{\sigma_{11}} \left\{ (\boldsymbol{\Sigma}_{12}, \boldsymbol{\Sigma}_{13}) \begin{pmatrix} \boldsymbol{\Sigma}_{22} & \boldsymbol{\Sigma}_{23} \\ \boldsymbol{\Sigma}_{32} & \boldsymbol{\Sigma}_{33} \end{pmatrix}^{-1} \begin{pmatrix} \boldsymbol{\Sigma}_{21} \\ \boldsymbol{\Sigma}_{31} \end{pmatrix} \right\},$$

see, for instance, Anderson (1958), the parametric spaces may be defined as follows:

$$\Omega_0 = \{(\boldsymbol{\mu}, \Sigma) \mid (\Sigma_{12}, \Sigma_{13}) = \mathbf{0}'\},$$
$$\Omega_1 = \{(\boldsymbol{\mu}, \Sigma) \mid (\Sigma_{12}, \Sigma_{13}) \neq \mathbf{0}'\}.$$

Let

$$\mathbf{S} = \begin{pmatrix} s_{11} & \mathbf{s}_{12} & \mathbf{s}_{13} \\ \mathbf{s}_{21} & \mathbf{s}_{22} & \mathbf{s}_{23} \\ \mathbf{s}_{31} & \mathbf{s}_{32} & \mathbf{s}_{33} \end{pmatrix}$$

$$= \begin{pmatrix} \sum_{i=1}^{N} (x_i - \overline{x})^2, & \sum_{i=1}^{N} (x_i - \overline{x})(\mathbf{y}_i - \overline{\mathbf{y}})', & \sum_{i=1}^{N} (x_i - \overline{x})(\mathbf{z}_i - \overline{\mathbf{z}})' \\ \mathbf{s}_{21} = \mathbf{s}_{12}', & \sum_{i=1}^{N} (\mathbf{y}_i - \overline{\mathbf{y}})(\mathbf{y}_i - \overline{\mathbf{y}})', & \sum_{i=1}^{N} (\mathbf{y}_i - \overline{\mathbf{y}})(\mathbf{z}_i - \overline{\mathbf{z}})' \\ \mathbf{s}_{31} = \mathbf{s}_{13}', & \mathbf{s}_{32} = \mathbf{s}_{23}', & \sum_{i=1}^{N} (\mathbf{z}_i - \overline{\mathbf{z}})(\mathbf{z}_i - \overline{\mathbf{z}})' \end{pmatrix},$$

$$\mathbf{T} = \begin{pmatrix} t_{11} & \mathbf{t}_{12} \\ \mathbf{t}_{21} & \mathbf{t}_{22} \end{pmatrix}$$

$$= \begin{pmatrix} \sum_{i=1}^{M} (u_i - \overline{u})^2, & \sum_{i=1}^{M} (u_i - \overline{u})(\mathbf{v}_i - \overline{\mathbf{v}})' \\ \mathbf{t}_{21} = \mathbf{t}_{12}', & \sum_{i=1}^{M} (\mathbf{v}_i - \overline{\mathbf{v}})(\mathbf{v}_i - \overline{\mathbf{v}})' \end{pmatrix},$$

$$\overline{\mathbf{ST}} = \begin{pmatrix} \overline{st}_{11} & \overline{\mathbf{st}}_{12} \\ \overline{\mathbf{st}}_{21} & \overline{\mathbf{st}}_{22} \end{pmatrix}$$

$$= \begin{pmatrix} \frac{MN}{M+N} (\overline{u} - \overline{x})^2, & \frac{MN}{M+N} (\overline{u} - \overline{x})(\overline{\mathbf{v}} - \overline{\mathbf{y}})' \\ \overline{\mathbf{st}}_{21} = \overline{\mathbf{st}}_{12}', & \frac{MN}{M+N} (\overline{\mathbf{v}} - \overline{\mathbf{y}})(\overline{\mathbf{v}} - \overline{\mathbf{y}})' \end{pmatrix}$$

and

$$\begin{pmatrix} \tau_{11} & \boldsymbol{\tau}_{12} \\ \boldsymbol{\tau}_{21} & \boldsymbol{\tau}_{22} \end{pmatrix} = \begin{pmatrix} s_{11} + t_{11} + \overline{st}_{11}, & \mathbf{s}_{12} + \mathbf{t}_{12} + \overline{\mathbf{st}}_{12} \\ \mathbf{s}_{21} + \mathbf{t}_{21} + \overline{\mathbf{st}}_{21}, & \mathbf{s}_{22} + \mathbf{t}_{22} + \overline{\mathbf{st}}_{22} \end{pmatrix}.$$

We will prove in the next section that the likelihood ratio test statistic for testing the null hypothesis given in (1) is

$$\lambda = \frac{\left[\left(\tau_{11} - \boldsymbol{\tau}_{12}\boldsymbol{\tau}_{22}^{-1}\boldsymbol{\tau}_{21}\right)/\tau_{11}\right]^{\frac{1}{2}(M+N)} \left(1 - r_{p+q,N}^2\right)^{\frac{1}{2}N}}{\left(1 - r_{p,N}^2\right)^{\frac{1}{2}N}},$$

where $r^2_{p+q,N}$ is the square of the sample multiple correlation coefficient between X and $(\mathbf{Y}', \mathbf{Z}')'$ for the sample on $\mathbf{W}$, and $r^2_{p,N}$ is the square of the sample multiple correlation coefficient between X and $\mathbf{Y}$ for the sample on $(X, \mathbf{Y}')'$, that is,

$$r^2_{p+q,N} = \frac{1}{s_{11}} \left\{ (\mathbf{s}_{12}, \mathbf{s}_{13}) \begin{pmatrix} \mathbf{s}_{22} & \mathbf{s}_{23} \\ \mathbf{s}_{32} & \mathbf{s}_{33} \end{pmatrix}^{-1} \begin{pmatrix} \mathbf{s}_{21} \\ \mathbf{s}_{31} \end{pmatrix} \right\}$$

and

$$r^2_{p,N} = \frac{\mathbf{s}_{12}\mathbf{s}_{22}^{-1}\mathbf{s}_{21}}{s_{11}}.$$

3. THE LIKELIHOOD RATIO STATISTIC

The likelihood function given $\mathbf{w}_1, \ldots, \mathbf{w}_N$ and $\{(u_i, \mathbf{v}_i)\}$, $i = 1, 2, \ldots, M$, is

$$L(\boldsymbol{\mu}, \boldsymbol{\Sigma}) = k \mid \boldsymbol{\Sigma}_{(11)} \mid^{-\frac{1}{2}M} \mid \boldsymbol{\Sigma} \mid^{-\frac{1}{2}N} \exp\left[-\frac{1}{2}\phi_1\right] \times \exp\left[-\frac{1}{2}\phi_2\right], \quad (2)$$

where

$$k = (2\pi)^{-\frac{1}{2}[(p+q+1)N+(p+1)M]},$$

$$\boldsymbol{\Sigma}_{(11)} = \begin{pmatrix} \sigma_{11} & \boldsymbol{\Sigma}_{12} \\ \boldsymbol{\Sigma}_{21} & \boldsymbol{\Sigma}_{22} \end{pmatrix},$$

$$\phi_1 = \sum_{i=1}^{N} (\mathbf{w}_i - \boldsymbol{\mu})' \boldsymbol{\Sigma}^{-1} (\mathbf{w}_i - \boldsymbol{\mu}),$$

$$\phi_2 = \sum_{i=1}^{M} \left(\mathbf{u}_i - \boldsymbol{\mu}_{(1)}\right)' \boldsymbol{\Sigma}_{(11)}^{-1} \left(\mathbf{u}_i - \boldsymbol{\mu}_{(1)}\right)$$

and

$$\mathbf{u}_i' = (u_i, \mathbf{v}_i') \text{ and } \boldsymbol{\mu}_{(1)}' = (\mu_1, \boldsymbol{\mu}_2').$$

Now letting

$$\mathbf{x}_i^{(1)} = (x_i, \mathbf{y}_i')', \ \boldsymbol{\Sigma}_{(1)3} = \begin{pmatrix} \boldsymbol{\Sigma}_{13} \\ \boldsymbol{\Sigma}_{23} \end{pmatrix} = \boldsymbol{\Sigma}_{3(1)}'$$

and

$$\boldsymbol{\Sigma}_{33.(1)} = \boldsymbol{\Sigma}_{33} - \boldsymbol{\Sigma}_{3(1)}\boldsymbol{\Sigma}_{(11)}^{-1}\boldsymbol{\Sigma}_{(1)3},$$

$$\begin{aligned}
\phi_1 &= \sum_{i=1}^{N} (\mathbf{w}_i - \boldsymbol{\mu})' \boldsymbol{\Sigma}^{-1} (\mathbf{w}_i - \boldsymbol{\mu}) \\
&= \operatorname{tr} \left\{ \sum_{i=1}^{N} \boldsymbol{\Sigma}^{-1} (\mathbf{w}_i - \boldsymbol{\mu}) (\mathbf{w}_i - \boldsymbol{\mu})' \right\} \\
&= \operatorname{tr} \left\{ \sum_{i=1}^{N} \boldsymbol{\Sigma}^{-1} (\mathbf{w}_i - \overline{\mathbf{w}}) (\mathbf{w}_i - \overline{\mathbf{w}})' + N \boldsymbol{\Sigma}^{-1} (\overline{\mathbf{w}} - \boldsymbol{\mu}) (\overline{\mathbf{w}} - \boldsymbol{\mu})' \right\} \\
&= \sum_{i=1}^{N} (\mathbf{w}_i - \overline{\mathbf{w}})' \boldsymbol{\Sigma}^{-1} (\mathbf{w}_i - \overline{\mathbf{w}}) + N \sum_{i=1}^{N} (\overline{\mathbf{w}} - \boldsymbol{\mu})' \boldsymbol{\Sigma}^{-1} (\overline{\mathbf{w}} - \boldsymbol{\mu}) \\
&= \sum_{i-1}^{N} \left\{ \left(\mathbf{x}_i^{(1)} - \overline{\mathbf{x}}^{(1)} \right)' \boldsymbol{\Sigma}_{(11)}^{-1} \left(\mathbf{x}_i^{(1)} - \overline{\mathbf{x}}^{(1)} \right) \right. \\
&\quad + \left(\mathbf{z}_i - \overline{\mathbf{z}} - \boldsymbol{\Sigma}_{3(1)} \boldsymbol{\Sigma}_{(11)}^{-1} \left(\mathbf{x}_i^{(1)} - \overline{\mathbf{x}}^{(1)} \right) \right)' \\
&\quad \left. \times \boldsymbol{\Sigma}_{33.(1)}^{-1} \left(\mathbf{z}_i - \overline{\mathbf{z}} - \boldsymbol{\Sigma}_{3(1)} \boldsymbol{\Sigma}_{(11)}^{-1} \left(\mathbf{x}_i^{(1)} - \overline{\mathbf{x}}^{(1)} \right) \right) \right\} \\
&\quad + N \left\{ \left(\overline{\mathbf{x}}^{(1)} - \boldsymbol{\mu}_{(1)} \right)' \boldsymbol{\Sigma}_{(11)}^{-1} \left(\overline{\mathbf{x}}^{(1)} - \boldsymbol{\mu}_{(1)} \right) \right. \\
&\quad + \left(\overline{\mathbf{z}} - \boldsymbol{\mu}_3 - \boldsymbol{\Sigma}_{3(1)} \boldsymbol{\Sigma}_{(11)}^{-1} \left(\overline{\mathbf{x}}^{(1)} - \boldsymbol{\mu}_{(1)} \right) \right)' \\
&\quad \left. \times \boldsymbol{\Sigma}_{33.(1)}^{-1} \left(\overline{\mathbf{z}} - \boldsymbol{\mu}_3 - \boldsymbol{\Sigma}_{3(1)} \boldsymbol{\Sigma}_{(11)}^{-1} \left(\overline{\mathbf{x}}^{(1)} - \boldsymbol{\mu}_{(1)} \right) \right) \right\} \\
&= \operatorname{tr} \left\{ \boldsymbol{\Sigma}_{(11)}^{-1} \mathbf{s}_{(11)} + \boldsymbol{\Sigma}_{33.(1)}^{-1} \left(\mathbf{s}_{33} - \mathbf{s}_{3(1)} \boldsymbol{\Sigma}_{(11)}^{-1} \boldsymbol{\Sigma}_{(1)3} \right. \right. \\
&\quad \left. \left. - \boldsymbol{\Sigma}_{3(1)} \boldsymbol{\Sigma}_{(11)}^{-1} \mathbf{s}_{(1)3} + \boldsymbol{\Sigma}_{3(1)} \boldsymbol{\Sigma}_{(11)}^{-1} \mathbf{s}_{(11)} \boldsymbol{\Sigma}_{(11)}^{-1} \boldsymbol{\Sigma}_{(1)3} \right) \right\} \\
&\quad + \operatorname{tr} \left\{ N \left[\boldsymbol{\Sigma}_{(11)}^{-1} \left(\overline{\mathbf{x}}^{(1)} - \boldsymbol{\mu}_{(1)} \right) \left(\overline{\mathbf{x}}^{(1)} - \boldsymbol{\mu}_{(1)} \right)' \right. \right. \\
&\quad + \boldsymbol{\Sigma}_{33.(1)}^{-1} \left(\overline{\mathbf{z}} - \boldsymbol{\mu}_3 - \boldsymbol{\Sigma}_{3(1)} \boldsymbol{\Sigma}_{(11)}^{-1} \left(\overline{\mathbf{x}}^{(1)} - \boldsymbol{\mu}_{(1)} \right) \right) \\
&\quad \left. \left. \times \left(\overline{\mathbf{z}} - \boldsymbol{\mu}_3 - \boldsymbol{\Sigma}_{3(1)} \boldsymbol{\Sigma}_{(11)}^{-1} \left(\overline{\mathbf{x}}^{(1)} - \boldsymbol{\mu}_{(1)} \right) \right)' \right] \right\}, \qquad (3)
\end{aligned}$$

where $\mathbf{s}_{3(1)} = (\mathbf{s}_{31}, \mathbf{s}_{32}) = \mathbf{s}_{(1)3}'$ and

$$\mathbf{s}_{(11)} = \begin{pmatrix} \mathbf{s}_{11} & \mathbf{s}_{12} \\ \mathbf{s}_{21} & \mathbf{s}_{22} \end{pmatrix}.$$

Letting $\bar{\mathbf{u}} = \sum_{i=1}^{M} \mathbf{u}_i / M$,

$$\begin{aligned}
\phi_2 &= \sum_{i=1}^{M} \left(\mathbf{u}_i - \boldsymbol{\mu}_{(1)}\right)' \boldsymbol{\Sigma}_{(11)}^{-1} \left(\mathbf{u}_i - \boldsymbol{\mu}_{(1)}\right) \\
&= \operatorname{tr} \left\{ \sum_{i=1}^{M} \boldsymbol{\Sigma}_{(11)}^{-1} \left(\mathbf{u}_i - \boldsymbol{\mu}_{(1)}\right) \left(\mathbf{u}_i - \boldsymbol{\mu}_{(1)}\right)' \right\} \\
&= \operatorname{tr} \left\{ \sum_{i=1}^{M} \boldsymbol{\Sigma}_{(11)}^{-1} (\mathbf{u}_i - \bar{\mathbf{u}}) (\mathbf{u}_i - \bar{\mathbf{u}})' \right. \\
&\qquad \left. + M \boldsymbol{\Sigma}_{(11)}^{-1} \left(\bar{\mathbf{u}} - \boldsymbol{\mu}_{(1)}\right) \left(\bar{\mathbf{u}} - \boldsymbol{\mu}_{(1)}\right)' \right\} \\
&= \operatorname{tr} \left\{ \boldsymbol{\Sigma}_{(11)}^{-1} \mathbf{T} \right\} + \operatorname{tr} \left\{ M \boldsymbol{\Sigma}_{(11)}^{-1} \left(\bar{\mathbf{u}} - \boldsymbol{\mu}_{(1)}\right) \left(\bar{\mathbf{u}} - \boldsymbol{\mu}_{(1)}\right)' \right\}. \qquad (4)
\end{aligned}$$

Therefore, the likelihood function in (2) where ϕ_1 and ϕ_2 are given in (3) and (4) respectively is maximized with respect to $\boldsymbol{\mu}_3$ when

$$\boldsymbol{\mu}_3 = \bar{\mathbf{z}} - \boldsymbol{\Sigma}_{3(1)} \boldsymbol{\Sigma}_{(11)}^{-1} \left(\bar{\mathbf{x}}^{(1)} - \boldsymbol{\mu}_{(1)}\right),$$

that is,

$$\begin{aligned}
\max_{\boldsymbol{\mu}_3} L(\boldsymbol{\mu}, \boldsymbol{\Sigma}) &= k \mid \boldsymbol{\Sigma}_{(11)} \mid^{-\frac{1}{2}M} \mid \boldsymbol{\Sigma} \mid^{-\frac{1}{2}N} \\
&\quad \times \exp\left[-\frac{1}{2} \operatorname{tr} \left\{ \boldsymbol{\Sigma}_{(11)}^{-1} \left(M \left(\bar{\mathbf{u}} - \boldsymbol{\mu}_{(1)}\right) \left(\bar{\mathbf{u}} - \boldsymbol{\mu}_{(1)}\right)' \right. \right. \right. \\
&\quad \left. \left. \left. + N \left(\bar{\mathbf{x}}^{(1)} - \boldsymbol{\mu}_{(1)}\right) \left(\bar{\mathbf{x}}^{(1)} - \boldsymbol{\mu}_{(1)}\right)' \right) \right\} \right] \\
&\quad \times \exp\left[-\frac{1}{2} \operatorname{tr} \left\{ \boldsymbol{\Sigma}_{(11)}^{-1} \mathbf{T} \right\} \right] \times \exp\left[-\frac{1}{2} \operatorname{tr} \left\{ \boldsymbol{\Sigma}_{(11)}^{-1} \mathbf{s}_{(11)} \right. \right. \\
&\quad + \boldsymbol{\Sigma}_{33.1}^{-1} \left(\mathbf{s}_{33} - \mathbf{s}_{3(1)} \boldsymbol{\Sigma}_{(11)}^{-1} \boldsymbol{\Sigma}_{(1)3} - \boldsymbol{\Sigma}_{3(1)} \boldsymbol{\Sigma}_{(11)}^{-1} \mathbf{s}_{(1)3} \right. \\
&\quad \left. \left. \left. + \boldsymbol{\Sigma}_{3(1)} \boldsymbol{\Sigma}_{(11)}^{-1} \mathbf{s}_{(11)} \boldsymbol{\Sigma}_{(11)}^{-1} \boldsymbol{\Sigma}_{(1)3} \right) \right\} \right].
\end{aligned}$$

In order to maximize this expression with respect to $\boldsymbol{\mu}_{(1)}$, we minimize

$$\begin{aligned}
\operatorname{tr} &\left\{ \boldsymbol{\Sigma}_{(11)}^{-1} \left[M \left(\bar{\mathbf{u}} - \boldsymbol{\mu}_{(1)}\right) \left(\bar{\mathbf{u}} - \boldsymbol{\mu}_{(1)}\right)' \right. \right. \\
&\quad \left. \left. + N \left(\bar{\mathbf{x}}^{(1)} - \boldsymbol{\mu}_{(1)}\right) \left(\bar{\mathbf{x}}^{(1)} - \boldsymbol{\mu}_{(1)}\right)' \right] \right\}
\end{aligned}$$

$$= \text{tr} \left\{ M \left(\overline{\mathbf{u}} - \boldsymbol{\mu}_{(1)} \right)' \boldsymbol{\Sigma}_{(11)}^{-1} \left(\overline{\mathbf{u}} - \boldsymbol{\mu}_{(1)} \right) + N \left(\overline{\mathbf{x}}^{(1)} - \boldsymbol{\mu}_{(1)} \right)' \boldsymbol{\Sigma}_{(11)}^{-1} \left(\overline{\mathbf{x}}^{(1)} - \boldsymbol{\mu}_{(1)} \right) \right\}$$

$$= M\overline{\mathbf{u}}'\boldsymbol{\Sigma}_{(11)}^{-1}\overline{\mathbf{u}} - 2M\boldsymbol{\mu}_{(1)}'\boldsymbol{\Sigma}_{(11)}^{-1}\overline{\mathbf{u}} + M\boldsymbol{\mu}_{(1)}'\boldsymbol{\Sigma}_{(11)}^{-1}\boldsymbol{\mu}_{(1)} + N\overline{\mathbf{x}}^{(1)'}\boldsymbol{\Sigma}_{(11)}^{-1}\overline{\mathbf{x}}^{(1)} - 2N\boldsymbol{\mu}_{(1)}'\boldsymbol{\Sigma}_{(11)}^{-1}\overline{\mathbf{x}}^{(1)} + N\boldsymbol{\mu}_{(1)}'\boldsymbol{\Sigma}_{(11)}^{-1}\boldsymbol{\mu}_{(1)}$$

$$\equiv h\left(\boldsymbol{\mu}_{(1)}\right). \tag{5}$$

Setting

$$\frac{\partial h\left(\boldsymbol{\mu}_{(1)}\right)}{\partial \boldsymbol{\mu}_{(1)}} = -2M\boldsymbol{\Sigma}_{(11)}^{-1}\overline{\mathbf{u}} + 2M\boldsymbol{\Sigma}_{(11)}^{-1}\boldsymbol{\mu}_{(1)} - 2N\boldsymbol{\Sigma}_{(11)}^{-1}\overline{\mathbf{x}}^{(1)} + 2N\boldsymbol{\Sigma}_{(11)}^{-1}\boldsymbol{\mu}_{(1)} = \mathbf{0},$$

we have that the maximum likelihood estimator of $\boldsymbol{\mu}_{(1)}$ is

$$\hat{\boldsymbol{\mu}}_{(1)} = \frac{M\overline{\boldsymbol{\mu}} + N\overline{\mathbf{x}}^{(1)}}{M+N}$$

so that maximum likelihood estimator of $\boldsymbol{\mu}_3$ is

$$\hat{\boldsymbol{\mu}}_3 = \overline{\mathbf{z}} - \frac{M}{M+N}\left\{\boldsymbol{\Sigma}_{3(1)}\boldsymbol{\Sigma}_{(11)}^{-1}\left(\overline{\mathbf{x}}^{(1)} - \overline{\mathbf{u}}\right)\right\}.$$

Putting $\boldsymbol{\mu}_{(1)} = \hat{\boldsymbol{\mu}}_{(1)}$ in (5), we have:

$$\text{tr}\left\{\boldsymbol{\Sigma}_{(11)}^{-1}\left[M\left(\overline{\mathbf{u}} - \frac{M\overline{\mathbf{u}} + N\overline{\mathbf{x}}^{(1)}}{M+N}\right)\left(\overline{\mathbf{u}} - \frac{M\overline{\mathbf{u}} + N\overline{\mathbf{x}}^{(1)}}{M+N}\right)' + N\left(\overline{\mathbf{x}}^{(1)} - \frac{M\overline{\mathbf{u}} + N\overline{\mathbf{x}}^{(1)}}{M+N}\right)\left(\overline{\mathbf{x}}^{(1)} - \frac{M\overline{\mathbf{u}} + N\overline{\mathbf{x}}^{(1)}}{M+N}\right)'\right]\right\}$$

$$= \text{tr}\left\{\boldsymbol{\Sigma}_{(11)}^{-1}\left[\frac{MN^2}{(M+N)^2}\left(\overline{\mathbf{u}} - \overline{\mathbf{x}}^{(1)}\right)\left(\overline{\mathbf{u}} - \overline{\mathbf{x}}^{(1)}\right)' + \frac{NM^2}{(M+N)^2}\left(\overline{\mathbf{x}}^{(1)} - \overline{\mathbf{u}}\right)\left(\overline{\mathbf{x}}^{(1)}\ \overline{\mathbf{u}}\right)'\right]\right\}$$

$$= \text{tr}\left\{\boldsymbol{\Sigma}_{(11)}^{-1}\frac{MN}{(M+N)}\left(\overline{\mathbf{u}} - \overline{\mathbf{x}}^{(1)}\right)\left(\overline{\mathbf{u}} - \overline{\mathbf{x}}^{(1)}\right)'\right\}$$

$$= \text{tr}\left\{\boldsymbol{\Sigma}_{(11)}^{-1}\overline{\mathbf{ST}}\right\}.$$

Then,

$$\max_{\boldsymbol{\mu}} L(\boldsymbol{\mu},\boldsymbol{\Sigma}) = k \mid \boldsymbol{\Sigma}_{(11)} \mid^{-\frac{1}{2}M} \mid \boldsymbol{\Sigma} \mid^{-\frac{1}{2}N}$$

$$\times \exp\left\{-\frac{1}{2}\ \mathrm{tr}\ \left(\boldsymbol{\Sigma}_{(11)}^{-1}\left(\mathbf{T}+\overline{\mathbf{ST}}\right)\right)\right\}$$

$$\times \exp\left\{-\frac{1}{2}\ \mathrm{tr}\ \left(\boldsymbol{\Sigma}_{(11)}^{-1}\mathbf{s}_{(11)} + \boldsymbol{\Sigma}_{33.(1)}^{-1}\mathbf{s}_{33}^{*}\right)\right\}$$

$$= k \mid \boldsymbol{\Sigma}_{(11)} \mid^{-\frac{(N+M)}{2}} \mid \boldsymbol{\Sigma}_{33.(1)} \mid^{-\frac{1}{2}N}$$

$$\times \exp\left\{-\frac{1}{2}\ \mathrm{tr}\ \left(\boldsymbol{\Sigma}_{(11)}^{-1}\left(\mathbf{T}+\overline{\mathbf{ST}}+\mathbf{s}_{(11)}\right)\right)\right\}$$

$$\times \exp\left\{-\frac{1}{2}\ \mathrm{tr}\ \left(\boldsymbol{\Sigma}_{33.(1)}^{-1}\mathbf{s}_{33}^{*}\right)\right\}$$

$$\equiv L(\boldsymbol{\Sigma}), \tag{6}$$

where

$$\mathbf{s}_{33}^{*} = \mathbf{s}_{33} - \mathbf{s}_{3(1)}\boldsymbol{\Sigma}_{(11)}^{-1}\boldsymbol{\Sigma}_{(1)3} - \boldsymbol{\Sigma}_{3(1)}\boldsymbol{\Sigma}_{(11)}^{-1}\mathbf{s}_{(1)3} + \boldsymbol{\Sigma}_{3(1)}\boldsymbol{\Sigma}_{(11)}^{-1}\mathbf{s}_{(11)}\boldsymbol{\Sigma}_{(11)}^{-1}\boldsymbol{\Sigma}_{(1)3}.$$

Under H_0,

$$\boldsymbol{\Sigma}_{(11)} = \begin{pmatrix} \sigma_{11} & \mathbf{0}' \\ \mathbf{0} & \boldsymbol{\Sigma}_{22} \end{pmatrix},\ \boldsymbol{\Sigma}_{3(1)} = (\mathbf{0},\boldsymbol{\Sigma}_{32}),$$

$$\boldsymbol{\Sigma}_{33.(1)} = \boldsymbol{\Sigma}_{33} - (\mathbf{0},\boldsymbol{\Sigma}_{32})\begin{pmatrix} \sigma_{11} & \mathbf{0}' \\ \mathbf{0} & \boldsymbol{\Sigma}_{22} \end{pmatrix}^{-1}\begin{pmatrix} \mathbf{0}' \\ \boldsymbol{\Sigma}_{23} \end{pmatrix}$$

$$= \boldsymbol{\Sigma}_{33} - \boldsymbol{\Sigma}_{32}\boldsymbol{\Sigma}_{22}^{-1}\boldsymbol{\Sigma}_{23} = \boldsymbol{\Sigma}_{33.2}\ ,$$

$\mathbf{s}_{33}^{*}$ becomes

$$\mathbf{s}_{33}^{*0} = \mathbf{s}_{33} - \mathbf{s}_{32}\boldsymbol{\Sigma}_{22}^{-1}\boldsymbol{\Sigma}_{23} - \boldsymbol{\Sigma}_{32}\boldsymbol{\Sigma}_{22}^{-1}\mathbf{s}_{23} + \boldsymbol{\Sigma}_{32}\boldsymbol{\Sigma}_{22}^{-1}\mathbf{s}_{22}\boldsymbol{\Sigma}_{22}^{-1}\boldsymbol{\Sigma}_{23}$$

and $L(\boldsymbol{\Sigma})$ becomes

$$L^{0}(\boldsymbol{\Sigma}) = k\sigma_{11}^{-\frac{1}{2}(N+M)} \mid \boldsymbol{\Sigma}_{22} \mid^{-\frac{1}{2}(N+M)} \mid \boldsymbol{\Sigma}_{33.2} \mid^{-\frac{1}{2}N}$$

$$\times \exp\left(-\frac{1}{2\sigma_{11}}\left(t_{11}+s_{11}+\overline{st}_{11}\right)\right)$$

$$\times \exp\left\{-\frac{1}{2}\ \mathrm{tr}\ \left[\boldsymbol{\Sigma}_{22}^{-1}\left(\mathbf{t}_{22}+\mathbf{s}_{22}+\overline{\mathbf{st}}_{22}\right)\right]\right\}$$

$$\times \exp\left\{-\frac{1}{2}\boldsymbol{\Sigma}_{33.2}^{-1}\mathbf{s}_{33}^{*0}\right\}.$$

Since the mapping

$$\left.\begin{matrix} \sigma_{11} \\ \mathbf{\Sigma}_{22} \\ \mathbf{\Sigma}_{32} \\ \mathbf{\Sigma}_{33} \end{matrix}\right\} \to \left\{\begin{matrix} \sigma_{11} \\ \mathbf{\Sigma}_{22} \\ \mathbf{\Sigma}_{32}\mathbf{\Sigma}_{22}^{-1} \equiv \mathbf{\Sigma}_{32}^{*} \\ \mathbf{\Sigma}_{33} - \mathbf{\Sigma}_{32}\mathbf{\Sigma}_{22}^{-1}\mathbf{\Sigma}_{23} \equiv \mathbf{\Sigma}_{33.2} \end{matrix}\right.$$

is one-to-one, we have that

$$\max_{(\sigma_{11},\mathbf{\Sigma}_{22},\mathbf{\Sigma}_{23},\mathbf{\Sigma}_{33})} L^0(\mathbf{\Sigma}) = \max_{\left(\sigma_{11},\mathbf{\Sigma}_{22},\mathbf{\Sigma}_{32}^{*},\mathbf{\Sigma}_{33.2}\right)} L^0(\mathbf{\Sigma})$$
$$= \max_{\mathbf{\Sigma}_{32}^{*}}\Big[\max_{(\sigma_{11},\mathbf{\Sigma}_{22},\mathbf{\Sigma}_{33.2})} L^0(\mathbf{\Sigma})\Big].$$

We need the following lemma proved by Giri (1977, p. 74) in order to maximize $L^0(\mathbf{\Sigma})$.

Let $\mathbf{A}_{p\times p}$ be any symmetric positive definite matrix and let $f(\mathbf{A}) = c(\det \mathbf{A})^{\frac{1}{2}N} \exp\left\{-\frac{1}{2} \operatorname{tr} \mathbf{A}\right\}$, then $f(\mathbf{A})$ has a maximum in the space of all positive definite matrices when $\mathbf{A} = N\mathbf{I}_{p\times p}$.

Letting

$$K_0 = k \mid N\mathbf{I}_{q\times q} \mid^{-\frac{1}{2}N} \mid (M+N)\mathbf{I}_{p\times p} \mid^{\frac{1}{2}(M+N)} (M+N)^{\frac{1}{2}(M+N)}$$
$$\times \exp\left\{-\frac{1}{2} \operatorname{tr} \left[N\mathbf{I}_{q\times q} + (M+N)\mathbf{I}_{p\times p} + (M+N)\right]\right\},$$

$$\max_{(\sigma_{11},\mathbf{\Sigma}_{22},\mathbf{\Sigma}_{33.2})} L^0(\mathbf{\Sigma}) = K_0 \cdot \left(t_{11} + s_{11} + \overline{st}_{11}\right)^{-\frac{1}{2}(M+N)}$$
$$\times \mid \mathbf{t}_{22} + \mathbf{s}_{22} + \overline{\mathbf{st}}_{22} \mid^{-\frac{1}{2}(M+N)}$$
$$\times \mid \mathbf{s}_{33} - \mathbf{\Sigma}_{32}^{*}\mathbf{s}_{23} - \mathbf{s}_{32}\mathbf{\Sigma}_{23}^{*} + \mathbf{\Sigma}_{32}^{*}\mathbf{s}_{22}\mathbf{\Sigma}_{23}^{*} \mid^{-\frac{1}{2}N} . \quad (7)$$

For example,

$$\mid \mathbf{\Sigma}_{33.2} \mid^{-\frac{1}{2}N} \exp\left\{-\frac{1}{2} \operatorname{tr} \left[\mathbf{\Sigma}_{33.2}^{-1}\mathbf{s}_{33}^{*0}\right]\right\}$$
$$= \mid \mathbf{s}_{33}^{*0} \mid^{-\frac{1}{2}N} \mid \mathbf{s}_{33}^{*0} \mid^{\frac{1}{2}N} \mid \mathbf{\Sigma}_{33.2} \mid^{-\frac{1}{2}N} \exp\left\{-\frac{1}{2} \operatorname{tr} \left[\mathbf{\Sigma}_{33.2}^{-1}\mathbf{s}_{33}^{*0}\right]\right\}$$

has a maximum when $\mathbf{\Sigma}_{33.2}^{-1}\mathbf{s}_{33}^{*0} = N\mathbf{I}_{q\times q}$.

Noting that

$$\mid \mathbf{s}_{33} - \mathbf{\Sigma}_{32}^{*}\mathbf{s}_{23} - \mathbf{s}_{32}\mathbf{\Sigma}_{23}^{*} + \mathbf{\Sigma}_{32}^{*}\mathbf{s}_{22}\mathbf{\Sigma}_{23}^{*} \mid$$
$$= \mid \mathbf{s}_{33} - \mathbf{s}_{32}\mathbf{s}_{22}^{-1}\mathbf{s}_{23} + \left(\mathbf{\Sigma}_{32}^{*} - \mathbf{s}_{32}\mathbf{s}_{22}^{-1}\right)\mathbf{s}_{22}\left(\mathbf{\Sigma}_{32}^{*} - \mathbf{s}_{32}\mathbf{s}_{22}^{-1}\right)' \mid,$$

we see that the R.H.S. of (7) is maximized when

$$\Sigma_{32}^{*} = s_{32}s_{22}^{-1}$$

and that

$$\max_{(\mu,\Sigma)\in\Omega_0} L(\mu,\Sigma) = K_0\left(t_{11}+s_{11}+\overline{st}_{11}\right)^{-\frac{1}{2}(M+N)}$$
$$\times \mid t_{22}+s_{22}+\overline{st}_{22}\mid^{-\frac{1}{2}(M+N)}$$
$$\times \mid s_{33}-s_{32}s_{22}^{-1}s_{23}\mid^{-\frac{1}{2}N}. \qquad (8)$$

Now, let us maximize $L(\Sigma)$ for all symmetric positive definite matrices of dimension $(p+q+1)\times(p+q+1)$. From (6),

$$\begin{aligned} L(\Sigma) &= \max_{\mu} L(\mu,\Sigma) \\ &= k \mid \Sigma_{(11)}\mid^{-\frac{1}{2}(N+M)} \mid \Sigma_{33.(1)}\mid^{-\frac{1}{2}N} \\ &\times \exp\left\{-\frac{1}{2}\,\mathrm{tr}\,\left[\Sigma_{(11)}^{-1}\left(\mathbf{T}+\overline{\mathbf{ST}}+s_{(11)}\right)\right]\right\} \\ &\times \exp\left\{-\frac{1}{2}\,\mathrm{tr}\,\left[\Sigma_{33.(1)}^{-1}s_{33}^{*}\right]\right\}. \end{aligned}$$

Since the mapping:

$$\left.\begin{array}{l}\Sigma_{(11)}\\ \Sigma_{(1)3}\\ \Sigma_{33}\end{array}\right\} \to \left\{\begin{array}{c}\Sigma_{(11)}\\ \Sigma_{3(1)}\Sigma_{(11)}^{-1}\equiv\Sigma_{31}^{*}\\ \Sigma_{33}-\Sigma_{3(1)}\Sigma_{(11)}^{-1}\Sigma_{(1)3}=\Sigma_{33.1}\end{array}\right.$$

is one-to-one, we have that

$$\begin{aligned} \max_{(\mu,\Sigma)\in\Omega_0\cup\Omega_1} L(\mu,\Sigma) &= \max_{\left(\Sigma_{(11)},\Sigma_{31}^{*},\Sigma_{33.1}\right)} L(\Sigma) \\ &= \max_{\Sigma_{31}^{*}}\Big[\max_{(\Sigma_{(11)},\Sigma_{33.1})} L(\Sigma)\Big] \\ &= K_1 \mid \mathbf{T}+s_{(11)}+\overline{\mathbf{ST}}\mid^{-\frac{1}{2}(M+N)} \\ &\times \mid s_{33}-s_{3(1)}s_{(11)}^{-1}s_{(1)3}\mid^{-\frac{1}{2}N}, \end{aligned} \qquad (9)$$

where

$$\begin{aligned} K_1 &= k \mid NI_{q\times q}\mid^{\frac{1}{2}N} \mid (M+N)I_{(p+1)\times(p+1)}\mid^{\frac{1}{2}(M+N)} \times \\ &\times \exp\left\{-\frac{1}{2}\,\mathrm{tr}\,\left[NI_{q\times q}+(M+N)I_{(p+1)\times(p+1)}\right]\right\} \\ &= K_0. \end{aligned}$$

From (8) and (9) the likelihood ratio is

$$
\begin{aligned}
\lambda &= \frac{\max_{(\boldsymbol{\mu},\Sigma)\in\boldsymbol{\Omega}_0} L(\boldsymbol{\mu},\Sigma)}{\max_{(\boldsymbol{\mu},\Sigma)\in\boldsymbol{\Omega}_0\cup\boldsymbol{\Omega}_1} L(\boldsymbol{\mu},\Sigma)} \\
&= \left[\left(t_{11}+s_{11}+\overline{st}_{11}\right)^{-\frac{1}{2}(M+N)} \mid \mathbf{t}_{22}+\mathbf{s}_{22}+\overline{\mathbf{st}}_{22} \mid^{-\frac{1}{2}(M+N)} \right.\\
&\quad \left. \times \mid \mathbf{s}_{33}-\mathbf{s}_{32}\mathbf{s}_{22}^{-1}\mathbf{s}_{23} \mid^{-\frac{1}{2}N}\right] \\
&\quad \times \left[\mid \mathbf{T}+\mathbf{s}_{(11)}+\overline{\mathbf{ST}} \mid^{\frac{1}{2}(N+M)} \mid \mathbf{s}_{33}-\mathbf{s}_{3(1)}\mathbf{s}_{(11)}^{-1}\mathbf{s}_{(1)3} \mid^{\frac{1}{2}N}\right].
\end{aligned}
$$

Letting $\mathbf{t}_{ij}+\mathbf{s}_{ij}+\overline{\mathbf{st}}_{ij}=\boldsymbol{\tau}_{ij}$, we have:

$$
\lambda = \frac{(\tau_{11})^{-\frac{1}{2}(N+M)} \mid \boldsymbol{\tau}_{22} \mid^{-\frac{1}{2}(N+M)} \left\{\begin{vmatrix} \mathbf{s}_{22} & \mathbf{s}_{23} \\ \mathbf{s}_{32} & \mathbf{s}_{33} \end{vmatrix} / \mid \mathbf{s}_{22} \mid\right\}^{-\frac{1}{2}N}}{\begin{vmatrix} \tau_{11} & \boldsymbol{\tau}_{12} \\ \boldsymbol{\tau}_{21} & \boldsymbol{\tau}_{22} \end{vmatrix}^{-\frac{1}{2}(M+N)} \left\{\mid \mathbf{S} \mid / \mid \mathbf{s}_{(11)} \mid\right\}^{-\frac{1}{2}N}},
$$

and noting that

$$
\begin{vmatrix} \tau_{11} & \boldsymbol{\tau}_{12} \\ \boldsymbol{\tau}_{21} & \boldsymbol{\tau}_{22} \end{vmatrix} = \mid \boldsymbol{\tau}_{22} \mid \left(\tau_{11}-\boldsymbol{\tau}_{12}\boldsymbol{\tau}_{22}^{-1}\boldsymbol{\tau}_{21}\right),
$$

we have

$$
\begin{aligned}
\lambda &= \frac{(\tau_{11})^{-\frac{1}{2}(N+M)}}{\left(\tau_{11}-\boldsymbol{\tau}_{12}\boldsymbol{\tau}_{22}^{-1}\boldsymbol{\tau}_{21}\right)^{-\frac{1}{2}(M+N)}} \cdot \left(\frac{\begin{vmatrix} \mathbf{s}_{23} & \mathbf{s}_{23} \\ \mathbf{s}_{32} & \mathbf{s}_{33} \end{vmatrix}}{\mid \mathbf{s}_{22} \mid} \cdot \frac{\begin{vmatrix} s_{11} & \mathbf{s}_{12} \\ \mathbf{s}_{21} & \mathbf{s}_{22} \end{vmatrix}}{\mid \mathbf{S} \mid} \right)^{-\frac{1}{2}N} \\
&= \left[\frac{(\tau_{11})^{-\frac{1}{2}(N+M)}}{\left(\tau_{11}-\boldsymbol{\tau}_{12}\boldsymbol{\tau}_{22}^{-1}\boldsymbol{\tau}_{21}\right)^{-\frac{1}{2}(N+M)}}\right] \\
&\quad \times \left(\frac{\begin{vmatrix} \mathbf{s}_{22} & \mathbf{s}_{23} \\ \mathbf{s}_{32} & \mathbf{s}_{33} \end{vmatrix} \cdot \mid \mathbf{s}_{22} \mid \cdot \left(s_{11}-\mathbf{s}_{12}\mathbf{s}_{22}^{-1}\mathbf{s}_{21}\right)}{\mid \mathbf{s}_{22} \mid \begin{vmatrix} \mathbf{s}_{22} & \mathbf{s}_{23} \\ \mathbf{s}_{32} & \mathbf{s}_{33} \end{vmatrix} \left(s_{11} - (\mathbf{s}_{12}, \mathbf{s}_{13}) \begin{pmatrix} \mathbf{s}_{22} & \mathbf{s}_{23} \\ \mathbf{s}_{32} & \mathbf{s}_{33} \end{pmatrix}^{-1} \begin{pmatrix} \mathbf{s}_{21} \\ \mathbf{s}_{31} \end{pmatrix} \right)} \right)^{-\frac{1}{2}N} \\
&= \frac{\left[\left(\tau_{11}-\boldsymbol{\tau}_{12}\boldsymbol{\tau}_{22}^{-1}\boldsymbol{\tau}_{21}\right)/\tau_{11}\right]^{\frac{1}{2}(M+N)} \left(1-r^2_{p+q,N}\right)^{\frac{1}{2}N}}{\left(1-r^2_{p,N}\right)^{\frac{1}{2}N}}. \qquad (10)
\end{aligned}
$$

We reject H_0 defined in (1) when $\lambda \leq K$, where λ is given in (10) and K is a constant depending on the significance level of the test. Some distributional aspects of the test statistic λ will be discussed in a later article.

REFERENCES

Afifi, A. A., and R. M. Elashoff (1966), "Missing observations in multivariate statistics I. Review of the literature." *Journal of the American Statistical Association* **61**, 595–604.

Afifi, A. A., and R. M. Elashoff (1967), "Missing observations in multivariate statistics II. Point estimation in simple linear regression." *Journal of the American Statistical Association* **62**, 10–29.

Afifi, A. A., and R. M. Elashoff (1969a), "Missing observations in multivariate statistics III: Large sample analysis of simple linear regression." *Journal of the American Statistical Association* **64**, 337–358.

Afifi, A. A., and R. M. Elashoff (1969b), "Missing observations in multivariate statistics—IV. A note on simple linear regression." *Journal of the American Statistical Association* **64**, 359–365.

Anderson, T. W. (1957), "Maximum likelihood estimates for a multivariate normal distribution when some observations are missing." *Journal of the American Statistical Association* **52**, 200–203.

Anderson, T. W. (1958), *An Introduction to Multivariate Analysis*. New York: Wiley.

Bhargava, R. P. (1975), "Some one-sample hypothesis testing problems when there is a monotone sample from a multivariate normal population." *Annals of the Institute of Statistical Mathematics* **27**, 327–339.

Giri, N. (1977), *Multivariate Statistical Inference*. New York: Academic Press.

Eaton, M. L., and T. Kariya (1974), "Testing for independence with additional information." Technical Report No. 238. University of Minnesota.

Eaton, M. L., and T. Kariya (1975), "Tests on means with additional information." Technical Report No. 243. University of Minnesota.

Edgett, G. L. (1956), "Multiple regression with missing observations among the independent variables." *Journal of the American Statistical Association* **51**, 122–131.

Little, R. J. A. (1976), "Inference about means from incomplete multivariate data." *Biometrika* **63**, 593–604.

Lord, F. M. (1955), "Estimation of parameters from incomplete data." *Journal of the American Statistical Association* **50**, 870–876.

Morrison, D. F. (1971), "Expectations and variances of maximum likelihood estimates of the multivariate normal distribution parameters with missing data." *Journal of the American Statistical Association* **66**, 602–604.

Radhakrishnan, R. (1982), "Inadmissibility of the maximum likelihood estimator for a multivariate normal distribution when some observations are missing." *Communications in Statistics A, Theory and Methods* **11**, 941–955.

Rao, C. R. (1956), "Analysis of dispersion with incomplete observations on one of the characters." *Journal of the Royal Statistical Society,* Series B **19**, 259–264.

Smith, W. B., and R. C. Pfaffenberger (1970), "Selection index estimation from partial multivariate normal data." *Biometrics* **26**, 625–639.

Trawinski, I. M., and R. E. Bargmann (1964), "Maximum likelihood estimation with incomplete multivariate data." *Annals of Mathematical Statistics* **35**, 647–657.

Wilks, S. S. (1932), "Moments and distributions of estimates of population parameters from fragmentary samples." *Annals of Mathematical Statistics* **3**, 163–195.

M. A. Ali [1]

MISSING VALUE PROBLEMS IN MULTIPLE LINEAR REGRESSION WITH TWO INDEPENDENT VARIABLES

ABSTRACT

The relative accuracies of frequently used estimators and an alternative estimator of regression coefficients are investigated for the case of two independent variables x_1 and x_2 with missing values on x_2 only. A Monte Carlo study of observations generated from a trivariate normal distribution is performed. It is found that for a wide class of conditions, in the sense of mean squared error, the alternative method is superior to the method of linear prediction and the method of maximum likelihood for estimating the coefficients of both completely and incompletely observed variables.

1. INTRODUCTION

1.1 Statement of the Problem

A problem frequently occurring in statistical practice is that of dealing with multiple regression when some of the values on the independent variables are missing. We discuss here the special case of regressing a dependent variable y on two independent variables x_1 and x_2, where there are missing values on x_2 only. Such a situation may arise in chemical experiments, for example, where destruction or removal of part of an experimental unit prevents the investigator from making complete measurements on x_2. The question then is how to deal properly with the incomplete sample.

We consider the linear model

$$y_i = \beta_0 + \beta_1 x_{1i} + \beta_2 x_{2i} + e_i, \quad i = 1, 2, \ldots, N, \tag{1.1.1}$$

where the e_i are i.i.d. random variables with mean 0 and variance σ^2, and $(N - n)$ values on x_2 are unobserved.

[1] Department of Statistics, Yarmouk University, Irbid, Jordan

I. B. MacNeill and G. J. Umphrey (eds.), Foundations of Statistical Inference, 163–182.

The simplest approach is to apply the usual least squares analysis to the subset of complete cases, an approach equivalent to applying the method of least squares to all available observations while regarding the missing values as additional parameters to be estimated (Yates, 1933). However, many alternatives to the complete case method have been proposed.

Wilks (1932) and Dear (1959) have studied the substitution of the sample means for missing observations, followed by the usual application of least squares. Walsh (1959) and Buck (1960) have investigated the substitution of predicted values from simple linear regressions among the observed values. Anderson (1957) and Beale and Little (1975) have applied the method of maximum likelihood to deal with the problem arising from missing data.

Properties of these procedures have been compared extensively, mainly through simulation studies by Beale and Little (1975), Donner and Rosner (1982), Haitovsky (1968), Heiberger (1977) and Timm (1970). Beale and Little (1975) have found that for multinormal distributions the method of maximum likelihood is preferable to the complete case method and the method of linear prediction in regards to minimizing the residual sum of squares of deviations of the observed and fitted values of the dependent variable. Donner and Rosner (1982) have shown that the method of maximum likelihood is superior to other methods in estimating β_1 and β_2 of the model (1.1.1).

Analytic comparisons have generally not been performed, largely because of the complexity of the problem that arises when missing values are scattered over several explanatory variables. However, Donner (1982) has derived the expressions for the bias and variance associated with procedures frequently used to deal with the problem. He has found that in terms of mean squared error (MSE), the mean substitution method is more effective for estimating β_2, the coefficients of incompletely observed variables when the correlations involving these variables are small and the proportion of missing cases is fairly high. In estimating β_1, the coefficients of completely observed variables, this method provides estimates which may have substantial bias unless the correlations involving these variables are small. Further, he has shown that the method of linear prediction offers advantages over the complete-case method for estimating β_1 while the two methods are equivalent for estimating β_2.

In this article an alternative estimator for the vector parameter $\boldsymbol{\beta} = (\beta_1, \beta_2)'$ has been proposed. Its properties are examined. It is observed that for a wide class of conditions, in the sense of MSE, the alternative method is superior to the method of linear prediction and the method of maximum likelihood for estimating the coefficient of both completely and incompletely observed variables.

1.2 Description of Procedures Investigated

The procedures considered are mentioned below with the resulting estimate of $\boldsymbol{\beta} = (\beta_1, \beta_2)'$ denoted in parentheses:

i) The complete case method in which the usual least squares analysis is applied to the cases for which both x_1 and x_2 are completely observed, $\hat{\boldsymbol{\beta}}_c = (\hat{\beta}_{1c}, \hat{\beta}_{2c})'$.

ii) The method of linear prediction in which a missing x_2 is replaced by its predicted value, obtained from the simple linear regression of x_2 on x_1 over the complete cases followed by the usual analysis, $\hat{\boldsymbol{\beta}}_\ell = (\hat{\beta}_{1\ell}, \hat{\beta}_{2\ell})'$.

iii) The mean substitution method in which a missing x_2 is replaced by the mean of x_2 over the complete cases followed by the usual analysis, $\hat{\boldsymbol{\beta}}_m = (\hat{\beta}_{1m}, \hat{\beta}_{2m})'$.

1.3 Notations Used

In presenting the results the following notations are used:

$f = n/N$ = proportion of cases observed on x_2, $\quad 0 < f \leq 1$.

r = simple correlation coefficient between x_1 and x_2 over the n complete cases.

r^* = simple correlation coefficient between x_1 and x_2 over all N cases, including those for which x_2 is not actually observed.

$R = (1 - r^2)\,(1 - f)$.

$D = (r^* - r)$.

$H = 1 - f(1 - r^2)$.

$H_1 = 1 - fr^2$.

Tr $\mathbf{Z}$ = Trace of a square matrix $\mathbf{Z}$.

$|\,\mathbf{Z}\,|$ = Determinant of a square matrix $\mathbf{Z}$.

$\|\,\boldsymbol{\beta}\,\|^2 = \boldsymbol{\beta}'\,\boldsymbol{\beta}$, for any vector $\boldsymbol{\beta}$.

$\text{Bias}^2(\hat{\boldsymbol{\beta}}_a) = \|\,E(\hat{\boldsymbol{\beta}}_a) - \boldsymbol{\beta}\,\|^2$, for any estimator $\hat{\boldsymbol{\beta}}_a$ of $\boldsymbol{\beta}$.

$D(\hat{\boldsymbol{\beta}}_a) = E\{[\hat{\boldsymbol{\beta}}_a - E(\hat{\boldsymbol{\beta}}_a)][\hat{\boldsymbol{\beta}}_a - E(\hat{\boldsymbol{\beta}}_a)]'\}$.

$\text{MSE}(\hat{\boldsymbol{\beta}}_a) = \text{Bias}^2(\hat{\boldsymbol{\beta}}_a) + \text{Tr}\,(D(\hat{\boldsymbol{\beta}}_a))$.

2. METHODS OF ESTIMATION

2.1 Assumptions

We are primarily concerned with estimating β_1 and β_2 under the assumptions that the $(N-n)$ values of x_2 are missing. We assume that the values (x_{1i}, x_{2i}) are fixed with zero means and unit variances (unadjusted for degrees of freedom) over all N cases, and over the n complete cases. We also assume that the y_i are centered about zero over all N cases and over the n complete cases. Writing the observations with missing x_{2i} last, we separate the complete cases from the incomplete cases in the linear model

$$\mathbf{Y} = \mathbf{X}\boldsymbol{\beta} + \mathbf{e}, \tag{2.1.1}$$

where

$\mathbf{Y}$: an $N \times 1$ vector of observations,

$\mathbf{X}$: an $N \times 2$ input matrix,

$\boldsymbol{\beta} = (\beta_1, \beta_2)'$,

and $\mathbf{e}$ = an $N \times 1$ vector of random errors.

We partition $\mathbf{Y}$ as follows:

$$\mathbf{Y} = \begin{pmatrix} \mathbf{Y}_C \\ \cdots \\ \mathbf{Y}_I \end{pmatrix}, \tag{2.1.2}$$

where

$\mathbf{Y}_C$ is the vector of observations on n complete cases,

$\mathbf{Y}_I$ is the vector of observations for incomplete cases.

The input matrix $\mathbf{X}$ and the vector of random errors, $\mathbf{e}$, are partitioned accordingly, i.e.,

$$\mathbf{X} = \begin{pmatrix} \mathbf{X}_c \\ \cdots \\ \mathbf{X}_I \end{pmatrix} = \begin{pmatrix} \mathbf{X}_{1c} & \mathbf{X}_{2c} \\ \cdots & \cdots \\ \mathbf{X}_{1I} & \mathbf{X}_{2I} \end{pmatrix}$$

with

$$\mathbf{X}_c = (\mathbf{X}_{1c}\ \mathbf{X}_{2c})$$
$$\mathbf{X}_I = (\mathbf{X}_{1I}\ \mathbf{X}_{2I})\,;\ \mathbf{X}_{2I} \text{ is missing},$$

and

$$\mathbf{e} = \begin{bmatrix} \mathbf{e}_c \\ \mathbf{e}_I \end{bmatrix}.$$

Our assumptions can be summarized as follows:

$$\begin{aligned}
\mathbf{1}'\mathbf{Y} &= \mathbf{1}'\mathbf{Y}_c = \mathbf{1}'\mathbf{Y}_I = 0, \quad \mathbf{1} \text{ is a column vector of ones,} \\
\mathbf{1}'\mathbf{X}_{1c} &= \mathbf{1}'\mathbf{X}_{2c} = \mathbf{1}'\mathbf{X}_{1I} = \mathbf{1}'\mathbf{X}_{2I} = 0, \\
n &= \mathbf{X}_{1c}'\mathbf{X}_{1c} = \mathbf{X}_{2c}'\mathbf{X}_{2c}, \\
N - n &= \mathbf{X}_{1I}'\mathbf{X}_{1I} = \mathbf{X}_{2I}'\mathbf{X}_{2I}, \\
rn &= \mathbf{X}_{1c}'\mathbf{X}_{2c}, \\
r^* N &= \mathbf{X}_{1c}'\mathbf{X}_{2c} + \mathbf{X}_{1I}'\mathbf{X}_{2I}.
\end{aligned} \tag{2.1.3}$$

2.2 Complete Case Method

In the complete case method the usual least squares analysis is applied to the model

$$\mathbf{Y}_c = \mathbf{X}_c\boldsymbol{\beta} + \mathbf{e}_c.$$

For the complete case, the estimator of $\boldsymbol{\beta}$ by $\hat{\boldsymbol{\beta}}_c$ is

$$\hat{\boldsymbol{\beta}}_c = (\mathbf{X}_c'\mathbf{X}_c)^{-1}\mathbf{X}_c'\mathbf{Y}_c, \tag{2.2.1}$$

where

$$(\mathbf{X}_c'\mathbf{X}_c) = \begin{bmatrix} n & rn \\ rn & n \end{bmatrix}$$

and

$$(\mathbf{X}_c'\mathbf{X}_c)^{-1} = \frac{1}{n(1-r^2)} \begin{bmatrix} 1 & -r \\ -r & 1 \end{bmatrix}.$$

$\hat{\boldsymbol{\beta}}_c$ is unbiased for $\boldsymbol{\beta}$ and its MSE is given by

$$\text{MSE}(\hat{\boldsymbol{\beta}}_c) = \sigma^2 \,\text{Tr}\,(\mathbf{X}_c'\mathbf{X}_c)^{-1} = \frac{2\sigma^2}{n(1-r^2)}. \tag{2.2.2}$$

2.3 Alternative Methods

In the linear prediction method we replace $\mathbf{X}_{2I}$ by its predicted value obtained by the relation

$$\mathbf{X}_{2I} = r\mathbf{X}_{1I}, \tag{2.3.1}$$

while in the mean substitution method we substitute $\mathbf{X}_{2I}$ by the mean of n available observations on x_2. Since x_{2i} are observed with zero mean, thus for this method we substitute

$$\mathbf{X}_{2I} = 0. \tag{2.3.2}$$

We consider, in general, a prediction model of the form

$$\mathbf{X}_{2I} = kr\mathbf{X}_{1I}, \ k \geq 0. \tag{2.3.3}$$

Note that the relations (2.3.1) and (2.3.2) can easily be obtained from the relation (2.3.3) by setting $k = 1$ and 0 respectively.
Let

$$\mathbf{X}_k = \begin{bmatrix} \mathbf{X}_{1c} & \mathbf{X}_{2c} \\ \mathbf{X}_{1I} & kr\mathbf{X}_{1I} \end{bmatrix}$$

and

$$\mathbf{S} = (\mathbf{X}_k'\mathbf{X}_k). \tag{2.3.4}$$

We define $\hat{\boldsymbol{\beta}}_k$, an estimate of $\boldsymbol{\beta}$ by

$$\hat{\boldsymbol{\beta}}_k = (\hat{\beta}_{1k}, \hat{\beta}_{2k})' = \mathbf{S}^{-1}\mathbf{X}_k'\mathbf{Y}. \tag{2.3.5}$$

The bias of $\hat{\boldsymbol{\beta}}_k$ is given by

$$\text{Bias}(\hat{\boldsymbol{\beta}}_k) = E[\hat{\boldsymbol{\beta}}_k - \boldsymbol{\beta}] = \frac{n\beta_2 G}{|\mathbf{S}|} \begin{bmatrix} (1 - kr^2) \\ r(k-1) \end{bmatrix}, \tag{2.3.6}$$

where

$$|\mathbf{S}| = Nn\left[(1 - r^2) + r^2(1-k)^2(1-f)\right]$$

and

$$G = N\left[D + r(1-k)(1-f)\right].$$

Therefore,

$$\text{Bias}^2(\hat{\boldsymbol{\beta}}_k) = \frac{n^2\beta_2^2 G^2}{|\mathbf{S}|^2}\left[\left(1 - kr^2\right)^2 + r^2(k-1)^2\right]. \tag{2.3.7}$$

The dispersion matrix of $\hat{\boldsymbol{\beta}}_k$ is given by

$$D(\hat{\boldsymbol{\beta}}_k) = \sigma^2\mathbf{S}^{-1}, \tag{2.3.8}$$

where

$$\mathbf{S}^{-1} = \frac{N}{|\mathbf{S}|}\begin{bmatrix} f + k^2r^2(1-f) & -rf - kr(1-f) \\ -rf - kr(1-f) & 1 \end{bmatrix}.$$

The MSE of $\hat{\boldsymbol{\beta}}_k$ is given by

$$\text{MSE}(\hat{\boldsymbol{\beta}}_k) = \sigma^2 \ \text{Tr}\ \mathbf{S}^{-1} + \ \text{Bias}^2(\hat{\boldsymbol{\beta}}_k). \tag{2.3.9}$$

Note from the relation (2.3.6) that for $k = 1/r^2$, the bias of $\hat{\beta}_k$ has the form

$$E\left[\hat{\boldsymbol{\beta}}_k - \boldsymbol{\beta}\right]_{k=1/r^2} = \frac{\beta_2(rr^* - H)}{H}\begin{bmatrix}0\\1\end{bmatrix}, \tag{2.3.10}$$

In sequel, for $k = 1/r^2$, we denote $\hat{\boldsymbol{\beta}}_k$ by $\hat{\boldsymbol{\beta}}_I = (\beta_{1I}, \beta_{2I})'$. Again note that the estimators $\hat{\boldsymbol{\beta}}_\ell$ and $\hat{\boldsymbol{\beta}}_m$ of $\boldsymbol{\beta}$ can be obtained from the relation (2.3.5) by setting $k = 1$ and 0 respectively.

We discuss the behaviour of $\hat{\boldsymbol{\beta}}_I, \hat{\boldsymbol{\beta}}_\ell$ and $\hat{\boldsymbol{\beta}}_m$ in the following section.

3. BEHAVIOUR OF $\hat{\boldsymbol{\beta}}_\ell, \hat{\boldsymbol{\beta}}_m$ AND $\hat{\boldsymbol{\beta}}_I$

The expressions for bias of $\hat{\boldsymbol{\beta}}_\ell, \hat{\boldsymbol{\beta}}_m$ and $\hat{\boldsymbol{\beta}}_I$ can easily be obtained from the relation (2.3.6) by setting $k = 1, 0$ and $1/r^2$ respectively. The dispersion matrices of $\hat{\boldsymbol{\beta}}_\ell, \hat{\boldsymbol{\beta}}_m$ and $\hat{\boldsymbol{\beta}}_I$ can be obtained from the relation (2.3.8) using $k = 1, 0$ and $1/r^2$ respectively. These results are presented in Table 1.

Donner (1982) has shown that for estimating β_1, the coefficient of a completely observed variable, $\hat{\beta}_{1\ell}$ offers advantages over $\hat{\beta}_{1c}$ and $\hat{\beta}_{1m}$. For estimating $\boldsymbol{\beta}_2$, the coefficient of an incompletely observed variable, the method of linear prediction and the complete-case method are equivalent, i.e., $\hat{\beta}_{2\ell} \equiv \hat{\beta}_{2c}$. Since $\hat{\beta}_{2\ell} \equiv \hat{\beta}_{2c}$, the only attractive alternative to $\hat{\beta}_{2c}$ in this case is the estimator $\hat{\beta}_{2m}$, which has lower MSE if r is small and proportion of missing cases is moderately high.

Replacing k in relation (2.3.5) by $1/r^2$ and using (2.1.3) we have

$$\begin{aligned}\hat{\boldsymbol{\beta}}_I &= \hat{\boldsymbol{\beta}}_{k=1/r^2} = (\mathbf{X}_2'\mathbf{X}_2)^{-1}\mathbf{X}_2'\mathbf{Y}\\ &= \mathbf{A}_2\mathbf{X}_2'\mathbf{Y}\\ &= \frac{1}{n(1-r^2)}\begin{bmatrix}1 & -r\\ -r & r^2/H\end{bmatrix}\begin{bmatrix}\mathbf{X}_{1c}'\mathbf{Y}_c + \mathbf{X}_{1I}'\mathbf{Y}_I\\ \mathbf{X}_{2c}'\mathbf{Y}_c + \mathbf{X}_{1I}'\mathbf{Y}_I/r\end{bmatrix},\end{aligned} \tag{3.1.1}$$

where

$$\mathbf{X}_2 = \begin{bmatrix}\mathbf{X}_{1c} & \mathbf{X}_{2c}\\ \mathbf{X}_{1I} & \mathbf{X}_{1I}/r\end{bmatrix}, \quad \mathbf{A}_2 = (\mathbf{X}_2'\mathbf{X}_2)^{-1}, \tag{3.1.2}$$

and $\mathbf{Y}_c$ and $\mathbf{Y}_I$ are defined in (2.1.2).

It can easily be shown from the relations (2.2.1) and (3.1.1) that

$$\hat{\beta}_{1I} \equiv \hat{\beta}_{1c}, \tag{3.1.3}$$

Table 1. *Bias and Dispersion Matrix of Estimators of $\boldsymbol{\beta}$.*

Estimator	Bias	Dispersion Matrix
$\hat{\boldsymbol{\beta}}_c = \begin{bmatrix} \hat{\beta}_{1c} \\ \hat{\beta}_{2c} \end{bmatrix}$	$\begin{bmatrix} 0 \\ 0 \end{bmatrix}$	$\frac{\sigma^2}{n(1-r^2)} \begin{bmatrix} 1 & -r \\ -r & 1 \end{bmatrix}$
$\hat{\boldsymbol{\beta}}_l = \begin{bmatrix} \hat{\beta}_{1l} \\ \hat{\beta}_{2l} \end{bmatrix}$	$\begin{bmatrix} \beta_2 D \\ 0 \end{bmatrix}$	$\frac{\sigma^2}{n(1-r^2)} \begin{bmatrix} f + r^2(1-f) & -r \\ -r & 1 \end{bmatrix}$
$\hat{\boldsymbol{\beta}}_m = \begin{bmatrix} \hat{\beta}_{1m} \\ \hat{\beta}_{2m} \end{bmatrix}$	$\frac{\beta_2(D+r(1-f))}{H_1} \begin{bmatrix} 1 \\ -r \end{bmatrix}$	$\frac{\sigma^2}{nH_1} \begin{bmatrix} f & -rf \\ -rf & 1 \end{bmatrix}$
$\hat{\boldsymbol{\beta}}_I = \begin{bmatrix} \hat{\beta}_{1I} \\ \hat{\beta}_{2I} \end{bmatrix}$	$\frac{\beta_2(rr^*-H)}{H} \begin{bmatrix} 0 \\ 1 \end{bmatrix}$	$\frac{\sigma^2}{n(1-r^2)} \begin{bmatrix} 1 & -r \\ -r & r^2/H \end{bmatrix}$

i.e., the two estimators are identical. Note from Table 1 that

$$\text{Bias}(\hat{\beta}_{2c}) = 0, \quad \text{Var}(\hat{\beta}_{2c}) = \sigma^2/n\left(1 - r^2\right),$$

$$\text{Bias}(\hat{\beta}_{2I}) = \beta_2(rr^* - H)/H, \quad \text{Var}(\hat{\beta}_{2I}) = \frac{\sigma^2 r^2}{n\left(1 - r^2\right)H},$$

$$\text{Bias}(\hat{\beta}_{2m}) = -\beta_2 r\left[D + r(1-f)\right]/H_1, \quad \text{Var}(\hat{\beta}_{2m}) = \sigma^2/nH_1.$$

Clearly, if $r^2 \leq \frac{1}{2}$, then

$$\text{Var}(\hat{\beta}_{2I}) \leq \text{Var}(\hat{\beta}_{2m}). \tag{3.1.4}$$

Further, if

$$r + \frac{1}{r}\left[R - \frac{(1-f)H}{H_1 + H}\right] \leq r^* \leq r + \frac{1}{r}\left[R + \frac{(1-f)H}{H_1 - H}\right]$$

then

$$\text{Bias}^2(\hat{\beta}_{2I}) \leq \text{Bias}^2(\hat{\beta}_{2m}). \tag{3.1.5}$$

Thus $\hat{\beta}_{2I}$ may be considered an effective alternative to $\hat{\beta}_{2m}$. From the relations (3.1.3)–(3.1.5) it follows that $\hat{\beta}_{2I}$ offers advantages in estimating β_2, while in estimating β_1 this method and the complete case method are equivalent.

4. AN ALTERNATIVE ESTIMATOR FOR β

It is interesting to note from Table 1 that $\hat{\boldsymbol{\beta}}_\ell$ provides a biased estimator for β_1 and an unbiased estimator for β_2, while $\hat{\boldsymbol{\beta}}_I$ gives us an unbiased estimator for β_1 and a biased estimator for β_2. This suggests that we can define another estimator $\hat{\boldsymbol{\beta}}_R = (\hat{\beta}_{1R}, \hat{\beta}_{2R})'$ for the vector parameter $\boldsymbol{\beta} = (\beta_1, \beta_2)'$ by

$$\hat{\boldsymbol{\beta}}_R = \begin{bmatrix} \hat{\beta}_{1R} \\ \hat{\beta}_{2R} \end{bmatrix} = \alpha\hat{\boldsymbol{\beta}}_\ell + (1-\alpha)\hat{\boldsymbol{\beta}}_I, \qquad \alpha \in [0,1]. \tag{4.1.1}$$

The bias of $\hat{\boldsymbol{\beta}}_R$ is given by

$$\begin{aligned} \text{Bias}(\hat{\boldsymbol{\beta}}_R) &= \alpha \ \text{Bias}(\hat{\boldsymbol{\beta}}_\ell) + (1-\alpha) \ \text{Bias}(\hat{\boldsymbol{\beta}}_I) \\ &= \beta_2 \begin{bmatrix} \alpha D \\ (1-\alpha)(rr^* - H)/H \end{bmatrix}. \end{aligned} \tag{4.1.2}$$

Thus for any α satisfying $0 \leq \alpha \leq 1$, we have

$$\begin{aligned} |\ \text{Bias}(\hat{\beta}_{1R})\ | &\leq |\ \text{Bias}(\hat{\beta}_{1\ell})\ |, \\ |\ \text{Bias}(\hat{\beta}_{2R})\ | &\leq |\ \text{Bias}(\hat{\beta}_{2I})\ |. \end{aligned} \tag{4.1.3}$$

From (4.1.2) we obtain

$$\begin{aligned} \text{Bias}(\hat{\boldsymbol{\beta}}_R) &= \beta_2^2 \left[\alpha^2 D^2 + (1-\alpha)^2 (rr^* - H)^2/H^2\right] \\ &= \alpha^2 \ \text{Bias}^2(\hat{\boldsymbol{\beta}}_\ell) + (1-\alpha)^2 \ \text{Bias}^2(\hat{\boldsymbol{\beta}}_I). \end{aligned} \tag{4.1.4}$$

Note from (4.1.4) that if

$$\text{Bias}^2(\hat{\boldsymbol{\beta}}_\ell) \geq \frac{1-\alpha}{1+\alpha} \ \text{Bias}^2(\hat{\boldsymbol{\beta}}_I),$$

then

$$\text{Bias}^2(\hat{\boldsymbol{\beta}}_R) \leq \text{Bias}^2(\hat{\boldsymbol{\beta}}_\ell).$$

Further, if

$$\text{Bias}^2(\hat{\boldsymbol{\beta}}_I) \geq \frac{\alpha}{2-\alpha} \text{Bias}^2(\hat{\boldsymbol{\beta}}_\ell),$$

then

$$\text{Bias}^2(\hat{\boldsymbol{\beta}}_R) \leq \text{Bias}^2(\hat{\boldsymbol{\beta}}_I).$$

Let

$$\mathbf{X}_1 = \begin{bmatrix} \mathbf{X}_{1c} & \mathbf{X}_{2c} \\ \mathbf{X}_{1I} & r\mathbf{X}_{1I} \end{bmatrix},$$

and

$$\mathbf{A}_1 = (\mathbf{X}_1'\mathbf{X}_1)^{-1}.$$

Setting $k = 1$ in relation (2.3.5) we get

$$\hat{\boldsymbol{\beta}}_\ell = \hat{\boldsymbol{\beta}}_{k=1} = \mathbf{A}_1\mathbf{X}_1'\mathbf{Y}. \tag{4.1.5}$$

Using relations (3.1.1) and (4.1.5) we may rewrite (4.1.1) as

$$\hat{\boldsymbol{\beta}}_R = [\alpha\mathbf{A}_1\mathbf{X}_1' + (1-\alpha)\mathbf{A}_2\mathbf{X}_2']\,\mathbf{Y}, \tag{4.1.6}$$

where $\mathbf{A}_2$ and $\mathbf{X}_2$ are defined in (3.1.2). The dispersion matrix of $\hat{\boldsymbol{\beta}}_R$ is given by

$$D\left(\hat{\boldsymbol{\beta}}_R\right) = \sigma^2\Big[\alpha^2\left\{\mathbf{A}_1 + \mathbf{A}_2 - \mathbf{A}_1\mathbf{X}_1'\mathbf{X}_2\mathbf{A}_2 - \mathbf{A}_2\mathbf{X}_2'\mathbf{X}_1\mathbf{A}_1\right\}$$
$$-2\alpha\left\{\mathbf{A}_2 - \frac{1}{2}\mathbf{A}_1\mathbf{X}_1'\mathbf{X}_2\mathbf{A}_2 - \frac{1}{2}\mathbf{A}_2\mathbf{X}_2'\mathbf{X}_1\mathbf{A}_1\right\} + \mathbf{A}_2\Big]$$

and

$$\text{Tr}(D(\hat{\boldsymbol{\beta}}_R)) = \sigma^2\left[\alpha^2\{\text{Tr}(\mathbf{A}_1 + \mathbf{A}_2) - 2\,\text{Tr}(\mathbf{A}_1\mathbf{X}_1'\mathbf{X}_2\mathbf{A}_2)\}\right.$$
$$\left.-2\alpha\{\text{Tr}(\mathbf{A}_2) - \text{Tr}(\mathbf{A}_1\mathbf{X}_1'\mathbf{X}_2\mathbf{A}_2)\} + \text{Tr}(\mathbf{A}_2)\right]; \tag{4.1.7}$$

we now determine the weight α in (4.1.1) in some optimal sense. An optimal value of α may be obtained by minimizing

$$\text{MSE}(\hat{\boldsymbol{\beta}}_R) = \text{Bias}^2(\hat{\boldsymbol{\beta}}_R) + \text{Tr}(D(\hat{\boldsymbol{\beta}}_R)).$$

Unfortunately, minimization of $\text{MSE}(\hat{\boldsymbol{\beta}}_R)$ requires knowledge about the unknown parameters β_2, σ^2 and r^*. Therefore, we determine α by minimizing

$\mathrm{Tr}\big(D(\hat{\boldsymbol{\beta}}_R)\big)$. The value of α that minimizes $\mathrm{Tr}\big(D(\hat{\boldsymbol{\beta}}_R)\big)$ can easily be shown to be

$$\alpha^* = \mathrm{Tr}\,[\mathbf{A}_2 - \mathbf{A}_1\mathbf{X}_1'\mathbf{X}_2\mathbf{A}_2] \,/\, \{\, \mathrm{Tr}\,[\mathbf{A}_1 + \mathbf{A}_2 - 2\mathbf{A}_1\mathbf{X}_1'\mathbf{X}_2\mathbf{A}_2]\}\,. \quad (4.1.8)$$

Note that the optimal value of α given by (4.1.8) is free from any unknown parameter. Substituting α by α^* from (4.1.8) in (4.1.7) we get

$$\mathrm{Tr}(D(\hat{\boldsymbol{\beta}}_R))\,|_{\alpha\,=\alpha^*} = \sigma^2[\,\mathrm{Tr}(\mathbf{A}_2) - \alpha^*\,\mathrm{Tr}(\mathbf{A}_2 - \mathbf{A}_1\mathbf{X}_1'\mathbf{X}_2\mathbf{A}_2)].$$

Since $\mathrm{Tr}\big(D(\hat{\boldsymbol{\beta}}_R)\big)$ is minimized at $\alpha = \alpha^*$, we obtain

$$\mathrm{Tr}(D(\hat{\boldsymbol{\beta}}_R))\,|_{\alpha=\alpha^*} \leq \mathrm{Tr}(D(\hat{\boldsymbol{\beta}}_R))\,|_{\alpha=0} = \mathrm{Tr}(D(\hat{\boldsymbol{\beta}}_I))$$

and

$$\mathrm{Tr}(D(\hat{\boldsymbol{\beta}}_R))\,|_{\alpha=\alpha^*} \leq \mathrm{Tr}(D(\hat{\boldsymbol{\beta}}_R))\,|_{\alpha=1} = \mathrm{Tr}(D(\hat{\boldsymbol{\beta}}_\ell)).$$

In other words, at $\alpha = \alpha^*$

$$\mathrm{Tr}(D(\hat{\boldsymbol{\beta}}_R)) \leq \mathrm{Tr}(D(\hat{\boldsymbol{\beta}}_\ell))$$

and

$$\mathrm{Tr}(D(\hat{\boldsymbol{\beta}}_R)) \leq \mathrm{Tr}(D(\hat{\boldsymbol{\beta}}_I)). \quad (4.1.9)$$

Thus $\hat{\boldsymbol{\beta}}_R$ is preferable to both $\hat{\boldsymbol{\beta}}_\ell$ and $\hat{\boldsymbol{\beta}}_I$. Unlike $\hat{\boldsymbol{\beta}}_\ell$, $\hat{\beta}_m$ and $\hat{\boldsymbol{\beta}}_I$, $\hat{\boldsymbol{\beta}}_R$ provides superior estimators for the coefficients of both completely and incompletely observed variables.

In the following section we perform simulation experiments to study the behaviour of $\hat{\boldsymbol{\beta}}_R = (\hat{\beta}_{1R}, \hat{\beta}_{2R})'$ in relation to the maximum likelihood estimator $\mathbf{b} = (b_{1m}, b_{2m})'$ of $\boldsymbol{\beta} = (\beta_1, \beta_2)'$. The expressions for b_{1m} and b_{2m} are derived by Donner and Rosner (1982). We present them in the Appendix.

5. THE ESTIMATED MEAN SQUARED ERRORS

The simulation experiments are used to estimate the mean squared errors for the two estimators, viz. the alternative estimator, given by the relation (4.1.1) and the maximum likelihood estimator, defined in (A.1.1) and (A.1.2) (see Appendix). The procedure is described below.

We generate N pairs (x_{1i}, x_{i2}), $i = 1, 2, \ldots, N$ from a bivariate normal distribution with means

$$\mu\,(x_1) = 0, \qquad \mu\,(x_2) = 0,$$

variances

$$\sigma^2(x_1) = 1, \qquad \sigma^2(x_2) = 1,$$

and specified correlation coefficient ρ_{x_1,x_2}. For each pair (x_{1i}, x_{2i}) and for specified values of β_1 and β_2 we generate an observation y_i according to

$$y_i = \beta_1 x_{1i} + \beta_2 x_{2i} + e_i, \qquad i = 1, 2, \ldots, N,$$

where

$$e_i \sim \text{NID}\left(0, \sigma^2\right), \text{ with } \sigma^2 = 0.25.$$

A fraction $(1-f)$ of the values on x_2 were then randomly deleted, for $f = 0.70$ and 0.90, and where the specific observations deleted were independently selected for each value of f. After deleting a fraction $(1-f)$ of the values on x_2, the remaining values of x_2 and the corresponding values of x_1 are fixed with zero means and unit variances (unadjusted for degrees of freedom). The values of y_i are also fixed with zero mean over all N cases and over the n complete cases. The estimators $\hat{\boldsymbol{\beta}}_\ell = (\hat{\beta}_{1\ell}, \hat{\beta}_{2\ell})'$ and $\hat{\boldsymbol{\beta}}_I = (\hat{\beta}_{1I}, \hat{\beta}_{2I})'$ are obtained from (2.3.5) by setting $k = 1$ and $1/r^2$ respectively. The value of α is computed using (4.1.8). Then the estimator $\hat{\boldsymbol{\beta}}_R = (\hat{\beta}_{1R}, \hat{\beta}_{2R})'$ of $\boldsymbol{\beta} = (\beta_1, \beta_2)'$ is obtained from (4.1.1). The maximum likelihood estimators b_{1m} and b_{2m} of β_1 and β_2 respectively are computed using the relations (A.1.1) and (A.1.2). Monte Carlo experiments are repeated 500 times. The average of the 500 sample values of each squared error, e.g. $(\hat{\beta}_{1R} - \beta_1)^2$, is taken as an estimate of the corresponding mean squared error which is denoted by MSE $(\cdot)$.

The estimates of the mean squared errors of $\hat{\beta}_{1R}, b_{1m}$ and $\hat{\beta}_{2R}, b_{2m}$ and the relative efficiencies,

$$R_1 = \text{MSE}(\hat{\beta}_{1R}) / \text{MSE}(b_{1m})$$

and

$$R_2 = \text{MSE}(\hat{\beta}_{2R}) / \text{MSE}(b_{2m}),$$

are calculated for $\beta_1 = 1.0,\ 5.0,\ \beta_2 = 1.0,\ 5.0,\ \rho_{x_1,x_2} = 0.30,\ 0.70,\ \sigma^2 = 0.25$ and $N = 10$, 20 and 40. The random numbers used in the study were generated on an IBM 370 computer using the algorithm of Lewis *et al.* (1969). Results of the simulation experiments are given in Tables 2–5.

Table 2. *Estimated Mean Squared Error and Relative Efficiency of* $\hat{\beta}_{1R}$ *and* b_{1m}, *for* $\rho_{x_1,x_2} = 0.30$ *and* $\sigma^2 = 0.25$.

Values of β_1 and β_2	Values of f	MSE	Values of N		
			10	20	40
$\beta_1 = 1.0$, $\beta_2 = 1.0$	0.90	MSE($\hat{\beta}_{1R}$)	0.029	0.017	0.012
		STD ERR	0.005	0.003	0.001
		MSE(b_{1m})	0.034	0.020	0.014
		STD ERR	0.006	0.003	0.001
		R_1	0.853	0.850	0.857
	0.70	MSE($\hat{\beta}_{1R}$)	0.032	0.020	0.015
		STD ERR	0.004	0.002	0.001
		MSE(b_{1m})	0.037	0.024	0.018
		STD ERR	0.005	0.003	0.002
		R_1	0.865	0.833	0.833
$\beta_1 = 1.0$, $\beta_2 = 5.0$	0.90	MSE($\hat{\beta}_{1R}$)	0.036	0.028	0.021
		STD ERR	0.007	0.004	0.009
		MSE(b_{1m})	0.042	0.033	0.023
		STD ERR	0.006	0.003	0.001
		R_1	0.857	0.848	0.913
	0.70	MSE($\hat{\beta}_{1R}$)	0.040	0.031	0.027
		STD ERR	0.008	0.004	0.003
		MSE(b_{1m})	0.046	0.035	0.029
		STD ERR	0.005	0.002	0.0008
		R_1	0.869	0.886	0.931
$\beta_1 = 5.0$, $\beta_2 = 1.0$	0.90	MSE($\hat{\beta}_{1R}$)	0.041	0.037	0.030
		STD ERR	0.015	0.008	0.004
		MSE(b_{1m})	0.048	0.042	0.033
		STD ERR	0.013	0.007	0.003
		R_1	0.854	0.881	0.909
	0.70	MSE($\hat{\beta}_{1R}$)	0.053	0.044	0.040
		STD ERR	0.017	0.011	0.005
		MSE(b_{1m})	0.059	0.048	0.042
		STD ERR	0.015	0.011	0.004
		R_1	0.898	0.917	0.952
$\beta_1 = 5.0$, $\beta_2 = 5.0$	0.90	MSE($\hat{\beta}_{1R}$)	0.066	0.055	0.042
		STD ERR	0.011	0.006	0.003
		MSE(b_{1m})	0.070	0.058	0.043
		STD ERR	0.010	0.006	0.003
		R_1	0.943	0.948	0.976
	0.70	MSE($\hat{\beta}_{1R}$)	0.081	0.067	0.053
		STD ERR	0.016	0.010	0.004
		MSE(b_{1m})	0.088	0.072	0.055
		STD ERR	0.018	0.009	0.005
		R_1	0.920	0.930	0.964

Table 3. *Estimated Mean Squared Error and Relative Efficiency of $\hat{\beta}_{1R}$ and b_{1m}, for $\rho_{x_1,x_2} = 0.30$ and $\sigma^2 = 0.25$.*

Values of β_1 and β_2	Values of f	MSE	Values of N		
			10	20	40
$\beta_1 = 1.0$, $\beta_2 = 1.0$	0.90	MSE($\hat{\beta}_{1R}$)	0.046	0.040	0.033
		STD ERR	0.009	0.008	0.007
		MSE(b_{1m})	0.052	0.045	0.036
		STD ERR	0.006	0.003	0.002
		R_1	0.885	0.889	0.917
	0.70	MSE($\hat{\beta}_{1R}$)	0.059	0.051	0.044
		STD ERR	0.010	0.006	0.003
		MSE(b_{1m})	0.065	0.055	0.047
		STD ERR	0.009	0.005	0.003
		R_1	0.908	0.927	0.936
$\beta_1 = 5.0$, $\beta_2 = 5.0$	0.90	MSE($\hat{\beta}_{1R}$)	0.049	0.043	0.035
		STD ERR	0.010	0.009	0.006
		MSE(b_{1m})	0.054	0.047	0.036
		STD ERR	0.007	0.004	0.003
		R_1	0.907	0.915	0.972
	0.70	MSE($\hat{\beta}_{1R}$)	0.064	0.056	0.047
		STD ERR	0.010	0.006	0.003
		MSE(b_{1m})	0.069	0.059	0.048
		STD ERR	0.010	0.007	0.004
		R_1	0.928	0.949	0.979
$\beta_1 = 5.0$, $\beta_2 = 1.0$	0.90	MSE($\hat{\beta}_{1R}$)	0.054	0.051	0.046
		STD ERR	0.012	0.009	0.005
		MSE(b_{1m})	0.060	0.055	0.048
		STD ERR	0.013	0.010	0.005
		R_1	0.900	0.927	0.958
	0.70	MSE($\hat{\beta}_{1R}$)	0.073	0.067	0.061
		STD ERR	0.014	0.011	0.008
		MSE(b_{1m})	0.078	0.070	0.062
		STD ERR	0.012	0.007	0.007
		R_1	0.936	0.957	0.984
$\beta_1 = 5.0$, $\beta_2 = 5.0$	0.90	MSE($\hat{\beta}_{1R}$)	0.079	0.074	0.069
		STD ERR	0.015	0.012	0.005
		MSE(b_{1m})	0.086	0.078	0.072
		STD ERR	0.012	0.007	0.003
		R_1	0.919	0.949	0.958
	0.70	MSE($\hat{\beta}_{1R}$)	0.088	0.081	0.077
		STD ERR	0.016	0.011	0.010
		MSE(b_{1m})	0.093	0.085	0.079
		STD ERR	0.011	0.010	0.009
		R_1	0.946	0.953	0.975

Table 4. *Estimated Mean Squared Error and Relative Efficiency of* $\hat{\beta}_{2R}$ *and* b_{2m}, *for* $\rho_{x_1,x_2} = 0.30$ *and* $\sigma^2 = 0.25$.

Values of β_1 and β_2	Values of f	MSE	Values of N: 10	20	40
$\beta_1 = 1.0$, $\beta_2 = 1.0$	0.90	MSE($\hat{\beta}_{2R}$)	0.421	0.405	0.386
		STD ERR	0.110	0.092	0.055
		MSE(b_{2m})	0.436	0.412	0.398
		STD ERR	0.114	0.095	0.060
		R_2	0.966	0.983	0.970
	0.70	MSE($\hat{\beta}_{2R}$)	0.364	0.357	0.333
		STD ERR	0.100	0.081	0.050
		MSE(b_{2m})	0.400	0.385	0.350
		STD ERR	0.113	0.090	0.005
		R_2	0.910	0.927	0.951
$\beta_1 = 5.0$, $\beta_2 = 5.0$	0.90	MSE($\hat{\beta}_{2R}$)	4.857	4.111	2.978
		STD ERR	0.256	0.208	0.169
		MSE(b_{2m})	4.432	3.681	2.623
		STD ERR	0.204	0.181	0.152
		R_2	1.096	1.117	1.135
	0.70	MSE($\hat{\beta}_{2R}$)	3.121	2.894	2.370
		STD ERR	0.207	0.163	0.121
		MSE(b_{2m})	2.859	2.344	2.102
		STD ERR	0.188	0.156	0.110
		R_2	1.092	1.233	1.127
$\beta_1 = 5.0$, $\beta_2 = 1.0$	0.90	MSE($\hat{\beta}_{2R}$)	0.432	0.415	0.390
		STD ERR	0.110	0.080	0.030
		MSE(b_{2m})	0.468	0.436	0.400
		STD ERR	0.115	0.110	0.070
		R_2	0.923	0.952	0.975
	0.70	MSE($\hat{\beta}_{2R}$)	0.395	0.358	0.342
		STD ERR	0.110	0.080	0.042
		MSE(b_{2m})	0.423	0.384	0.353
		STD ERR	0.112	0.091	0.005
		R_2	0.934	0.935	0.969
$\beta_1 = 5.0$, $\beta_2 = 5.0$	0.90	MSE($\hat{\beta}_{2R}$)	4.387	4.019	2.111
		STD ERR	0.200	0.165	0.114
		MSE(b_{2m})	4.232	3.840	2.001
		STD ERR	0.189	0.150	0.110
		R_2	1.037	1.047	1.055
	0.70	MSE($\hat{\beta}_{2R}$)	3.851	3.109	2.100
		STD ERR	0.160	0.134	0.009
		MSE(b_{2m})	3.563	2.914	1.973
		STD ERR	0.120	0.110	0.007
		R_2	1.081	1.067	1.064

Table 5. *Estimated Mean Squared Error and Relative Efficiency of $\hat{\beta}_{2R}$ and b_{2m}, for $\rho_{x_1,x_2} = 0.70$ and $\sigma^2 = 0.25$.*

Values of β_1 and β_2	Values of f	MSE	Values of N 10	20	40
$\beta_1 = 1.0$, $\beta_2 = 1.0$	0.90	MSE($\hat{\beta}_{2R}$)	0.016	0.012	0.009
		STD ERR	0.004	0.002	0.0008
		MSE(b_{2m})	0.021	0.015	0.011
		STD ERR	0.005	0.002	0.0009
		R_2	0.762	0.800	0.818
	0.70	MSE($\hat{\beta}_{2R}$)	0.019	0.017	0.012
		STD ERR	0.009	0.008	0.005
		MSE(b_{2m})	0.025	0.021	0.014
		STD ERR	0.005	0.003	0.009
		R_2	0.760	0.810	0.857
$\beta_1 = 5.0$, $\beta_2 = 5.0$	0.90	MSE($\hat{\beta}_{2R}$)	0.106	0.095	0.059
		STD ERR	0.006	0.003	0.001
		MSE(b_{2m})	0.138	0.110	0.067
		STD ERR	0.008	0.007	0.003
		R_2	0.768	0.864	0.881
	0.70	MSE($\hat{\beta}_{2R}$)	0.290	0.272	0.214
		STD ERR	0.114	0.101	0.070
		MSE(b_{2m})	0.320	0.298	0.230
		STD ERR	0.124	0.105	0.051
		R_2	0.906	0.913	0.930
$\beta_1 = 5.0$, $\beta_2 = 1.0$	0.90	MSE($\hat{\beta}_{2R}$)	0.019	0.016	0.011
		STD ERR	0.005	0.003	0.001
		MSE(b_{2m})	0.024	0.019	0.013
		STD ERR	0.007	0.003	0.001
		R_2	0.792	0.842	0.846
	0.70	MSE($\hat{\beta}_{2R}$)	0.023	0.018	0.014
		STD ERR	0.005	0.003	0.001
		MSE(b_{2m})	0.028	0.022	0.017
		STD ERR	0.009	0.005	0.005
		R_2	0.821	0.818	0.823
$\beta_1 = 5.0$, $\beta_2 = 5.0$	0.90	MSE($\hat{\beta}_{2R}$)	0.146	0.117	0.083
		STD ERR	0.009	0.009	0.007
		MSE(b_{2m})	0.163	0.128	0.090
		STD ERR	0.010	0.005	0.004
		R_2	0.896	0.914	0.922
	0.70	MSE($\hat{\beta}_{2R}$)	0.333	0.304	0.251
		STD ERR	0.014	0.011	0.006
		MSE(b_{2m})	0.365	0.329	0.267
		STD ERR	0.015	0.010	0.005
		R_2	0.912	0.924	0.940

6. CONCLUSIONS

Although the results derived above apply strictly to only very limited cases, they are suggestive of some general conclusions regarding the relative efficiencies of the alternative estimator given by (4.1.1) and the maximum likelihood estimator (MLE) defined in (A.1.1) and (A.1.2). We note from Tables 2 and 3 that the MSE of $\hat{\beta}_{IR}$ is always smaller than the MSE of b_{1m}. It seems that for estimating β_1 the method of maximum likelihood does not offer any advantage over the alternative method. The advantage of the alternative method is most marked when N is small.

Further, the comparison statistics in Tables 4 and 5 show that the MSE of $\hat{\beta}_{2R}$ is smaller than or equal to the MSE of b_{2m} except for cases when the correlations between x_1 and x_2 are small and the values of β_2 are large. Thus the alternative method behaves better than the method of maximum likelihood for a wide class of conditions. The expressions for the MLE's are very cumbersome. The alternative method is simpler and it does not require any knowledge about the form of the distribution of the variables under study. In conclusion, therefore, we recommend the use of the alternative method to deal with the problem arising from missing data.

APPENDIX: THE MAXIMUM LIKELIHOOD ESTIMATORS OF β_1 AND β_2

From Donner and Rosner (1982), the maximum likelihood estimators b_{1m} and b_{2m} of β_1 and β_2, respectively, under the assumption of an underlying trivariate normal distribution are given by:

$$b_{1m} = \frac{P_{y1} - Y_{y2}P_{21}}{(1 - P_{21}^2)} \left(S_y^2/S_1^2\right)^{\frac{1}{2}} \tag{A.1.1}$$

and

$$b_{2m} = \frac{P_{y2} - P_{y1}P_{21}}{(1 - P_{21}^2)} \left(S_y^2/S_2^2\right)^{\frac{1}{2}}, \tag{A.1.2}$$

where

$$S_y^2 = \sum_{i=1}^{N} (y_i - \bar{y}_N)^2 /N, \qquad \bar{y}_N = \sum_{i=1}^{N} y_i/N,$$

$$S_1^2 = \sum_{i=1}^{N} (x_{1i} - \bar{x}_{1N})^2 /N, \qquad \bar{x}_{1N} = \sum_{i=1}^{N} x_{1i}/N,$$

$$
\begin{aligned}
S_2^2 &= \left[K + A^2T_2 + A_2^2T_y + 2A_2A_1T_{y1}^1 + A_1^2T_1^1\right], \\
P_{y1} &= S_{y1}/\left(S_y^2S_1^2\right)^{\frac{1}{2}}, \\
P_{y2} &= S_{y2}/\left(S_y^2S_2^2\right)^{\frac{1}{2}}, \\
P_{21} &= S_{21}/\left(S_2^2S_1^2\right)^{\frac{1}{2}}, \\
S_{y1} &= \sum_{i=1}^{N}\left(x_{1i} - \bar{x}_{1N}\right)\left(y_i - \bar{y}_N\right), \\
A &= \left(S_y^2S_1^2 - S_{y1}^2\right), \\
A_2 &= \frac{c\left(g_1a_1 - g_2T_1\right) - g\left(c_1a_1 - c_2T_2\right) + T_{y1}\left(c_1g_2 - c_2g_1\right)}{c\left(T_{y1}a_1 - aT_1\right) - T\left(c_1a_1 - c_2T_1\right) + T_{y1}\left(c_1a - c_2T_{y1}\right)}, \\
A_1 &= \frac{c\left(T_{y1}g_2 - ag_1\right) - T\left(c_1g_2 - c_2g_1\right) + g\left(c_1a - c_2T_{y1}\right)}{c\left(T_{y1}a_1 - aT_1\right) - T\left(c_1a_1 - c_2T_1\right) + T_{y1}\left(c_1a - c_2T_{y1}\right)}, \\
T_1 &= \sum_{i=1}^{N}\left(x_{1i} - \bar{x}_{1N}\right)^2, \\
T_{y1} &= \sum_{i=1}^{N}\left(y_i - \bar{y}_N\right)\left(x_{1i} - \bar{x}_{1N}\right), \\
T &= \sum_{i=1}^{N}\left(y_i - \bar{y}_N\right), \\
K &= (N-n)A\left(A_2S_{y2} + A_1S_{21}\right), \\
S_{y2} &= \left[-A_2S_y^2 - A_1S_{y1}\right]/A, \\
S_{y1} &= \left[S_1^2A_1 + A_2S_{y1}\right]/(-A), \\
a &= n\bar{x}_N - \sum_{i=1}^{N} y_i, \\
a_1 &= n\,\bar{x}_{1N} - \sum_{i=1}^{n} x_{2i}, \\
g &= A\left(\bar{y}_N\sum_{i=1}^{n} x_{2i} - \sum_{i=1}^{n} y_ix_{2i}\right), \\
g_1 &= A\left(\bar{x}_{1N}\sum_{i=1}^{n} x_{2i} - \sum x_{1i}x_{2i}\right),
\end{aligned}
$$

$$g_2 = A\sum_{i=1}^{n} x_{2i},$$

$$c = Aa, \qquad c_1 = Aa_1, \qquad c_2 = nA,$$

$$T_2 = \sum_{i=1}^{n} (x_{2i} - m_2)^2,$$

$$T_y = \sum_{i=1}^{N} (y_i - \bar{y}_N)^2,$$

$$T_1^1 = \sum_{i=1}^{N} (x_{1i} - \bar{x}_{1N})^2,$$

$$T_{y1}^1 = \sum_{i=1}^{N} (x_{1i} - \bar{x}_{1N})(y_i - \bar{y}_N),$$

and

$$m_2 = \frac{g(T_{y1}a_1 - T_1 a) - T(a_1 g_1 - T_1 g_2) + T_{y1}(g_1 a - T_{y1} g_2)}{c(T_{y1}a_1 - T_1 a) - T(a_1 c_1 - T_1 c_2) + T_{y1}(c_1 a - T_{y1} c_2)}.$$

REFERENCES

Anderson, T. W. (1957), "Maximum likelihood estimation of the multivariate normal distribution when some observations are missing." *Journal of the American Statistical Association* **52**, 200–203.

Beale, E. M. L., and R. J. A. Little (1975), "Missing values in multivariate analysis." *Journal of the Royal Statistical Society*, Series B **37**, 129–145.

Buck, S. F. (1960), "A method of estimation of missing values in multivariate data, suitable for use with an electronic computer." *Journal of the Royal Statistical Society*, Series B **22**, 302–307.

Dear, R. E. (1959), "A principal component missing data method for multiple regression models." SP-86, Santa Monica, California: System Development Corporation.

Donner, A. (1982), "The relative effectiveness of procedures commonly used in multiple regression analysis for dealing with missing values." *The American Statistician* **36**, 378–381.

Donner, A., and B. Rosner (1982), "Missing value problems in multiple linear regression with two independent variables." *Communications in Statistics A, Theory and Methods* **11**, 127–140.

Haitovsky, Y. (1968), "Missing data in regression analysis." *Journal of the Royal Statistical Society*, Series B **30**, 67–82.

Heiberger, R. M. (1977), "Regression with the pairwise present covariance matrix: a dangerous practice." Technical Report No. 19, Department of Statistics, University of Pennsylvania.

Lewis, P. A. W., A. S. Goodman, and T. M. Miller (1969), "A pseudo-random number generator." *IBM Systems Journal* **8**, 136–196.

Timm, N. H. (1970), "The estimation of variance covariance matrices from incomplete data." *Psychometrika* **35**, 417–437.

Walsh, J. F. (1959), "Computer feasible general method for fitting and using regression functions when data is incomplete." SP-71, Santa Monica, California: System Development Corporation.

Wilks, S. S. (1932), "Moments and distributions of estimates of population parameters from fragmentary samples." *Annals of Mathematical Statistics* **3**, 163–195.

Yates, F. (1933), "The analysis of replicated experiments when the field results are incomplete." *The Empire Journal of Experimental Agriculture* **1**, 129–142.

M. M. Ali [1]

A BOUND FOR THE TAIL AREA OF THE t DISTRIBUTION FOR SAMPLES FROM A SYMMETRICALLY TRUNCATED NORMAL POPULATION

ABSTRACT

It is shown in this paper that the tail probability of the usual Student's t statistic under the assumption that the sample arises from a symmetrically truncated normal population, is greater than the corresponding tail probability of the usual t-distribution under normality assumptions for the population for sufficiently large values of t.

1. INTRODUCTION

When testing hypotheses by means of Student's t statistic, one uses the tail probabilities of the distribution. The behaviour of the tail probabilities of the distribution of the t-statistic under non-normal situations sheds light on the robustness of the t-statistic under non-normal assumptions for the population distribution.

Hotelling (1961) studied the limiting ratio of the tail probability of the usual t-distribution (under normality assumptions) to that of the t-statistic under several specified non-normal population distributions.

Mukherjee (1971) studied the behaviour of the t-distribution under the assumption that the underlying population is a symmetrically truncated normal distribution. This work involved extensive computations by numerical integration for several sufficiently large values of t and for several different degrees of symmetric truncation of the normal population. Mukherjee concluded that in all cases studied in the thesis, the tail probability for large values of t is greater than the corresponding tail probability for the usual t-distribution.

---

[1] Department of Statistical and Actuarial Sciences, The University of Western Ontario, London, Ontario N6A 5B9

I. B. MacNeill and G. J. Umphrey (eds.), Foundations of Statistical Inference, 183–187.

In this paper we show that the conclusion arrived at by Mukherjee (1971) for several specific cases is true for the general case. An analytic general proof is given by employing arguments of a geometrical nature. More precisely, we show that the tail probability of the Student's t-distribution based on a sample from a symmetrically truncated normal distribution is heavier than the corresponding tail probability of the usual Student's t-distribution based on a sample from an untruncated normal population.

2. GEOMETRY OF THE STUDENT'S t-DISTRIBUTION

The geometrical interpretation of Student's t-distribution in terms of the probability mass contained inside a right circular cone is very well known. We sketch this geomerical interpretation in order to develop appropriate notations and also because our arguments are heavily based on geometrical considerations.

In the n-dimensional Euclidean space, consider a general point P with coordinates $(x_1, \ldots, x_n)$. Let OQ denote the equiangular line having direction cosines $(1/\sqrt{n}, \ldots, 1/\sqrt{n})$ passing through the origin O. Let PM denote the perpendicular from P onto OQ. Then the length OM is equal to $\sqrt{n}\bar{x}$, while OP has length $\{\sum x_1^2\}^{1/2}$ and the length of PM is $\{\sum(x_i - \bar{x})^2\}^{1/2} = \sqrt{n-1}s$; $\bar{x}$ and s^2 are respectively the sample mean and the sample variance.

Let θ denote the angle POM, so that

$$\cot \theta = \frac{OM}{PM} = \frac{\sqrt{n}\bar{x}}{\sqrt{(n-1)}s} = T/\sqrt{(n-1)},$$

where

$$T = \frac{\bar{x}}{s/\sqrt{n}} = \sqrt{n-1} \cot \theta.$$

We further assume that the distribution of $(x_1, \ldots, x_n)$ is centered at O, i.e., $E(x_1) = E(x_2) = \cdots = E(x_n) = 0$. Hence

$$G(t) = \Pr\{T \geq t\} = \Pr\{\sqrt{n-1} \cot \theta \geq t\}.$$

Hence

$$G(t) = \Pr\{T \geq t\} = \Pr\left\{\theta \leq \text{ arc cot } \frac{t}{\sqrt{n-1}}\right\},$$

which is the probability content of a right circular cone whose axis passes through the general mean of the distribution along the equiangular line having a semivertical angle $\theta = \text{arc cot}(t/\sqrt{(n-1)})$.

3. DISTRIBUTION OF t FOR SPHERICAL DISTRIBUTIONS

We now consider a special case where the random variable $(x_1, \ldots, x_n)$ has a p.d.f. whose contours of constant density are concentric spheres with centre at the origin. This class of p.d.f. is invariant under rotation of axes around the centre of the probability distribution and is variously termed the class of rotationally invariant distributions, the class of spherically symmetric distributions, or simply as the class of spherical distributions. Obvious geometric considerations show that, for this class, if the probability mass is projected onto the unit sphere centered around the mean of the distribution, the projected mass is uniformly distributed on the surface of the unit sphere and

$$G(t) = \Pr\{T \geq t\}$$

is the probability content of the arc of the spherical cap of the unit sphere contained in a cone of semivertical angle arc cot whose axis is $(t/\sqrt{n-1})$ any line passing through the centre; hence $G(t)$ has the same value as any given t for any member of the class of spherical distributions. The p.d.f. $f(x_1, \ldots x_n) = (2\pi)^{-n/2} \exp\left(-\frac{1}{2}\sum x_i^2\right)$ is a member of the class of spherical distributions. Hence, it is readily seen that the distribution of t is the same as that of the usual t under the normality assumption for any member having a p.d.f. belonging to the spherical class—an observation first made by Efron (1969).

4. THE BOUND

Consider a sample $x_1, x_2, \ldots, x_n$ of size n from a symmetrically truncated normal population whose p.d.f. is given by

$$f(x) = \begin{cases} ke^{-x^2/2} & \text{for } |x| \leq \alpha \\ 0 & \text{elsewhere,} \end{cases}$$

where k is the normalizing constant given by

$$k^{-1} = \int_{-\alpha}^{+\alpha} e^{-t^2/2} dt,$$

so that

$$\begin{aligned} f(x_1, \ldots, x_n) &= k^n \exp\left(-\frac{1}{2}\sum x_i^2\right) && \text{inside the } n \text{ cube } C_n(\alpha) \\ &= 0 && \text{otherwise,} \end{aligned} \tag{1}$$

where $C_n(\alpha)$ is the n-cube given by $| x_i | \le \alpha$ for $i = 1, 2, \ldots, n$.

Let $S_n(\rho)$ define an n sphere of radius ρ, i.e.,

$$S_n(\rho) = \left\{(x_1, \ldots, x_n) \mid \sum x_i^2 \le \rho^2\right\} \tag{2}$$

and let ρ be so determined that for the same k as in (1),

$$k^{-n} \int_{S_n(\rho)} \exp\left(-\frac{1}{2}\sum x_i^2\right) \prod dx_i = 1.$$

This defines a symmetrically *spherically* truncated sample from a normal population having the joint distribution given by

$$\begin{aligned} f(x_1, \ldots, x_n) &= k^n \exp\left(-\frac{1}{2}\sum x_i^2\right) && \text{inside the sphere } S_n(\rho) \\ &= 0 && \text{elsewhere.} \end{aligned}$$

The p.d.f.s of $(x_1, \ldots, x_n)$ under regular symmetric truncation, $| x_i | \le \alpha$, and under spherical truncation, $\sum x_i^2 \le \rho^2$, agree over the common part

$$C_n(\alpha) \cap S_n(\rho). \tag{3}$$

Since both $C_n(\alpha)$ and $S_n(\rho)$ contain unit mass, $S_n(\rho)$ could not be inscribed inside $C_n(\alpha)$ nor could $S_n(\rho)$ circumscribe $C_n(\alpha)$. Therefore $S_n(\rho)$ will cut off the corners of the n-cube $C_n(\alpha)$ while the lost mass will be made-up by picking up mass in $S_n(\rho) - S_n(\rho) \cap C_n(\alpha)$.

Let us denote by $G(t)$ and $H(t)$ the following probabilities:

$$G(t) = \Pr\{T \ge t \mid (x_1, \ldots x_n) \text{ is } \alpha\text{-truncated}\}$$

and

$$H(t) = \Pr\{T \ge t \mid (x_1, \ldots, x_n) \text{ is } \rho\text{-truncated}\}.$$

For sufficiently large t, the angle arc cot $(t/\sqrt{n-1})$ will be small. Also the equiangular line passes through $(0, \ldots, 0)$ and $(\alpha, \ldots, \alpha)$, which is one of the vertices of the n-cube $| x_i | \le \alpha$ for $i = 1, \ldots, n$. Hence ρ will be less than $\sqrt{n}\alpha$.

Let $K_n(t)$ denote the cone with axis along the equiangular line passing through the origin and having a semivertical angle arc cot $(t/\sqrt{n-1})$. Then it is easy to see that for sufficiently large values of t,

$$S_n(\rho) \cap K_n(t) \subset C_n(\rho) \cap K_n(t).$$

Now

$$\begin{aligned} G(t) &= \Pr\{T \geq t \mid (x_1, \ldots, x_n) \text{ has an } \alpha\text{-truncated distribution}\} \\ &= \text{Pr. Content of } [C_n(\alpha) \cap K_n(t) \mid \text{an } \alpha\text{-truncated distribution}] \\ &\geq \text{Pr. Content of } [S_n(\rho) \cap K_n(t) \mid \text{an } \alpha\text{-truncated distribution}], \quad (4) \end{aligned}$$

since the sphere $S_n(\rho)$ cuts off the corner of the n-cube $C_n(\rho)$. By virtue of (3),

$$S_n(\rho) \cap K_n(t)$$

has a common density under either α-truncation or ρ-truncation. Hence

$$G(t) \geq \text{Pr. Content of } [S_n(\rho) \cap K_n(t) \mid \rho\text{-truncation}].$$

We remark at this point that the p.d.f. (2) under ρ-truncation belongs to the class of *spherically* symmetric distribution and hence the distribution of its t statistic is the same as the usual t distribution under the normality assumption. Therefore:

$$\begin{aligned} &\text{Pr. Content of } [S_n(\rho) \cap K_n(\rho) \mid \rho\text{-truncation}] \\ &= \Pr\Bigg\{T \geq t \mid \text{ when } (x_1, \ldots, x_n) \text{ has p.d.f.} \\ &\left(\frac{1}{\sqrt{2\pi}}\right)^n e^{-\frac{1}{2}\sum x_i^2} \text{ for } -\infty < x_i < +\infty, \quad i = 1, \ldots, n\Bigg\}. \end{aligned}$$

From (4) it follows that we have the bound for sufficiently large values of t

$$G(t) \geq \Pr\{T \geq t \mid \rho\text{-truncation}\} = 1 - F(t),$$

where $F(t)$ represents the distribution function of the Student's t distribution under the usual normality assumption.

Hence we have shown that, for sufficiently large values of t, the tail ends of the t-distribution for the case when the sample arises from *any symmetric* truncated normal distribution is heavier than the corresponding tail probability of the usual t-distribution.

REFERENCES

Efron, B. (1969), "Student's t-test under symmetry conditions". *Journal of the American Statistical Association,* **64,** 1278–1302.

Hotelling, H. (1961), "The behavior of some standard tests under non-standard conditions". *Proceedings of the Fourth Berkeley Symposium on Mathematical Statistics and Probability* **1,** 319–360.

Mukherjee, G. K. (1971), "On Student's t-test based on truncated normal distribution". Ph.D. thesis, The University of Western Ontario.

Chu-In Charles Lee [1]

MAXIMUM LIKELIHOOD ESTIMATES FOR STOCHASTICALLY ORDERED MULTINOMIAL POPULATIONS WITH FIXED AND RANDOM ZEROS

ABSTRACT

Likelihood ratio tests for and against a stochastic ordering between two multinomial populations are considered by Robertson and Wright as discrete analogues for the one-sided Mann-Whitney-Wilcoxon and Kolmogorov-Smirnov tests. They obtained closed form expressions for the maximum likelihood estimates when all frequencies are positive. The purpose of this article is to provide closed form expressions for the maximum likelihood estimates with fixed and random zeros.

1. INTRODUCTION

Let X and Y be two random variables with distribution functions F and G respectively. The random variable X is said to be stochastically no larger than Y if $F(t) \geq G(t)$ for every real number t, which we denote symbolically $X < Y$. If $X < Y$ and $Y < X$ then X and Y are identically distributed. We are interested in the case that both X and Y are discrete with $t_1 < t_2 < \ldots < t_k$ as all possible values that the two random variables may take. Let $p_i = P(X = t_i)$ and $q_i = P(Y = t_i)$, $i = 1, 2, \ldots, k$. Then $X < Y$ is equivalent to

$$\sum_{j=1}^{i} p_j \geq \sum_{j=1}^{i} q_j \quad \text{for} \quad i = 1, 2, \ldots, k-1,$$

which we denote symbolically $\mathbf{p} >> \mathbf{q}$.

[1] Department of Mathematics and Statistics, Memorial University of Newfoundland, St. John's, Newfoundland A1C 5S7

I. B. MacNeill and G. J. Umphrey (eds.), Foundations of Statistical Inference, 189–197.

We are interested in the following hypotheses:

$$H_0 : \mathbf{p} = \mathbf{q} \text{ and } H_1 : \mathbf{p} >> \mathbf{q}.$$

It is well known that one-sided procedures are more powerful than their two-sided counterparts. If we are interested in testing the hypothesis H_0 and if $\mathbf{p} >> \mathbf{q}$ is our firm belief, then it is to our advantage to restrict the alternatives to $H_1 - H_0$. On the other hand, we may be interested in testing the hypothesis H_1 against all alternatives which may embody our intuition about the ordering of the two underlying distributions.

The two-sample problem is the one in which two independent random samples are taken from two separate multinomial distributions with unknown parameters $\mathbf{p} = (p_1, p_2, \ldots, p_k)$ and $\mathbf{q} = (q_1, q_2, \ldots, q_k)$ respectively, while the one-sample problem is the one in which we have a known standard $\mathbf{q}$ and a random sample is taken from a multinomial distribution with unknown parameter $\mathbf{p}$. The fixed-zero problem is the one for which at least one of $p_1, p_2, \ldots, p_k;\ q_1, q_2, \ldots, q_k$ is zero and we know where these zeros are located. While the random-zero problem is the one for which $p_1, p_2, \ldots, p_k;\ q_1, q_2, \ldots, q_k$ are all positive but some of their relative frequencies in the sample (or the two samples) take zero values.

Robertson and Wright (1981) studied likelihood ratio tests for testing H_0 against $H_1 - H_0$ and for testing H_1 against all alternatives for the one-sample and the two-sample problems under the condition that every cell count is at least one. However, in practice we may encounter the case of fixed zeros and random zeros in certain cells. The purpose of this article is to provide closed form expressions for the maximum likelihood estimates under this general setting.

2. RESTRICTED MAXIMUM LIKELIHOOD ESTIMATES

We shall consider the following three problems: (a) the two-sample problem with random zeros only, (b) the one-sample problem with random zeros only, and (c) both the one-sample and the two-sample problems with fixed zeros. The parameters $\mathbf{p} = (p_1, p_2, \ldots, p_k)$ and $\mathbf{q} = (q_1, q_2, \ldots, q_k)$ are in the triangular region $\Delta = \{\boldsymbol{\theta} \in R^k : \theta_i \geq 0,\ i = 1, \ldots, k,\ \sum_{i=1}^{k} \theta_i = 1\}$. We first consider the problem (a) in which the components of $\mathbf{p}$ and $\mathbf{q}$ are all positive but their values are unknown. Two independent random samples of size m and n are taken from two separate populations associated with $\mathbf{p}$ and $\mathbf{q}$ respectively. Let $\hat{\mathbf{p}} = (\hat{p}_1, \hat{p}_2, \ldots, \hat{p}_k)$ and $\hat{\mathbf{q}} = (\hat{q}_1, \hat{q}_2, \ldots, \hat{q}_k)$ be the corresponding vectors of relative frequencies, so that $m\hat{\mathbf{p}}$ and $n\hat{\mathbf{q}}$ are independent multinomial random vectors with parameters $(m, \mathbf{p})$ and $(n, \mathbf{q})$

respectively. A restricted m.l.e. $(\overline{\mathbf{p}}, \overline{\mathbf{q}})$ is a $2k$-component vector in $\Delta \times \Delta$ which solves the following optimization problem:

$$\max \sum_{i=1}^{k} \{m\hat{p}_i \ln(p_i) + n\hat{q}_i \ln(q_i)\} \quad \text{subject to} \quad \mathbf{p} >> \mathbf{q}, \tag{2.1}$$

under the convention $0 \ln(0) = 0$.

The approach uses the theory given by Barlow *et al.* (1972, Chapter 1) which requires the following notation. We have a vector $\mathbf{g} = (g_1, g_2, \ldots, g_\ell)$ and a vector $\mathbf{w} = (w_1, w_2, \ldots, w_\ell)$ of positive components. Let

$$D^\ell = \{\mathbf{x} \in R^\ell : x_1 \geq x_2 \geq \ldots \geq x_\ell\}.$$

The solution to the weighted least-squares problem:

$$\min \sum_{r=1}^{\ell} (g_r - f_r)^2 w_r \quad \text{subject to} \quad \mathbf{f} \in D^\ell$$

exists and it is unique. We denote this solution by $E_{\mathbf{w}}(\mathbf{g} \mid D^\ell)$ with its rth component given by $E_{\mathbf{w}}(\mathbf{g} \mid D^\ell)_r$. We can compute this solution by the Pool-Adjacent-Violators algorithm.

Let $j_0 = 0$ and let

$$i_r = \min\{i : \hat{p}_i > 0, \ i > j_{r-1}\} \tag{2.2}$$

and

$$j_r = \min\{j : \hat{q}_j > 0, \ j \geq i_r\} \tag{2.3}$$

be defined successively so that we have a finite sequence of indices $i_1 \leq j_1 < i_2 \leq j_2 < \ldots$ which may be ended in j_t if $j_t = k$ or $\hat{p}_i = 0$ for each $i > j_t$, or otherwise ended in i_t when $\hat{q}_j = 0$ for each $j \geq i_t$. If the sequence ends in j_t we let $\ell = t$ and $i_{\ell+1} = k + 1$. Otherwise, we let $\ell = t - 1$ so that $i_{\ell+1} = i_t \leq k$. We define

$$\hat{P}_r = \sum_{h=i_r}^{i_{r+1}-1} \hat{p}_h \quad \text{and} \quad \hat{Q}_r = \sum_{h=i_r}^{i_{r+1}-1} \hat{q}_h$$

for $r = 1, 2, \ldots, \ell$. Then the ℓ components of $\hat{P}$ and those of $\hat{\mathbf{Q}}$ are all positive and the vector $\hat{\mathbf{Q}}/\hat{\mathbf{P}} \equiv (\hat{Q}_1/\hat{P}_1, \hat{Q}_2/\hat{P}_2, \ldots, \hat{Q}_\ell/\hat{P}_\ell)$ is meaningful. A restricted m.l.e. $(\overline{\mathbf{p}}, \overline{\mathbf{q}})$ is provided by the following theorem.

Theorem 2.1. Let $N = m + n$ and let $(\overline{\mathbf{p}}, \overline{\mathbf{q}})$ be defined by

(a) $\overline{p}_i = \overline{q}_i = n\hat{q}_i/N$ if $i \geq 1$ and $i < i_1$,
(b) $\overline{p}_i = \hat{p}_i\{m + nE_{\hat{\mathbf{P}}}(\hat{\mathbf{Q}}/\hat{\mathbf{P}} \mid D^\ell)_r\}/N$ and $\overline{q}_i = \hat{q}_i\{mE_{\hat{\mathbf{P}}}(\hat{\mathbf{Q}}/\hat{\mathbf{P}} \mid D^\ell)_r^{-1} + n\}/N$ if $i_r \leq i < i_{r+1}$, $r = 1, 2, \ldots, \ell$, and
(c) $\overline{p}_i = \overline{q}_i = m\hat{p}_i/N$ if $i \leq k$ and $i \geq i_{\ell+1}$.

Then $(\overline{\mathbf{p}}, \overline{\mathbf{q}})$ is a restricted m.l.e.

Once we obtain a restricted m.l.e., the likelihood ratio statistics T_{01} for testing H_0 against $H_1 - H_0$ and T_{12} for testing H_1 against all alternatives can be evaluated. Their asymptotic distributions under the corresponding null hypotheses (H_0 for T_{01} and H_1 for T_{12}) can be found in Theorem 4.1 and Theorem 4.2 of Robertson and Wright (1981).

We next consider the problem (b) in which the components of $\mathbf{p}$ and $\mathbf{q}$ are all positive, $\mathbf{q}$ is known but $\mathbf{p}$ is unknown. A random sample of size m is taken from the population associated with $\mathbf{p}$. Let $\hat{\mathbf{p}} = (\hat{p}_1, \hat{p}_2, \ldots, \hat{p}_k)$ be the vector of relative frequencies, so that $m\hat{\mathbf{p}}$ is a multinomial random vector with parameters $(m, \mathbf{p})$. A restricted m.l.e. $\overline{\mathbf{p}}$ is a k-component vector in Δ which solves the following optimization problem:

$$\max \sum_{i=1}^{k} m\hat{p}_k \ln(p_i) \quad \text{subject to} \quad \mathbf{p} >> \mathbf{q}$$

under the convention $0 \ln(0) = 0$.

Suppose the components of $\hat{\mathbf{p}}$ are positive at $i_1 < i_2 < \ldots < i_\ell$ and zero elsewhere. We define

$$\hat{P}_r = \sum_{h=i_r}^{i_{r+1}-1} \hat{p}_h = \hat{p}_{i_r} \quad \text{and} \quad Q_r = \sum_{h=i_r}^{i_{r+1}-1} q_h \tag{2}$$

for $r = 1, 2, \ldots, \ell$, where $i_{\ell+1} = k+1$. Then the ℓ-components of $\hat{\mathbf{P}}$ and those of $\mathbf{Q}$ are all positive and the vector $\mathbf{Q}/\hat{\mathbf{P}} \equiv (Q_1/\hat{P}_1, Q_2/\hat{P}_2, \ldots, Q_\ell/\hat{P}_\ell)$ is meaningful. A restricted m.l.e. $\overline{\mathbf{p}}$ is provided by the following corollary.

Corollary 2.2. Let $\overline{\mathbf{p}}$ be defined by
(a) $\overline{p}_i = q_i$ if $i \geq 1$ and $i < i_1$,
(b) $\overline{p}_i = \hat{p}_i E_{\hat{\mathbf{P}}}(\mathbf{Q}/\hat{\mathbf{P}} \mid D^\ell)_r$ if $i_r \leq i < i_{r+1}$, $r = 1, 2, \ldots, \ell$.

Then $\overline{\mathbf{p}}$ is a restricted m.l.e.

Once we obtain a restricted m.l.e., the likelihood ratio statistics S_{01} for testing H_0 against $H_1 - H_0$ and S_{12} for testing H_1 against all alternatives can be evaluated. Their asymptotic distributions under the corresponding null hypotheses (H_0 for S_{01} and H_1 for S_{12}) can be found in Theorem 3.1 and Theorem 3.2 of Robertson and Wright (1981).

We finally consider the problem (c) in which there are two subsets Λ and Γ of $\{1, 2, \ldots, k\}$ so that $p_i > 0$ if $i \in \Lambda$, $q_j > 0$ if $j \in \Gamma$ and the remaining p_i and q_j are fixed zeros. Let $i_1 \leq j_1 < i_2 \leq j_2 < \ldots$ be defined by (2.2) and (2.3) with $\hat{p}_i$ and $\hat{q}_j$ replaced by p_i and q_j respectively. The finite monotone increasing sequence of indices may be ended in j_t if $j_t = k$ or $p_i = 0$ for each $i > j_t$ or otherwise ended in i_t when $q_j = 0$ for each $j \geq i_t$. However, if the sequence ends in i_t or if $q_j > 0$ for some j, $1 \leq j < i_1$, then the hypothesis $H_1 : \mathbf{p} >> \mathbf{q}$ can never be the case. Therefore, the problem for testing H_0 against $H_1 - H_0$ and that for testing H_1 against all alternatives are considered only if we have the sequence $i_1 \leq j_1 < i_2 \leq j_2 < \ldots < i_t \leq j_t$ ended in j_t and $q_j = 0$ for each j, $1 \leq j < i_1$. We let $k' = t$, $i_{k'+1} = k + 1$, and

$$\theta_r = \sum_{h=i_r}^{i_{r+1}-1} p_h \quad \text{and} \quad \phi_r = \sum_{h=i_r}^{i_{r+1}-1} q_h$$

for $r = 1, 2, \ldots, k'$. Then $\theta_1, \theta_2, \ldots, \theta_{k'}, \phi_1, \phi_2, \ldots, \phi_{k'}$ are all positive and $\sum_{r=1}^{k'} \theta_r = \sum_{r=1}^{k'} \phi_r = 1$. It is trivial that the restriction $\mathbf{p} >> \mathbf{q}$ over the two k-component vectors can be summarized by the restriction $\boldsymbol{\theta} >> \boldsymbol{\phi}$ over the two k'-component vectors.

Once the dimensionality is reduced from k to k', we can compute a restricted m.l.e. $(\bar{\boldsymbol{\theta}}, \bar{\boldsymbol{\phi}})$ by Theorem 2.1 for the two-sample problem and a restricted m.l.e. $\bar{\boldsymbol{\theta}}$ by Corollary 2.2 for the one-sample problem. Asymptotic distributions under the corresponding null hypotheses are applied to the reduced problem with parameters $\boldsymbol{\theta}$ and $\boldsymbol{\phi}$.

APPENDIX: PROOFS AND REMARKS

Proof of Theorem 2.1. The proof for the case that $\ell = 0$ is trivial. The proof is given in the sequel for the case when ℓ is positive. For convenience, we shall let $g_r = \hat{Q}_r / \hat{P}_r$ and $g_r^* = E_{\hat{\mathbf{P}}}(\hat{\mathbf{Q}}/\hat{\mathbf{P}} \mid D^\ell)_r$ for $r = 1, \ldots, \ell$. It follows from the max-min formula, see (1.9) of Barlow *et al.* (1972),

$$g_r^* \geq g_\ell^* = \min_{h \leq \ell} \left\{ \sum_{s=h}^{\ell} \hat{Q}_s \Big/ \sum_{s=h}^{\ell} \hat{P}_s \right\} \geq \hat{Q}_\ell > 0$$

for $r = 1, \ldots, \ell$. Therefore, $\bar{p}_i \geq 0$ and $\bar{q}_i \geq 0$ for $i = 1, \ldots, k$.

Consider a fixed index i, $i_1 \leq i < i_{\ell+1}$. There is a corresponding r so that $j_r \leq i < j_{r+1}$ where $j_{\ell+1} = i_{\ell+1}$ and $j_0 = i_1$ if $i_1 < j_1$. Let $\boldsymbol{\delta} = (\delta_1, \ldots, \delta_\ell)$ be a corresponding indicator function so that $\delta_h = 1$ if $h \leq r$

and $\delta_h = 0$ elsewhere. By (2.2), (2.3) and (b) of the theorem, we have

$$N\sum_{h=i_1}^{i}\overline{p}_h \geq N\sum_{h=i_1}^{i_{r+1}-1}\overline{p}_h = m\sum_{h=1}^{r}\hat{P}_h + n\sum_{h=1}^{r}g_h^*\hat{P}_h$$

and

$$N\sum_{h=i_1}^{i}\overline{q}_h \leq N\sum_{h=i_1}^{i_{r+1}-1}\overline{q}_h = m\sum_{h=1}^{r}\hat{Q}_h/g_h^* + n\sum_{h=1}^{r}\hat{Q}_h.$$

By (1.21) of Barlow *et al.* (1972),

$$\sum_{h=1}^{r}g_h^*\hat{P}_h = \sum_{h=1}^{\ell}g_h^*\delta_h\hat{P}_h \geq \sum_{h=1}^{\ell}g_h\delta_h\hat{P}_h = \sum_{h=1}^{r}\hat{Q}_h$$

and by Theorem 1.8 of Barlow *et al.* (1972),

$$\sum_{h=1}^{r}\hat{Q}_h/g_h^* = \sum_{h=1}^{r}g_h(g_h^*)^{-1}\hat{P}_h = \sum_{h=1}^{\ell}g_h(g_h^*)^{-1}\delta_h\hat{P}_h$$

$$\leq \sum_{h=1}^{\ell}g_h^*(g_h^*)^{-1}\delta_h\hat{P}_h = \sum_{h=1}^{r}\hat{P}_h.$$

It follows that

$$N\sum_{h=i_1}^{i}\overline{p}_h \geq m\sum_{h=1}^{r}\hat{P}_h + n\sum_{h=1}^{r}\hat{Q}_h \geq N\sum_{h=i_1}^{i}\overline{q}_h$$

for $i = i_1, i_1+1, \ldots, i_{\ell+1}-1$. The above inequality holds as an equality when $i = i_{\ell+1}-1$; see (1.22) and Theorem 1.7 of Barlow *et al.* (1972). By (a) and (c) of the theorem,

$$\sum_{h=1}^{i}\overline{p}_h \geq \sum_{h=1}^{i}\overline{q}_h$$

for $i = 1, \ldots, k-1$ and $\sum_{h=1}^{k}\overline{p}_h = \sum_{h=1}^{k}\overline{q}_h = \sum_{h=1}^{k}(m\hat{p}_h + n\hat{q}_h)/N = 1$. Therefore, $\overline{\mathbf{p}}$ and $\overline{\mathbf{q}}$ are in Δ and $\overline{\mathbf{p}} >> \overline{\mathbf{q}}$.

Let $\mathbf{p}$ and $\mathbf{q}$ be two multinomial parameters in Δ with $\mathbf{p} >> \mathbf{q}$. For each λ between 0 and 1, let $\mathbf{p}(\lambda)$ and $\mathbf{q}(\lambda)$ be two vectors defined by

$$p_i(\lambda) = \lambda p_i + (1-\lambda)\overline{p}_i \quad \text{and} \quad q_i(\lambda) = \lambda q_i + (1-\lambda)\overline{q}_i$$

for $i = 1, \ldots, k$. Then $\mathbf{p}(\lambda)$ and $\mathbf{q}(\lambda)$ are in Δ and $\mathbf{p}(\lambda) >> \mathbf{q}(\lambda)$ for each λ between 0 and 1. Let $f(\lambda)$ be the objective function in (2.1) with p_i and q_i been replaced by $p_i(\lambda)$ and $q_i(\lambda)$ respectively. Since f is concave, if we can establish the inequality that $f'(0) \leq 0$ then by the concave inequality

$$f(\lambda) \leq f(0) + \lambda f'(0),$$

we have that $f(1) \leq f(0)$ and hence $(\overline{\mathbf{p}}, \overline{\mathbf{q}})$ is a desired solution.

Let

$$P_0 = \sum_{i=1}^{i_1-1} p_i, \quad Q_0 = \sum_{i=1}^{i_1-1} q_i, \quad P_{\ell+1} = \sum_{i=i_{\ell+1}}^{k} p_i, \quad Q_{\ell+1} = \sum_{i=i_{\ell+1}}^{k} q_i$$

and for $r = 1, \ldots, \ell$ let

$$P_r = \sum_{i=i_r}^{i_{r+1}-1} p_i, \quad Q_r = \sum_{i=i_r}^{i_{r+1}-1} q_i, \qquad \alpha_r = m/(m + ng_r^*)$$

and $\beta_r = ng_r^*/(m + ng_r^*)$. It can be shown that

$$\begin{aligned} f'(0) &= m \sum_{i \in \Lambda} \hat{p}_i p_i / \overline{p}_i + n \sum_{i \in \Gamma} \hat{q}_i q_i / \overline{q}_i - N \\ &\leq N \left\{ \sum_{r=1}^{\ell} \alpha_r P_r + P_{\ell+1} + Q_0 + \sum_{r=1}^{\ell} \beta_r Q_r - 1 \right\}, \end{aligned}$$

where $\Lambda = \{i : \hat{p}_i > 0\}$ and $\Gamma = \{i : \hat{q}_i > 0\}$. It is trivial that

$$Q_0 = \alpha_1 Q_0 + \beta_1 Q_0 \leq \alpha_1 P_0 + \beta_1 Q_0,$$

$$P_{\ell+1} = \alpha_\ell P_{\ell+1} + \beta_\ell P_{\ell+1} \leq \alpha_\ell P_{\ell+1} + \beta_\ell Q_{\ell+1}$$

and

$$\alpha_r - \alpha_{r+1} = \beta_{r+1} - \beta_r = n\alpha_r\alpha_{r+1}(g_{r+1}^* - g_r^*)/m \leq 0$$

for each $r = 1, \ldots, \ell - 1$. By the Abel method of summation we have that

$$\begin{aligned} f'(0)/N &\le \alpha_1 P_0 + \sum_{r=1}^{\ell} \alpha_r P_r + \alpha_\ell P_{\ell+1} + \beta_1 Q_0 \\ &+ \sum_{r=1}^{\ell} \beta_r Q_r + \beta_\ell Q_{\ell+1} - 1 \\ &= \sum_{r=1}^{\ell-1} (\alpha_r - \alpha_{r+1}) \sum_{r=0}^{r} P_h + \sum_{r=1}^{\ell-1} (\beta_r - \beta_{r+1}) \sum_{h=0}^{r} Q_h \\ &+ \alpha_\ell + \beta_\ell - 1 \\ &= \sum_{r=1}^{\ell-1} (\alpha_r - \alpha_{r+1}) \left(\sum_{i=1}^{i_r} p_i - \sum_{i=1}^{i_r} q_i \right) \le 0. \end{aligned}$$

This completes the proof.

Proof of Corollary 2.2. The proof follows similar lines as that of Theorem 2.1 above. In fact, the corollary can be regarded as a special case of Theorem 2.1 by letting m be fixed and $n \to \infty$.

Remark 1. Let ϵ be any arbitrary positive real number and let $\hat{\mathbf{p}}_\epsilon$ and $\hat{\mathbf{q}}_\epsilon$ be two vectors defined by

$$\hat{p}_{\epsilon,i} = (\hat{p}_i + \epsilon)/(1 + k\epsilon) \quad \text{and} \quad \hat{q}_{\epsilon,i} = (\hat{q}_i + \epsilon)/(1 + k\epsilon)$$

for $i = 1, \ldots, k$ and let

$$\overline{p}_{\epsilon,i} = \hat{p}_{\epsilon,i}\{m + nE_{\hat{\mathbf{p}}_\epsilon}(\hat{\mathbf{q}}_\epsilon/\hat{\mathbf{p}}_\epsilon \mid D^k)_i\}/N$$

and

$$\overline{q}_{\epsilon,i} = \hat{q}_{\epsilon,i}\{mE_{\hat{\mathbf{p}}_\epsilon}(\hat{\mathbf{q}}_\epsilon/\hat{\mathbf{p}}_\epsilon \mid D^k)_i^{-1} + n\}/N$$

for $i = 1, \ldots, k$. It can be shown that as $\epsilon \to 0$ the values of $\overline{p}_{\epsilon,i}$ and $\overline{q}_{\epsilon,i}$, $i_1 \le i < i_{\ell+1}$, are the same as those of (b) in Theorem 2.1.

Remark 2. It has been shown (see Robertson, 1965) that

$$E_{\hat{\mathbf{P}}}(\hat{\mathbf{Q}}/\hat{\mathbf{P}} \mid D^\ell)^{-1} = E_{\hat{\mathbf{Q}}}(\hat{\mathbf{P}}/\hat{\mathbf{Q}} \mid -D^\ell).$$

This identity provides a symmetric expression for $(\overline{\mathbf{p}}, \overline{\mathbf{q}})$ in (b) of Theorem 2.1.

ACKNOWLEDGMENTS

This research was partially supported by NSERC of Canada and ONR contract N00014-80-C-0321.

REFERENCES

Barlow, R. E., D. J. Bartholomew, J. M. Bremner, and H. D. Brunk (1972), *Statistical Inference Under Order Restrictions*. New York: Wiley and Sons.

Robertson, T. (1965), "A note on the reciprocal of the conditional expectation of a positive random variable". *Annals of Mathematical Statistics* **36**, 1302–1305.

Robertson, T., and F. T. Wright (1981), "Likelihood ratio tests for and against a stochastic ordering between multinomial populations". *Annals of Statistics* **9**, 1248–1257.

Masafumi Akahira [1] and Kei Takeuchi [2]

ON THE DEFINITION OF ASYMPTOTIC EXPECTATION

ABSTRACT

In this paper a definition of asymptotic expectation is given and its fundamental properties are discussed.

1. INTRODUCTION

In the asymptotic theory of estimation, the concept of asymptotic expectation is widely used (e.g., Akahira and Takeuchi, 1981; Ibragimov and Has'minskii, 1981; Lehmann, 1982), and it is usually remarked that it can be different from the asymptotic value of expectation. It is, however, not sufficiently accurately defined in the literature, especially when the asymptotic distribution does not exist.

In the paper we shall give a definition of the asymptotic expectation and show its properties, e.g., its linearity and a Markov type inequality. We shall also obtain the necessary and sufficient conditions for the convergences in probability and distribution.

2. RESULTS

Let $\{X_n\}$ be a sequence of non-negative random variables. For any sequence $\{A_n\}$ of positive numbers we define an A_n-censored sequence as

$$X_n^*(A_n) = \min\{X_n, A_n\}, \qquad n = 1, 2, \ldots.$$

For $\{X_n\}$ we denote by $\mathcal{A}(X_n)$ a set of all the sequences $\{A_n\}$ of positive

[1] Department of Mathematics, University of Electro-Communications, Chofu, Tokyo 182, Japan

[2] Faculty of Economics, University of Tokyo, Hongo, Bunkyo-ku, Tokyo 113, Japan.

I. B. MacNeill and G. J. Umphrey (eds.), Foundations of Statistical Inference, 199–208.

numbers satisfying

$$\Pr\{X_n > A_n\} \to 0 \quad \text{as} \quad n \to \infty.$$

We define the upper and the lower asymptotic expectations of $\{X_n\}$ as

$$\overline{As}.E(X_n) = \inf_{\{A_n\} \in \mathcal{A}(X_n)} \limsup_{n\to\infty} E(X_n^*(A_n))$$

and

$$\underline{As}.E(X_n) = \inf_{\{A_n\} \in \mathcal{A}(X_n)} \liminf_{n\to\infty} E(X_n^*(A_n)),$$

respectively. Note that they can be infinity.

Definition 2.1. If $\overline{As}.E(X_n) = \underline{As}.E(X_n) < \infty$, then we call it the asymptotic expectation of $\{X_n\}$ and denote it by $As.E(X_n)$.

The following theorem establishes that the above definition is reduced to the usual concept when X_n has an asymptotic distribution.

Theorem 2.1. If X_n has a proper asymptotic distribution F, that is

$$\Pr\{X_n \le x\} \to F(x) \quad \text{as} \quad n \to \infty$$

for every continuity point x of F, then

$$As.E(X_n) = \int_0^\infty x dF(x),$$

provided that the right-hand side is finite.

Proof. Denote by $F_n(x)$ the distribution of X_n. We put

$$\mu = \int_0^\infty x dF(x).$$

Since for any fixed positive number A

$$|E(X_n^*(A)) - E(X^*(A))| \le \left| \int_0^A x dF_n(x) - \int_0^A x dF(x) \right| + A|F_n(A) - F(A)|,$$

we have

$$|E(X_n^*(A)) - E(X^*(A))| \to 0 \quad \text{as} \quad n \to \infty. \tag{2.1}$$

Since μ is finite, it follows that

$$|E(X^*(A)) - \mu| \to 0 \quad \text{as} \quad A \to \infty. \tag{2.2}$$

Then it follows from (2.1) and (2.2) that for any sequence $\{\varepsilon_m\}$ of positive numbers there exists a sequence $\{A_m\}$ of positive numbers such that

$$|E(X^*(A_m)) - \mu| < \varepsilon_m \qquad (m = 1, 2, \ldots). \tag{2.3}$$

For the sequence $\{A_m\}$ there is a monotone increasing sequence $\{N(m)\}$ of positive integers such that for $n \geq N(m)$

$$|E(X_n^*(A_m)) - E(X^*(A_m))| < \varepsilon_m. \tag{2.4}$$

For n satisfying $N(m) \leq n < N(m+1)$ we define $m = N^{-1}(n)$. Denoting $A_{N^{-1}(n)}$ and $\varepsilon_{N^{-1}(n)}$ by A_n and ε_n, respectively, we have by (2.3) and (2.4)

$$|E(X_n^*(A_n)) - \mu| < 2\varepsilon_n$$

for sufficiently large n.

Letting $\varepsilon_n \to 0$ as $n \to \infty$, we obtain

$$|E(X_n^*(A_n)) - \mu| \to 0 \quad \text{as} \quad n \to \infty.$$

This completes the proof.

In order to obtain some properties on the asymptotic expectation we have to prove some lemmas.

Lemma 2.1. If $\overline{As}.E(X_n) < \infty$, then for any sequence $\{A_n\}$ of positive numbers tending to infinity as $n \to \infty$, we have that $\{A_n\} \in \mathcal{A}(X_n)$.

Proof. Assume that there exists a sequence $\{A'_n\}$ of positive numbers such that $\lim_{n\to\infty} A'_n = \infty$ and $\{A'_n\} \notin \mathcal{A}(X_n)$. For some $\varepsilon > 0$ there exists a subsequence $\{A'_{n_j}\}$ of $\{A'_n\}$ such that

$$\Pr\{X_{n_j} > A'_{n_j}\} > \varepsilon.$$

Since for any $\{A_n\} \in \mathcal{A}(X_n)$, $\Pr\{X_n > A_n\} \to 0$ as $n \to \infty$, it follows that for sufficiently large j

$$\Pr\{X_{n_j} > A_{n_j}\} < \varepsilon,$$

hence $A_{n_j} > A'_{n_j}$. And we have for sufficiently large j

$$\begin{aligned} E(X^*_{n_j}(A_{n_j})) &\geq E(X^*_{n_j}(A'_{n_j})) \\ &\geq A'_{n_j} \Pr\{X_{n_j} \geq A'_{n_j}\} \\ &> A'_{n_j} \varepsilon, \end{aligned}$$

hence the last term tends to infinity as $j \to \infty$. Hence we have

$$\limsup_{n\to\infty} E(X^*_n(A_n)) = \infty,$$

which contradicts the condition of the lemma. This completes the proof.

We denote by $\mathcal{A}_\infty(X_n)$ a subset of $\mathcal{A}(X_n)$ whose element $\{A_n\}$ satisfies $\lim_{n\to\infty} A_n = \infty$.

Lemma 2.2. The following hold:

$$\overline{As}.E(X_n) = \inf_{\{A_n\}\in\mathcal{A}_\infty(X_n)} \limsup_{n\to\infty} E(X^*_n(A_n));$$

$$\underline{As}.E(X_n) = \inf_{\{A_n\}\in\mathcal{A}_\infty(X_n)} \liminf_{n\to\infty} E(X^*_n(A_n)).$$

Proof. If the left-hand side of either of the two equalities above is infinity, it is obvious that the right-hand side is also infinity, hence we may assume that the right-hand side is finite.

For any $\{A_n\} \in \mathcal{A}(X_n)$ we put $p_n = \Pr\{X_n > A_n\}$. Since $p_n \to 0$ as $n \to \infty$, putting $A'_n = \max\{A_n, p_n^{-1/2}\}$ we have $A'_n \to \infty$ as $n \to \infty$, hence $\{A'_n\} \in \mathcal{A}_\infty(X_n)$. Since $A'_n \geq A_n$, we obtain

$$E(X^*_n(A'_n)) \geq E(X^*_n(A_n)). \tag{2.5}$$

Since

$$\begin{aligned} &E(X^*_n(A'_n)) - E(X^*_n(A_n)) \\ &\leq (A'_n - A_n)\Pr\{X_n > A_n\} \\ &\leq p_n^{1/2}, \end{aligned}$$

it follows from (2.5) that

$$\lim_{n\to\infty} \{E(X^*_n(A'_n)) - E(X^*_n(A_n))\} = 0.$$

Hence for any $\{A_n\} \in \mathcal{A}(X_n)$ there exists a sequence $\{A'_n\} \in \mathcal{A}_\infty(X_n)$ such that

$$\lim_{n\to\infty} \{E(X^*_n(A'_n)) - E(X^*_n(A_n))\} = 0.$$

This fact leads to the conclusion of the lemma.

Lemma 2.3. If $\overline{As}.E(X_n) < \infty$, then there exists a sequence $\{A_n^*\} \in \mathcal{A}_\infty(X_n)$ such that

$$\limsup_{n\to\infty} E(X_n^*(A_n^*)) = \overline{As}.E(X_n);$$
$$\liminf_{n\to\infty} E(X_n^*(A_n^*)) = \underline{As}.E(X_n).$$

Proof. For a sequence $\{\varepsilon_m\}$ of positive numbers, such that $\varepsilon_m \downarrow 0$, there exists a sequence $\{A_{n,m}\}$ of positive numbers such that

$$\limsup_{n\to\infty} E(X_n^*(A_{n,m})) \leq \overline{As}.E(X_n) + \varepsilon_m.$$

Without loss of generality we assume that, for each n, $A_{n,m}$ is monotone decreasing in m. We put $p_{n,m} = \Pr\{X_n > A_{n,m}\}$. Then for each n, $p_{n,m}$ is monotone increasing in m and for each m, $\lim_{n\to\infty} p_{n,m} = 0$. For any m there exists $n(m)$ such that $p_{n,m} < \varepsilon_m$ for $n > n(m)$. For any n let $m = m^{-1}(n)$ be a maximum value satisfying $n(m) \geq n$. Since $m^{-1}(n) \to \infty$ as $n \to \infty$, putting $A_n^{**} = A_{n,m^{-1}(n)}$ we have

$$\Pr\{X_n > A_n^{**}\} < \varepsilon_{m^{-1}(n)},$$

of which the right-hand side tends to 0 as $n \to \infty$. For any fixed n_0 we obtain

$$\begin{aligned}\limsup_{n\to\infty} E(X_n^*(A_n^{**})) &\leq \limsup_{n\to\infty} E(X_n^*(A_{n,m^{-1}(n_0)}))\\ &\leq \overline{As}.E(X_n) + \varepsilon_{m^{-1}(n_0)}.\end{aligned}$$

Since $\varepsilon_{m^{-1}(n_0)} \to 0$ as $n_0 \to \infty$, we have

$$\limsup_{n\to\infty} E(X_n^*(A_n^{**})) \leq \overline{As}.E(X_n). \tag{2.6}$$

On the other hand it follows from the definition of $\overline{As}.E(X_n)$ that the inverse inequality of (2.6) holds. Hence it is seen that the equality in (2.6) holds.

In a way similar to the above, it is shown that there exists a sequence $\{A_n^{***}\} \in \mathcal{A}_\infty(X_n)$ such that

$$\liminf_{n\to\infty} E(X_n^*(A_n^{***})) = \underline{As}.E(X_n).$$

We put $A_n^* = \min\{A_n^{**}, A_n^{***}\}$. Then we have

$$\Pr\{X_n > A_n^*\} \to 0 \quad \text{as} \quad n \to \infty$$

and

$$\limsup_{n\to\infty} E(X_n^*(A_n^*)) \le \limsup_{n\to\infty} E(X_n^*(A_n^{**})) = \overline{As}.E(X_n); \tag{2.7}$$

$$\liminf_{n\to\infty} E(X_n^*(A_n^*)) \le \liminf_{n\to\infty} E(X_n^*(A_n^{***})) = \underline{As}.E(X_n). \tag{2.8}$$

Returning to the definitions of $\overline{As}.E(X_n)$ and $\underline{As}.E(X_n)$ we see that the inverse inequalities of (2.7) and (2.8) hold. This completes the proof.

Lemma 2.4. Assume that $\overline{As}.E(X_n) < \infty$. Let $\{A_n^*\}$ be a sequence satisfying the condition of Lemma 2.3. Let $\{B_n^*\}$ be any sequence such that $B_n^* \to \infty$ as $n \to \infty$ and $B_n^* \le A_n^*$ for all $n \ge n_0$ with some n_0. Then $\{B_n^*\}$ satisfies the condition of Lemma 2.3.

Proof. By Lemmas 2.1 and 2.2 we have $\{B_n^*\} \in \mathcal{A}(X_n)$ and

$$\limsup_{n\to\infty} E(X_n^*(B_n^*)) \le \limsup_{n\to\infty} E(X_n^*(A_n^*)); \tag{2.9}$$

$$\liminf_{n\to\infty} E(X_n^*(B_n^*)) \le \liminf_{n\to\infty} E(X_n^*(A_n^*)). \tag{2.10}$$

From the definitions of $\overline{As}.E(X_n)$ and $\underline{As}.E(X_n)$ we see that the inverse inequalities of (2.9) and (2.10) hold. This completes the proof.

Lemma 2.5. Assume that $\overline{As}.E(X_n) < \infty$ and $\overline{As}.E(Y_n) < \infty$. Then

(i) $\overline{As}.E(X_n) + \overline{As}.E(Y_n) \ge \overline{As}.E(X_n + Y_n)$;

(ii) $\underline{As}.E(X_n) + \underline{As}.E(Y_n) \le \underline{As}.E(X_n + Y_n)$;

(iii) $\underline{As}.E(X_n + Y_n) \le \overline{As}.E(X_n) + \underline{As}.E(Y_n) \le \overline{As}.E(X_n + Y_n)$.

Further if $X_n \ge Y_n$, then

$$\overline{As}.E(X_n - Y_n) \le \overline{As}.E(X_n) - \underline{As}.E(Y_n);$$

$$\underline{As}.E(X_n - Y_n) \ge \underline{As}.E(X_n) - \overline{As}.E(Y_n).$$

Proof. (i) By Lemma 2.3 we take $\{A_n\} \in \mathcal{A}_\infty(X_n)$ and $\{B_n\} \in \mathcal{A}_\infty(Y_n)$ such that

$$\overline{As}.E(X_n) = \limsup_{n\to\infty} E(X_n^*(A_n));$$

$$\underline{As}.E(Y_n) = \limsup_{n\to\infty} E(Y_n^*(B_n)).$$

Let $c_n = \min\{A_n, B_n\}$. Since

$$E((X_n + Y_n)^*(c_n)) \le E(X_n^*(A_n)) + E(Y_n^*(B_n)),$$

it follows from Lemmas 2.3 and 2.4 that

$$\limsup_{n\to\infty} E((X_n+Y_n)^*(c_n)) \le \overline{As}.E(X_n) + \overline{As}.E(Y_n).$$

By Lemma 2.2 we obtain

$$\overline{As}.E(X_n+Y_n) \le \overline{As}.E(X_n) + \overline{As}.E(Y_n).$$

(ii) By Lemma 2.3 we take a sequence $\{c_n\} \in \mathcal{A}_\infty(X_n+Y_n)$ such that

$$\overline{As}.E(X_n+Y_n) = \limsup_{n\to\infty} E((X_n+Y_n)^*(c_n)).$$

Since

$$E\left(X_n^*\left(\frac{c_n}{2}\right)\right) + E\left(Y_n^*\left(\frac{c_n}{2}\right)\right) \le E((X_n+Y_n)^*(c_n)),$$

it follows from Lemmas 2.3 and 2.4 that

$$\liminf_{n\to\infty} E\left(X_n^*\left(\frac{c_n}{2}\right)\right) + \liminf_{n\to\infty} E\left(Y_n^*\left(\frac{c_n}{2}\right)\right) \le \underline{As}.E(X_n+Y_n).$$

By Lemma 2.2 we have

$$\underline{As}.E(X_n) + \underline{As}.E(Y_n) \le \underline{As}.E(X_n+Y_n).$$

(iii) By Lemma 2.3 we take $\{c_n\} \in \mathcal{A}_\infty(X_n+Y_n)$ such that

$$\overline{As}.E(X_n+Y_n) = \limsup_{n\to\infty} E((X_n+Y_n)^*(c_n)).$$

Since

$$E((X_n+Y_n)^*(c_n)) \ge E\left(X_n^*\left(\frac{c_n}{2}\right)\right) + E\left(Y_n^*\left(\frac{c_n}{2}\right)\right),$$

it follows that

$$\limsup_{n\to\infty} E((X_n+Y_n)^*(c_n)) \ge \limsup_{n\to\infty} E\left(X_n^*\left(\frac{c_n}{2}\right)\right) + \liminf_{n\to\infty} E\left(Y_n^*\left(\frac{c_n}{2}\right)\right).$$

By Lemmas 2.3 and 2.4 we have

$$\overline{As}.E(X_n+Y_n) \ge \overline{As}.E(X_n) + \underline{As}.E(Y_n).$$

Similarly we obtain

$$\underline{As}.E(X_n+Y_n) \le \underline{As}.E(X_n) + \overline{As}.E(Y_n).$$

The proof of the rest follows directly from (iii). This completes the proof.

From the above lemmas we have the following theorems.

Theorem 2.2. Assume that $As.E(X_n)$ and $As.E(Y_n)$ exist. Then

$$As.E(X_n+Y_n)=As.E(X_n)+As.E(Y_n).$$

Further if $X_n \geq Y_n$, then

$$As.E(X_n-Y_n)=As.E(X_n)-As.E(Y_n).$$

The proof of the theorem is directly derived from Lemma 2.5.

Theorem 2.3. (Markov type inequality.) If $\overline{As}.E(X_n)<\infty$, then for any $c>0$

$$\limsup_{n\to\infty} \Pr\{X_n \geq c\} \leq \frac{\overline{As}.E(X_n)}{c};$$

$$\liminf_{n\to\infty} \Pr\{X_n \geq c\} \leq \frac{\underline{As}.E(X_n)}{c}.$$

Proof. There exists a sequence $\{A_n\}$ such that

$$\overline{As}.E(X_n)=\limsup_{n\to\infty} E(X_n^*(A_n))$$

and $A_n \to \infty$ as $n\to\infty$. Then we can take n_0 such that for $n \geq n_0$, $A_n > c$ and

$$\Pr\{X_n \geq c\}=\Pr\{X_n^*(A_n)\geq c\} \leq \frac{E(X_n^*(A_n))}{c}.$$

By Lemma 2.3 we have

$$\limsup_{n\to\infty} \Pr\{X_n \geq c\} \leq \frac{\overline{As}.E(X_n)}{c}$$

The other inequality is obtained similarly. This completes the proof.

Theorem 2.4. X_n converges in probability to zero if and only if $As.E(X_n)=0$.

Proof. The proof of necessity is clear. If X_n does not converge in probability to zero, then there exists a positive number c such that $\lim_{n\to\infty}\Pr\{X_n > c\} > 0$. It follows from Theorem 2.3 that this contradicts the condition. Thus the proof is completed.

Theorem 2.5. Assume that random variables X_n and X have the distribution functions F_n and F, respectively, and the moment generating function (m.g.f.) of X,

$$g(\theta) = \int e^{\theta x} dF(x),$$

exists for all θ in some open interval I including the origin. Then X_n converges in distribution to X as $n \to \infty$, i.e., $X_n \xrightarrow{D} X$ if and only if $As.E(e^{\theta X_n}) = g(\theta)$ for all $\theta \in I$.

Proof. If $X_n \xrightarrow{D} X$, then $e^{\theta X_n} \xrightarrow{D} e^{\theta X}$, hence from Theorem 2.1

$$As.E(e^{\theta X_n}) = g(\theta) \quad \text{for all} \quad \theta \in I.$$

If $As.E(e^{\theta X_n}) = g(\theta)$ for all $\theta \in I$, then it follows from Theorem 2.3 that the sequence $\{F_n\}$ is tight, hence for any $\varepsilon > 0$ there exist $n_0 > 0$ and $K > 0$ such that for $n \geq n_0$

$$\Pr\{X_n > K\} < \varepsilon.$$

Then there exists a subsequence $\{n_i\}$ of $\{n\}$ such that F_{n_i} converges in distribution to some distribution F_0 as $i \to \infty$ and

$$As.E(X_{n_i}) = \lim_{i \to \infty} \int e^{\theta x} dF_{n_i}(x) = g(\theta) \quad \text{for all} \quad \theta \in I.$$

If a subsequence $\{F_{m_i}\}$ of $\{F_n\}$ does not converge to F_0, then there is some positive number δ such that a set $A_\delta = \{n : \lambda(F_n, F_0) > \delta\}$ is infinite, where $\lambda(F, G)$ denotes the Levy distance between two one-dimensional distributions. We also have a subsequence $\{F_{m'_i}\}$ of $\{F_{m_i}\}$ such that $m'_i \in A_\delta$ $(i = 1, 2, \ldots)$ and $F_{m'_i}$ converges in distribution to some distribution F'_0 as $i \to \infty$. By the first part of the proof we have $F'_0 = F_0$, which contradicts the assumption. This completes the proof.

Further, if Y_n converges in probability to zero as $n \to \infty$ and a function $f(x, y)$ is continuous in y, then it may be shown that $As.E(f(X_n, Y_n)) = As.E(f(X_n, 0))$.

For general real random variables X_n we have $X_n = X_n^+ - X_n^-$, where $X_n^+ = \max(X_n, 0)$ and $X_n^- = \max(-X_n, 0)$. Then we define the upper and the lower asymptotic expectations of $\{X_n\}$ as

$$\overline{As}.E(X_n) = \overline{As}.E(X_n^+) - \underline{As}.E(X_n^-);$$
$$\underline{As}.E(X_n) = \underline{As}.E(X_n^+) - \overline{As}.E(X_n^-),$$

respectively, and if $-\infty < \overline{As}.E(X_n) = \underline{As}.E(X_n) < \infty$ we call it the asymptotic expectation of $\{X_n\}$ and denote it by $As.E(X_n)$.

In a way similar to the case of the non-negative random variable it may be shown that

$$\overline{As}.E(-X_n) = -\underline{As}.E(X_n);$$
$$\underline{As}.E(-X_n) = -\overline{As}.E(X_n);$$
$$\overline{As}.E(X_n + Y_n) \leq \overline{As}.E(X_n) + \overline{As}.E(Y_n);$$
$$\underline{As}.E(X_n + Y_n) \geq \underline{As}.E(X_n) + \underline{As}.E(Y_n),$$

provided that $\overline{As}.E(|X_n|) < \infty$ and $\overline{As}.E(|Y_n|) < \infty$. Hence

$$As.E(X_n + Y_n) = As.E(X_n) + As.E(Y_n),$$

provided that $As.E(X_n)$ and $As.E(Y_n)$ exist.

ACKNOWLEDGMENTS

The paper was written while the first author was at Queen's University in Canada as a visiting Professor. The visit was supported by a grant of the Natural Sciences and Engineering Research Council of Canada. He is grateful to Professor Colin R. Blyth for inviting him.

REFERENCES

Akahira, M., and K. Takeuchi (1981). *Asymptotic Efficiency of Statistical Estimators: Concepts and Higher Order Asymptotic Efficiency.* Lecture Notes in Statistics 7. New York: Springer.

Ibragimov, I. A., and R. Z. Has'minskii (1981). *Statistical Estimation: Asymptotic Theory.* New York: Springer.

Lehmann, E. L. (1982). *Theory of Point Estimation.* New York: Wiley and Sons.

Shizuhiko Nishisato [1]

ROBUST TECHNIQUES FOR QUANTIFYING CATEGORICAL DATA

ABSTRACT

This paper reviews three previously developed methods for handling optimal quantification of categorical data with some outlier responses, and then proposes two methods which are distinct from the previous three. The two proposed methods are based, respectively, on the principle of forced classification and on the projection of data onto a subspace without outlier responses. Some numerical examples are presented to show how resistant some of the procedures are to outlier responses.

1. INTRODUCTION

To quantify categorical data, there is a well-known technique that has been referred to by many different names. These names include: the method of reciprocal averages (Richardson and Kuder, 1933; Horst, 1935), appropriate scoring (Fisher, 1948), optimal scaling (Bock, 1960), ***analyse des correspondances*** (Benzécri, 1968), dual scaling (Nishisato, 1980) and homogeneity analysis (Gifi, 1981). Given the $N \times m$ matrix of frequencies or the incidence matrix F, this technique assigns a vector of weights for the rows, $\mathbf{y}$, and a vector of weights for the columns, $\mathbf{x}$, of F in such a way that, for a maximum value of constant ρ $(0 < \rho < 1)$, the following relations hold:

$$\begin{aligned} \rho\mathbf{y} &= D_N^{-1}F\mathbf{x} \\ \rho\mathbf{x} &= D^{-1}F'\mathbf{y} \end{aligned} \tag{1}$$

where D_N and D are diagonal matrices of row totals and column totals of F, respectively, and $\mathbf{y}'F\mathbf{1} = \mathbf{1}'F\mathbf{x} = 0$. This set of equations is referred to

[1] Department of Measurement, Evaluation and Computer Applications, The Ontario Institute for Studies in Education, 252 Bloor Street West, Toronto, Ontario M5S 1V6

I. B. MacNeill and G. J. Umphrey (eds.), Foundations of Statistical Inference, 209–217.

as *formules de transition* (Benzécri *et al.*, 1973), or dual relations (Nishisato, 1980). It is nothing but the singular-value decomposition of matrix F, because (1) implies $\mathbf{y}'F\mathbf{x} = c\rho$, where $c = \mathbf{y}'D_N\mathbf{y} = \mathbf{x}'D\mathbf{x}$ is a scaling constant. The equations in (1) are of particular interest for data analysis, for they lead to expressions for the product-moment correlation coefficient,

$$\rho = \mathbf{y}'F\mathbf{x}/c, \tag{2}$$

squared correlation ratio,

$$\rho^2 = \mathbf{y}'FD^{-1}F'\mathbf{y} = \mathbf{x}'F'D_N^{-1}F\mathbf{x}/c \tag{3}$$

and the internal consistency reliability,

$$\alpha = 1 - (1 - \rho^2)/[(n - 1)\rho^2], \tag{4}$$

where n is the number of categorical variables when F is an incidence matrix. Thus, when (1) holds for a maximum value of ρ, $\mathbf{x}$ and $\mathbf{y}$ maximize such statistics as (2), (3) and (4).

This quantification method, hereafter referred to as dual scaling, has been widely used; see, for example, Benzécri *et al.* (1973), Nishisato (1975, 1980, 1982), Gifi, (1981), Lebart *et al.* (1976), Komazawa (1982), Gauch (1982), and Greenacre (1984). In determining so-called optimal vectors of weights $\mathbf{x}$ and $\mathbf{y}$, dual scaling employs the least-squares principle; hence any "outliers" in response data are likely to have substantial influence on the optimal weights. This fear has been substantiated by several investigators, such as Chan (1978), Nishisato (1980), and Meulman (1982), in relation to quantification of multiple-choice data, in which the task of dual scaling is to determine weights for multiple-choice options so as to produce maximally reliable scores for respondents. Tanaka (1984) used an influence function to identify optimal weights which fluctuate widely when a datum is discarded.

With an increasing number of applications of dual scaling to multiple-choice data, the problem of unstable weights has become an issue that can no longer be ignored. The present study discusses a few practical ways to arrive at weights which are resistant to pecularities of a particular data set.

2. NOTATION

We define the following notation:

N = the number of subjects (respondents)
n = the number of multiple-choice questions
m_j = the number of response options for item (question) j

$m = \sum_{j=1}^{n} m_j$
$F = [F_1, F_2, \ldots, F_n]$, where F_j is the $N \times m_j$ incidence matrix for item j, and $F_j 1 = 1$
$\mathbf{g} = (g_i)$ is the $N \times 1$ vector of row totals of F
$D_N = \text{diag}(g_i)$
$\mathbf{f} = (f_j)$ is the $m \times 1$ vector of column totals of F
$D = \text{diag}(f_j)$
$\mathbf{y}$ is an $N \times 1$ vector of scores for N subjects, scaled as $\mathbf{y}'\mathbf{g} = 0$
$\mathbf{x}_j$ is an $m_j \times 1$ vector of weights for m_j options for item j
$\mathbf{x}' = (\mathbf{x}_1, \mathbf{x}_2, \ldots \mathbf{x}_n)$ is an $1 \times m$ vector of weights for m options of n items, scaled as $\mathbf{x}'\mathbf{f} = 0$.

3. SPURIOUS RESULTS

When the optimal weight vector $\mathbf{x}$ is determined, it is known (Nishisato, 1982) that the product-moment correlation between item j, $F_j\mathbf{x}_j$, and the total questionnaire score, $F\mathbf{x}$, can be expressed as follows:

$$r_{jt} = \mathbf{x}'_j D_j \mathbf{x}_j \frac{n_{\rho^2}}{\mathbf{x}'D\mathbf{x}} \propto \mathbf{x}'_j D_j \mathbf{x}_j. \tag{5}$$

Thus, r_{jt} is strictly proportional to $\mathbf{x}'_j D_j \mathbf{x}_j$, suggesting that a single extreme value in $\mathbf{x}_j$ may increase the value of r_{jt} spuriously. Similarly, the product-moment correlation between item scores is likely to be drastically increased or decreased by extreme weights. Tables 1 and 2 show examples of inter-item correlations which are drastically altered by a few outliers. Table 1 is obtained by dual scaling from fifty subjects answering eleven questions with the total of 550 responses. Table 2 is obtained from dual scaling of the same data except for the fact that three responses from one subject, out of 550 responses, were omitted. One can see how much effect this small fraction of responses (i.e., 3/550) had on the correlation structure.

Note that responses to items 4, 6 and 7 of one of the fifty subjects were omitted to obtain the results in Table 2, and that large discrepancies between the two tables can be found with respect to r_{46}, r_{47} and r_{67}.

4. MORE RESISTANT ALTERNATIVES

Equation (1) indicates that $\mathbf{y}$ and $\mathbf{x}$ are proportional to averages of responses weighted respectively by $\mathbf{x}$ and $\mathbf{y}$. Let $\rho^* = 1/\rho$. Then (1) can be written as follows:

$$\begin{aligned} \mathbf{y} &= \rho^* D_N^{-1} F\mathbf{x}, \\ \mathbf{x} &= \rho^* D^{-1} F'\mathbf{y}, \end{aligned} \tag{6}$$

Table 1. *Inter-item Correlation Matrix of Optimally Quantified 550 Responses of Fifty Subjects (Complete Data)*

		Items										
		1	2	3	4	5	6	7	8	9	10	11
Items	1	100*										
	2	20	100									
	3	13	00	100								
	4	11	07	14	100							
	5	05	12	-06	02	100						
	6	21	13	23	<u>65</u>	12	100					
	7	15	04	19	<u>92</u>	12	<u>69</u>	100				
	8	09	02	-20	10	11	08	10	100			
	9	35	45	02	28	-02	25	19	17	100		
	10	19	-06	-22	38	12	26	38	03	20	100	
	11	30	40	04	26	-12	19	15	26	66	14	100

*The decimal point is omitted.

From (6), one can consider a few alternatives which are more resistant to pecularities in the data than (1).

[A] *Method of Reciprocal Medians*

Nishisato (1984a) proposed the following expressions:

$$\begin{aligned} y_i &= \mathrm{Mdn}_j[f_{ij}x_j] + \varepsilon_i, \\ x_j &= \mathrm{Mdn}_i[f_{ij}y_i] + \varepsilon_j, \end{aligned} \tag{7}$$

where ε_i and ε_j are as negligible as possible, and "Mdn" is the median operator. Nishisato added ε_i and ε_j because his empirical investigation has shown that the algorithm for simple reciprocal medians does not always converge to a stable point.

[B] *Alternating Reciprocal Averaging and Reciprocal Medians*

In the same paper as above, Nishisato proposed alternating a mathematically convergent process of reciprocal averaging and a resistant process

Table 2. *Inter-item Correlation Matrix of Optimally Quantified 547 Responses of Fifty Subjects (Responses to Items 4, 6, and 7 of One Subject were Omitted)*

		Items										
		1	2	3	4	5	6	7	8	9	10	11
	1	100*										
	2	36	100									
	3	13	00	100								
	4	03	32	03	100							
	5	-11	15	-06	38	100						
Items	6	38	22	20	17	-02	100					
	7	18	34	15	20	12	31	100				
	8	09	16	-20	35	03	17	25	100			
	9	38	44	01	31	-03	19	06	15	100		
	10	-09	-03	-22	15	12	03	04	06	03	100	
	11	30	51	04	45	12	20	11	26	66	12	100

*The decimal point is omitted.

of reciprocal medians, which converges, at least empirically, to two stable points corresponding to the two processes.

[C] *Method of Trimmed Reciprocal Averages*

As a more promising alternative than Methods A and B, Nishisato (1984b) proposed this third method, in which the q largest and the q smallest values in each set of $[f_{ij}y_i]$ and $[f_{ij}x_j]$ are omitted from averaging. The value of q may change from 0 (the method of reciprocal averages) up to half the number of elements in $[f_{ij}y_i]$ and $[f_{ij}x_j]$ (the method of reciprocal medians). Empirically, it is known that $q = 1$ leads to convergence of the reciprocal process and comparatively stable results.

Methods A and B require further modifications in their algorithms before they can be useful, while Method C seems to be promising. The present study offers two more procedures which seem to be at least as good as Method C:

[D] *Method of Generalized Forced Classification,*

and

[E] *Method by Projection onto a Subspace.*

These two procedures are asymptotically similar. In the procedure of forced classification (Nishisato, 1984c), it is known that the singular structure of matrix $[F_1, F_2, \ldots, k\ F_j, \ldots, F_n]$ approaches, as the weight k for matrix F_j increases, the singular structure of matrix $[P_jF_1, P_jF_2, \ldots, P_jF_n]$, where $P_j = F_j(F_j'F_j)^{-1}F_j'$ is a projection operator.

Method D generalizes the original formulation of the matrix $[F_1, F_2, \ldots, kF_j, \ldots, F_n]$ into the form:

$$E = [Q_1F_1, Q_2F_2, \ldots Q_jF_j, \ldots Q_nF_n], \tag{8}$$

where Q_j, $j = 1, 2, \ldots, n$, are diagonal matrices of weights. The weights are all equal to one, except for the options which contain spuriously influential responses (i.e., outliers) which receive weights smaller than one and greater than zero; the weight of one and that of zero correspond respectively to the analysis of observed data and that of data with influential responses discarded. Empirical investigation has shown that weights of 0.1 up to 0.5 on outliers have the effect of suppressing spuriousness of inter-item correlation, and that spuriousness increases quadratically when the weight increases from 0.6 to one. Therefore, a tentative suggestion here is to use a weight of 0.3 to 0.5 on each option containing an outlier.

In Method E, those response options which contain outliers are first identified, as in the case with Method D, by dual scaling of the data matrix F. The options containing outliers are then deleted from F to form the reduced matrix F_r, and the projection operator P_r for a subspace spanned by the rows of F_r, is calculated as follows:

$$P_r = F_r(F_r'F_r)^{-1}F_r'. \tag{9}$$

A weight vector is obtained by dual scaling of the matrix P_rF; that is, the vector $\mathbf{x}$ that maximizes the between-row sum of squares of the matrix P_rF. Once $\mathbf{x}$ is obtained $\mathbf{y}$ can be calculated by $D_N^{-1}F\mathbf{x}$ with an appropriate choice of units.

5. SAMPLE RESULTS

Out of the five procedures described above, Methods A and B require further refinement, and Method E is computationally cumbersome. Therefore, Table 3 shows results only for Methods C ($q = 1$ and $q = 2$) and Method

Table 3. *Comparisons of Methods C* $(q = 1, q = 2)$ *and D* $(k = 0.5)$ *with Dual Scaling of Original Matrix F and Reduced Matrix* F_r*. Options 4, 2 and 2 of Items 4, 7 and 6, Respectively, Contain Outliers* $(N = 50, n = 11)$

		Optimal Weights		C	C	D
Item	Option	F	F_r	$(q = 1)$	$(q = 2)$	$(k = 0.5)$
	1	.15	.94	.87	.99	.92
	2	−.24	−.27	−.60	−.55	−.33
4	3	−1.14	−2.41	−1.97	−1.40	−2.40
	(4)	10.05	0	3.33	3.29	1.92
	1	−.07	.30	.22	.41	.27
7	(2)	10.05	0	3.43	3.02	1.92
	3	−1.38	−2.75	−2.66	−3.00	−2.72
	1	−.13	.37	.38	.64	.35
6	(2)	6.65	2.40	2.46	1.34	1.96
	3	−.93	−1.92	−2.18	−2.47	−1.92

D $(k = 0.5)$ in comparison with the original data used for Table 1 and the reduced data used for Table 2.

One can see that Method D provides results very similar to those obtained from F_r and quite different from those of F. This can also be seen from comparisons of r_{jt}.

(F)	$r_{4t} = .83,$	$r_{7t} = .81,$	$r_{6t} = .75$
(F_r)	$r_{4t} = .62,$	$r_{7t} = .50,$	$r_{6t} = .50$
(D)	$r_{4t} = .63,$	$r_{7t} = .49,$	$r_{6t} = .50$

Although further work is needed, a preliminary evaluation of these procedures indicates, among other things, that:

(i) Method D is computationally fast and the most promising of the procedures investigated. Choosing the appropriate value of k has yet to be studied, although $k = 0.5$ appears to be adequate.

(ii) Method C looks interesting, but seems to present computational problems when some options are chosen only by one or two subjects; hence the method cannot provide a general solution to the problem under investigation.

(iii) Methods A and B also require further refinement, before they can be put into routine use.

(iv) Finally, it seems reasonable to consider that results by Method E are what investigators want, and that they are similar to the results obtained by Method D with the value of k being 0.5.

REFERENCES

Benzécri, J. P. (1969), "Statistical analysis as a tool to make patterns emerge from data". In *Methodologies of Pattern Recognition*, ed. S. Watanabe, pp. 35–74. New York: Academic Press.

Benzécri, J. P. *et al.* (1973), *L'Analyse des données: II. L'Analyse des correspondances.* Paris: Dunod.

Bock, R. D. (1960), *Methods of Applications of Optimal Scaling.* The University of North Carolina Psychometric Laboratory Research Memorandum No. 25.

Chan, D. (1978), "Treatment of missing data by optimal scaling". M.A. Thesis, University of Toronto.

Fisher, R. A. (1948), *Statistical Methods for Research Workers,* 10th edition. London: Oliver and Boyd.

Gauch, H. G. (1982), *Multivariate Analysis in Community Ecology.* Cambridge, MA: Cambridge University Press.

Gifi, A. (1981), *Nonlinear Multivariate Analysis.* Leiden: DSWO Press.

Greenacre, M. J. (1984), *Theory and Application of Correspondence Analysis.* London: Academic Press.

Horst, P. (1935), "Measuring complex attitudes". *Journal of Social Psychology* **6**, 369–374.

Lebart, L., A. Morineau, and N. Tabard (1977), *Technique de la Description Statistique: Methodes et Logiciels pour l'Analyse des Grands Tableaux.* Paris: Dunod.

Meulman, J. (1982), *Homogeneity Analysis of Incomplete Data.* Leiden: DSWO Press.

Nishisato, S. (1975), *Oyô shinri shakudohô: Shitsuteki data no bunseki to kaishaku (Applied Psychological Scaling: Analysis and Interpretation of Qualitative Data).* Tokyo: Sheishin Shobo Publishers.

Nishisato, S. (1980), *Analysis of Categorical Data: Dual Scaling and Its Applications.* Toronto: University of Toronto Press.

Nishisato, S. (1982), *Shitsuteki Data no Suryoka (Quantifying Qualitative Data)*. Tokyo, Japan: Asakura Shoten.

Nishisato, S. (1984a), "Dual scaling by reciprocal medians". In *Proceedings of the 32nd Scientific Conference of the Italian Statistical Society*, Sorrento, Italy, 141–147; handout, 1984b (4 pages).

Nishisato, S. (1984c), "Forced classification: a simple application of a quantification method". *Psychometrika* **49**, 25–36.

Richardson, M., and G. F. Kuder (1933), "Making a rating scale that measures". *Personnel Journal* **12**, 36–40.

Tanaka, Y. (1984), "Sensitivity analysis in Hayashi's third method of quantification". *Behaviormetrika* **16**, 31–44.

Henry E. Kyburg, Jr. [1]

THE BASIC BAYESIAN BLUNDER

ABSTRACT

Bayesian statistical inference appears to offer a way to solve the problems of inferring general statistical distributions in populations from observed distributions in samples. Early in the development of mathematical statistics this form of inference was referred to as "inverse" inference, in contrast with "direct" inference, the inference from knowledge of a distribution in a population to a distribution in a sample. It was felt that direct inference was reasonably well understood, and that inverse inference was a problem. This paper argues that Bayesian techniques are based on principles that actually conflict with direct inference. It is concluded that we should hold fast to direct inference based on our knowledge of frequencies or chances, and accept Bayesian procedures only when they can be put into the framework of direct inference.

1. DIRECT INFERENCE AND INVERSE INFERENCE

If any distinction in the realm of statistics or inductive logic deserves to be called "classical" the distinction between direct and inverse inference does. In philosophy, it is the classic distinction between inductive and deductive argument. Inferring that Socrates the man is mortal from the premise that all men are mortal, is an instance of direct inference. The corresponding inverse inference is that which proceeds from premises, Socrates the man is mortal, Plato the man is mortal,...Churchill the man is mortal, to the general conclusion that all men are mortal. Inverse inference proceeds from the particular to the general, direct inference from the general to the particular. Inverse inference is characterized by inductive logic; direct inference by deductive logic.

[1] Department of Philosophy, The University of Rochester, River Campus Station, Rochester, New York 14627

I. B. MacNeill and G. J. Umphrey (eds.), Foundations of Statistical Inference, 219–232.

In statistics the distinction is even more straight-forward: it is the distinction between inferences that take knowledge of a distribution in a population as a premise, and infer the probable character of a particular sample—this is direct inference—and inferences that take knowledge of a sample as a premise, and infer the probable character of the population from which the sample comes—this is inverse inference.

Inverse inference is characterized by what in computer science (or anyhow artificial intelligence) is called non-monotonicity. This means that, in contrast to deductive inference, an increase in the premises may undermine a conclusion already reached. This was recognized explicitly by R. A. Fisher in 1936. He writes, "There is one peculiarity of uncertain inference which often presents a difficulty to mathematicians trained only in the technique of rigorous deductive argument, namely that our conclusions are arbitrary, and therefore invalid, unless all the data, exhaustively, are taken into account. In rigorous deductive reasoning we may make any selection from the data, and any certain conclusions which may be deduced from this selection will be valid, whatever additional data we have at our disposal." (Fisher, 1936, p. 254.)

In general, whether in logic or in statistics, direct inference has been regarded as relatively unproblematic. Basic, first order, deductive logic is almost universally accepted as being all right as far as it goes, though there are some people who think it does not go far enough. (It does not capture modal arguments, for example.)

In a similar way, early on in the history of probability theory agreement was achieved concerning the inferences that were warranted whose premises concerned general distributions, and whose conclusions concerned samples. Given as a premise that heads among coin tosses are distributed binomially, with a parameter of 1/2, we all easily calculate that the probability of four heads in succession is 1/16. There is no uncertainty in the argument here. But there are also direct inferences that embody uncertainty: We infer—or, as Neyman for example might prefer to put it, we *behave* as if—the next ten tosses of this coin will not yield ten heads.

Given as premises however, the distribution of heads in a quite large sample of tosses, subjected to whatever constraints concerning randomness you wish, there is inevitably controversy concerning what conclusion is warranted and to what degree.

Note that in statistics both the direct and inverse inference may be non-monotonic: to augment the premises may undermine direct uncertain inference as well as the inverse inference. To learn that not only are the ten tosses we are concerned with the *next* ten tosses, but that nine of them have already yielded heads undermines our inference from the general distribution to the conclusion that we won't get ten heads in a row. Similarly, given the results of n tosses, the inference (whatever it may be) to a general distribution of heads among tosses will be undermined (rendered epistemically irrelevant) by knowledge of the outcomes of an additional m tosses. This fact is of interest, particularly when it comes to constructing a logic that will reflect the realities of uncertain inference. But it is not essential to the distinction between direct and inverse inference. What is essential is that in direct inference the movement is from the population to an actual or hypothetical sample, while in inverse inference the movement is from the sample to some statement concerning the parent population.

2. INVERSE INFERENCE AND BAYES' THEOREM

What is special about inverse inference is *not* the use of Bayes' theorem. When Neyman (1941, p. 131) writes "...persons who would like to deal only with classical probabilities, having their counterparts in the really observable frequencies, are forced to look for a solution of the problem of estimation other than by means of the theorem of Bayes," we must understand him to be emphasizing the phrase "solution to the problem of estimation," since Bayes' theorem is, after all, a theorem.

What this means is that, as R. A. Fisher (1930) saw clearly, there are many situations in which Bayes' theorem is applicable that can easily be construed in terms of direct inference. In 1930 (p. 530) he notes that drawing from a super-population in which the parameter of interest (say theta) has a known distribution F, and then getting a posterior distribution for theta is "...a perfectly direct argument...", though of course it uses Bayes' theorem. In the same way the famous example described by Laplace, concerning $n+1$ urns, each containing n black and white balls in each possible combination involves Bayes' theorem, but makes no use of inverse inference proper. (The application of this model to sampling does.)

For inverse inference *proper*—that is, inference whose uncertainty is *not* based on known frequences—Fisher (1931, pp. 528–529) has nothing but contempt: "In fact, the argument runs somewhat as follows: a number of useful but uncertain judgments can be expressed with exactitude in terms of probability; our judgments respecting causes or hypotheses are uncertain, therefore our rational attitude towards them is expressible in terms of probability." Neyman's attitude is even less tolerant.

Fisher and Neyman were, of course, reacting against the use of the so-called "axiom" of Bayes that gave uniform priors, against Laplace's principle of indifference, and the like. Since that time, however, inverse inference has become respectable again. It gained respectability by admitting what de Finetti called its "subjective sources" and claiming nevertheless to provide a rationale for inferences from a sample to a population, thus completing, in a sense, the theory of statistically uncertain inference. Direct inference governs the inference from the population to the sample; indirect inference governs the inference from the sample to the population.

I claim that there is a serious blunder involved here—not quite so obvious as the fallacy Fisher offers us, but a blunder nevertheless. It lies in the fact that direct inference and inverse inference do not coexist happily. Historically, we were all confident and happy with the use of direct inference. A number of people had *philosophical* qualms about its application to specific objects and events: "the next toss," "the next sample of a thousand balls to be drawn," etc. Nobody had formulated careful rules of application for direct inference; but few people doubted that direct inference was in principle sound: if you know that a coin is fair, you can infer that the probability that the next toss will yield heads is a half, that the probability is a sixteenth that the next sample of four consecutive tosses will have the structure HHTH, etc. Those who had serious qualms replaced talk of probability with talk of confidence, inductive behavior, rules, etc.

What we wanted, and didn't have, was a generally acceptable rationale for inverse inference. That is what Thomas Bayes sought, and what the modern philosophical Bayesian seeks. But inverse inference *proper* undermines direct inference. In order to have inverse inference, to "complete" our theory of uncertain inference, we find we must abandon direct inference in many of its classical applications.

3. INDEPENDENCE

The most elementary example of this conflict can be seen in the case of a simple binomial distribution. If we know that a coin is fair and that its tosses are independent, we have no difficulty in calculating the probability of, say, ten heads on the next fifteen trials. This is our old, unproblematic, direct inference.

But where does this knowledge come from? Inverse inference, so the story goes. That is, we suppose that the way we got our binomial hypothesis was by looking at a lot of coin tosses. So let H be the hypothesis that the coin is fair. If we "get" H by inverse inference, that cannot mean that we assign it probability one: inverse inference via conditionalization can't raise a probability to one that doesn't start there. But if the probability of H is not one, then all our conventional direct inferences are undermined. In particular, we can no longer regard the tosses as independent, since every toss will change (by conditionalization) the probability we assign to H. H says that the tosses are independent, but this just means that any permutation of a specified sequence occurs just as often as that same sequence. The *dependence* among tosses is epistemic, and depends on our changing epistemic state with regard to H. We can no longer just say that the probability of ten heads if $(1/2)^{10}$—for (a) the very first head will change the probability we assign to H itself, and (b) we must also take account of the vague hypotheses not-H and what it assigns to ten heads.

4. THE UNIFORM PRIOR

It has been claimed that both R. A. Fisher's fiducial inference, in the case of estimating the mean for a normal distribution, and Neyman's method of confidence intervals for estimating the mean of a binomial distribution, require a "flat" or "uninformative" prior distribution for their validity. As Fisher and Neyman, respectively, have pointed out, this is untrue. Whatever be the mean μ of the normal population, the quantity $(x - \mu)/\sigma$ will be normally distributed with unit variance and mean 0. For confidence interval estimation, whatever be the binomial parameter p, the frequency with which a sample will fall in the confidence region will be at least as great as the confidence level.

The germ of falsehood—or better, irrelevance—in this observation lies in the fact that *if* we had some prior distribution that was *not* the flat or uninformative distribution, *then* the fiducial or confidence argument would *not* be valid. But one should distinguish between knowing that p is uniformly distributed between 0 and 1, and knowing that p has some (totally unknown) value in that interval. This is the classic—but not always helpful—distinction between an unknown constant and a random quantity.

(Many writers insist that one can make probability statements *only* about random quantities and *not* about constants. But, for example, $(\mathbf{x} - \mu)/\sigma$ is a perfectly good random quantity, and yet to say that its absolute value is less than $\mathbf{e}$ is exactly to say that $\mathbf{x} - \sigma\mathbf{e} < \mu < \mathbf{x} + \sigma\mathbf{e}$, which is a statement "about" the constant μ. Of course there are epistemological problems to be dealt with before we can go around substituting observed values for the quantity $\mathbf{x}$, but that's another matter.)

Now if the prior distribution is a frequency-like distribution in some super-population, then it is merely that a *different* direct inference is called for (as Fisher and Neyman both say), and we aren't talking about inverse *proper*. But if the prior distribution is a *priori* or subjective, as it must be for an inverse inference proper, then there *is* conflict between the inverse inference and the fiducial or confidence argument based on direct inference.

5. RECONCILING DIRECT INFERENCE AND INVERSE INFERENCE

Inverse inference and direct inference will agree if the prior distribution provided by the inverse inference happens to be the uninformative prior. I have argued that this fortunate situation is relatively rare. In a previous paper (Kyburg, 1968) I used a procedure of de Finetti to show that it is very easy for general empirical hypotheses to achieve impressively high probabilities in the absence of any evidence in their favor. This generates just the sort of bias that cannot be tolerated by arguments that depend on direct inference.

If the sequence *Ti* is a sequence of exchangeable trials (in fact all we need stipulate is that the probability of a success followed by a failure is the same as the probability of a failure followed by a success—much less than full exchangeability) and the prior probability of a success is .01 and the conditional probability of a success on a second trial, given a success on the first trial is .02, *then* we must assign a probability of *at least* .9996 that no more than half the trials in the arbitrarily long run will yield successes. Or, we can calculate that the probability that less than 80% of the trials in the long run will yield successes is *at least* (this is very conservative—we use only Tchebycheff's inequality) $.999844 = 1 - .000156$.

Now let us perform 16 trials. Suppose they all yield successes. Neyman has taught us that there *is* frequency information bearing on hypotheses about the long-run frequency of success. Specifically, whatever the actual frequency of success may be, at least 90% of the performances of 16 trials will yield results falling in what Neyman (1939) refers to as the confidence belt, and the bounds of the confidence belt in this case are .80 and 1.00.

Neyman would say that we can be 90% confident that the long run success rate is in the interval $[.80, 1.00]$. This is *not* a probability for Neyman, but it *does* correspond to a *before trial* relative frequency or probability. In fact, he writes (1934, p. 123): "If the confidence belt is constructed we may affirm that the point *will* [Neyman's emphasis!] lie inside the belt. This statement may be erroneous, but the probability of error is either equal to or less than 1–*e*—thus is as small as desired."

But the gap between a frequency in a general class and the probability of a specific occurrence was exactly the gap that direct inference was supposed to be capable of crossing. Leave aside sophisticated philosophical doubts about the meaning of probability, and nothing could be more natural than for the holder of a specific ticket in a thousand ticket lottery to say that the probability of his winning the first prize is 1/1000. Similarly, holding a sample comprising sixteen successes, nothing could be more natural than to say, since at least 90% of such trials yield Neyman-representative results, that the probability is at least 0.9 that *our* trial yielded a Neyman-representative result—i.e., a result in the confidence belt. And it did this if and only if the success rate in the population is in the interval $[.80, 1.00]$.

Note that I haven't spelled out a principle of direct inference that gives this result and doesn't lead to difficulties. This isn't easy, thought I think that after some 25 years, I've gotten close to it. But this is *exactly* the sort of thing that everybody took for granted when the problems of inverse inference were first raised. This is the sort of uncertain inference that seemed unproblematic. It is furthermore a kind of uncertain inference that Bayesians seem to be rediscovering.

(For example, David Lewis, "A Subjectivists' Guide to Objective Chance", in R. C. Jeffrey (ed.), *Studies in Inductive Logic and Probability,* University of California Press, 1980, pp. 263–294, proposes a "principal principle" that is none other than a newly discovered principle of direct inference.)

Suppose the direct inference does go through. How does it relate to the previous result? Writing the appropriate form of Bayes' theorem, we have:

$$P(r > .8 \mid E(16,16)) = \frac{P(r > .8)P(E(16,16) \mid r > .8)}{P(r > .8)P(E(16,16 \mid r > .8) + P(r \leq .8)P(E(16,16) \mid r \leq .8)}$$

where r is the long run success rate and $E(16,16)$ is our evidence.

For consistency with direct inference, we require $P(r > .8 \mid E(16,16)) \geq 0.9$, or

$$P(r > .8)P(E(16,16) \mid r > .8) \geq .9(P(r > .8)P(E(16,16) \mid r > .8) + P(r \leq .8)P(E(16,16) \mid r \leq .8)).$$

Simplifying, and taking account of the fact that $P(r > .8) < .000156$ and $P(r \leq .8)/.999844$, we require:

$$P(E(16,16) \mid r > .8) > 1.40\ 10^5 P(E(16,16) \mid r \leq .8).$$

This doesn't seem very plausible, but perhaps it just shows that our original intuitions about the frequency of successes were not as plausible as they seemed.

6. ERRORS OF OBSERVATION

When we formulate a principle of direct inference that allows for imprecise knowledge, even this sort of retroactive adjustment is impossible.

I assume that we want to apply direct inference even when we do not know *exactly* the relevant frequencies in our reference classes. This is obviously important pragmatically, but it raises a theoretical question. Suppose that we know the relevant frequency in a large class quite precisely, but that our knowledge concerning a subclass is rather vague. If there is no *conflict* between the two items of knowledge—if the interval corresponding to the larger class is a subinterval of the interval corresponding to the smaller class, it seems appropriate to take the smaller interval, based on the larger class, as legislative for probability.

For example, I know that very nearly a half of coin-tosses in general yield heads. I know much less about the frequency with which this particular 1980 quarter lands heads—perhaps, by an inductive inference I could say that I know that between 40% and 60% of its tosses yield heads. But if you ask for the probability of heads on the next toss of this 1980 quarter, since there is no conflict between what I know of it and what I know of tosses in general, I shall take the narrow interval corresponding to my knowledge of tosses in general to give that probability.

This, in my systematic treatment of direct inference (Kyburg, 1974), is called "the strength rule". I shall now describe an example that shows that direct inference, if it incorporates the strength rule, is flatly inconsistent with inverse inference. The example is due to Isaac Levi (1977, 1980)who draws the conclusion that the strength rule is unacceptable. I shall alter the example slightly, but I shall keep the numbers roughtly the same. And I shall draw the opposite conclusion: that we should hold fast to the strength rule, and let inverse inference and the form of conditionalization that requires it go hang.

Suppose we measure lengths with one of three instruments, A, B, and C. A gives results accurate within a margin of error m between 88.5% and 88.9% of the time; of type B between 90.5% and 90.9% of the time; and of type C between 91.5% and 91.9% of the time. To fix our ideas, suppose "within a margin of error m" amounts to having a reading within .001 of the true value being measured. We may suppose that these frequencies are reported by the manufacturer of the three types of instrument. In general, though, we know that the combined results of all three kinds of measurement are accurate within a margin of error m between 89.9% and 90.1% of the time. Put otherwise: we have a population of measurements of which between 89.9% and 90.1% are accurate; this population is partitioned into three subpopulations, A, B, and C, characterized by the error rates mentioned.

A particular measurement is made. We don't know what instrument was used. It seems natural to say that the probability of its being accurate within the margin of error m is (about) .90—more exactly, the interval [.899, .901] seems to capture what we know.

We also know that an instrument yielding an error rate between 88.5% (the minimum for instrument A) and 91.9% (the maximum for instrument C) was used, so one might be tempted to think that the appropriate interval was [.885, .919]. The strength principle argues against this; if we have more accurate information we should use it. We should use the most exact statistical knowledge we have for direct inference, provided that it is not in conflict with other knowledge that we have.

But this position is in flat-out contradiction to inverse inference construed as conditionalization on a prior probability—i.e., as inverse inference proper. To see this, suppose that P is a Bayesian belief function, and that direct inference, as I have described it holds. Following Levi, we show that this leads, in combination with other plausible assumptions, to a contradiction.

First off, if all we know is that the measurement we have made was made with an instrument manufactured by the firm in question, we should accept the general frequency of error as constraining our epistemic probability *Prob*:

$$Prob(\text{“}S \in G_m\text{”}, K_{A\cup B\cup C}) \in [.899, .901]$$

where S is the particular measurement at issue, G_m is the set of measurements accurate within a margin of error m, and $K_{A\cup B\cup C}$ is the body of knowledge embodying merely the information that S was made with one of the types of instrument A, B, or C.

Clearly, if we know which subset of the population of measurements we are in, the error rate in that subset is indicated as the appropriate basis for a direct inference.

Let A, B, C be the sets of all trials, past, present, and future, with instruments of *types* A, B, and C, respectively.

We are warranted in accepting

$$\%(A, G_m) \in [.885, .889] \tag{1}$$

$$\%(B, G_m) \in [.905, .909] \tag{2}$$

$$\%(C, G_m) \in [.915, .919] \tag{3}$$

$$\text{Let } K_A = K \cup \{\text{“}S \in A\text{”}\}; K_B, K_C \text{ similarly };$$
$$K_{K\cup B} = K \cup \{\text{“}S \in A \cup B\text{”}\}; K_{B\cup C}, K_{A\cup C} \text{ similarly}.$$

From (1), (2), and (3) if follows that

$$\%(A \cup B, G_m) \in [.885, .909] \tag{4}$$

$$\%(A \cup C, G_m) \in [.885, .919] \tag{5}$$

$$\%(B \cup C, G_m) \in [.905, .919] \tag{6}$$

It *may* be the case that we have more precise knowledge of these disjunctive reference sets, as we do of $A \cup B \cup C$—but that *need* not be the case. We may have lost the data; we may (reasonably) be depending on what the manufacturer tells us; in general we cannot suppose that we know everything that anybody else knows or ever knew.

For the same reason we have, by direct inference, using the strength rule,

$$\begin{aligned}
&Prob(\text{“}S \in G_m\text{”}, K_A) \in [.885, .889]\\
&Prob(\text{“}S \in G_m\text{”}, K_B) \in [.905, .909]\\
&Prob(\text{“}S \in G_m\text{”}, K_C) \in [.915, .919]\\
&Prob(\text{“}S \in G_m\text{”}, K_{A\cup B}) \in [.899, .901]\\
&Prob(\text{“}S \in G_m\text{”}, K_{A\cup C}) \in [.899, .901]\\
&Prob(\text{“}S \in G_m\text{”}, K_{B\cup C}) \in [.905, .919]
\end{aligned}$$

There is no function P that's a conditional belief function such that in general

$$P(\text{“}S \in G_m\text{”}/K_X) \in Prob(\text{“}S \in G_m\text{”}, K_X).$$

To see this, suppose that P is such a belief function. By conditionalization and "total probability" we have

$$\begin{aligned}
P(\text{“}S \in G_m\text{”}/K) = {} & P(\text{“}S \in A \cup B\text{”}/K)P(\text{“}S \in G_m\text{”}/K_{A\cup B})\\
& + P(\text{“}S \in C\text{”}/K)P(\text{“}S \in G_m/K_C) \in [.899, .901].
\end{aligned} \tag{7}$$

Similarly,

$$\begin{aligned}
P(\text{“}S \in G_m\text{”}/K) = {} & P(S \in_{A\cup C} \text{”}/K)P(\text{“}S \in G_m\text{”}/K_{A\cup C})\\
& + P(S \in B\text{”}/K)P(S \in G_m\text{”}/K_B) \in [.899, .901].
\end{aligned} \tag{8}$$

Let $\alpha = P(\text{“}S \in A\text{”}/K)$; $\beta = P(\text{“}S \in B\text{”}/K)$; $\gamma = P(\text{“}S \in C\text{”}/K)$.
From (7) and (8), together with the principle that beliefs should be constrained by probabilities, we obtain:

$$\begin{aligned}
&(1-\beta)P(\text{“}S \in G_m\text{”}/K_{A\cup C}) + \beta P(\text{“}S \in G_m\text{”}/K_B \in [.899, .901],\\
&(1-\gamma)P\text{“}S \in G_m\text{”}/K_{A\cup B}) + \gamma P(S \in G_m\text{”}/K_C) \in [.899, .901],
\end{aligned}$$

from which it follows that

$$\begin{aligned}
&\gamma \le \frac{2}{17} = .1176,\\
&\beta \le \frac{2}{7} = .2857,\\
&\alpha \ge .5967.
\end{aligned}$$

Note that α, β, and γ are *not* probabilities, based on known frequencies, but mere degrees of belief, based on the principle of direct inference *together with* the probability calculus.

Given these constraints on α, β, and γ, we may derive a constraint on $P(\text{“}S \in G_m\text{”}/K)$.

$$\begin{aligned} P(\text{“}S \in G_m\text{”}/K) = {} & P(\text{“}S \in A\text{”}/K) \cdot P(\text{“}S \in G_m\text{”}/K_A) \\ & + P(\text{“}S \in B\text{”}/K) \cdot P(\text{“}S \in G_m\text{”}/K_B) \\ & + P(\text{“}S \in C\text{”}/K) \cdot P(\text{“}S \in G_m\text{”}/K_C). \end{aligned}$$

The *maximum* possible value for $P(\text{“}S \in G_m\text{”}/K)$ given these constraints is

$$\begin{aligned} & \alpha_{\min} P(\text{“}S \in G_m\text{”}/K_A) + \beta_{\max} P(\text{“}S \in G_m\text{”}\, K_B) \\ & + \gamma_{\max} \cdot P(\text{“}S \in G_m\text{”}/K_C) \\ & \leq .5967(.899) + .2857(.909) + .1176(.919) \\ & = .5305 + .2597 + .1801 = .8982. \end{aligned}$$

But this does not fall within the constraints imposed by the principle of direct inference, *viz.*, $[.899, .901]$.

It might be suggested that these are just not plausible statistics for us to know. The response is, first, that these may just be the statistics we have to work with. The second response is that even if we must get the statistics from our own data, we can generate the problem.

Pick a level of acceptance—e.g., $1 - \alpha = 0.99$. Look up a number n such that a .99 confidence interval based on a sample of size n, with observed relative frequency of G_m about 0.917, is included in $[915, .919]$. Since the .99 confidence interval corresponds to about 2.575 standard deviations (using a normal approximation and $1/2\sqrt{n}$ as the upper bound of the standard deviation, n is about 414414. Similarly, for A and for B, about the same n will do. To obtain an overall confidence interval of $[.899, .901]$, we may suppose a further, undifferentiated sample of 413395, of which 367498 are G_m. (Not all possible data is recorded; not all recorded data is kept.)

Even if we try to be realistic about the data, we encounter the conflict. But we have no reason to suppose we *have* the data—the error rates may just be reported in this form by the manufacturer.

7. DIRECT INFERENCE AND DUTCH BOOK

This just exhibits one more conflict between direct inference and inverse inference. What do we do about it? One answer is to circumscribe direct inference enough so that it can be reconciled with inverse inference. One way to do this is to obtain probabilities from statistical knowledge only when they concern objects (or events or whatever) that are *random members* of their appropriate reference classes. This is Levi's (1977, 1980) suggestion. But to construe randomness in this way is, as I see it, to abandon direct inference. We do not obtain the probability of accuracy of our measurement from knowledge of the frequencies of error in $A \cup B \cup C$ and its subsets A, B, and C, but from that statistical knowledge *combined* with non-statistical "probabilistic" knowledge about how the measurement was generated. (Clearly if knowledge about how the measurement was generated is statistical, we face no problem; but then all probabilities can come from direct inference.)

The most important philosophical counterargument to sacrificing conditionalization to direct inference is the Dutch Book argument. Just as it is alleged that one's degrees of belief should satisfy the axioms of the probability calculus, else one could have a book made against one, so, it is argued, if conditional bets are allowed, one's *conditional* degrees of belief must satisfy the principle of conditionalization. More explicitly, suppose that the interval of probability for S is $[.3, .4]$ and that for $S\&T$ is $[.1, .2]$. Then the Bayesian conditional probability of T *given* S should be constrained by the interval $[.26, .67]$. Every classical probability function P such that $P(S) \in [.3, .4]$ and $P(S\&T) \in [.1, .2]$ is such that $P(T/S) \in [.25, .67]$.

Clearly the interval $[.25, .67]$ should constrain the odds of conditional bets on T given S. It is claimed that the same interval should constrain my bets on T after I have added S to my body of knowledge. This principle is one that Levi has called "confirmational conditionalization". As was first pointed out by him, and as we have just seen, confirmational conditionalization is in conflict with at least some forms of direct inference.

Suppose we abandon confirmational conditionalization, as I have suggested. Then after observing S the (new) probability of T need not be the interval $[.25, .67]$—or any subinterval of it—but might be (say) $[.70, .80]$. The cunning bettor, knowing that I will modify my probabilities in this way, offers a bet at odds of 4 to 6 against S, and also a conditional bet at even money on T given S for a stake of 11. Then, knowing how I will modify my odds on learning S, he plans to make a *new* bet after S has occured (if it does) against T, at 6 units to 18. Here is what happens:

If S fails to occur, the bettor gains 6 units and no other bets are activated.

If S does occur, there are two cases. If T occurs the bettor loses 4 from the first bet; gains 11 from the conditional bet and loses 6 from the third bet, for a net gain of 1. If T fails to occur, he loses 4 from the first bet; loses 11 from the conditional bet, but gains 18 from the third bet for a net gain of 3. In any case, the bettor wins. I have been Dutch booked!

By giving up confirmational conditionalization, have I not laid myself open to a sure loss? Of course not. Even in the elementary case, I could be willing to offer odds of 2 to 1 on S and 2 to 1 against S without being willing to make both bets at once. But even if I *must* post odds, and *must* take any bets consistent with that posting, so that the posted odds must be coherent, on pain of sure loss, that is no argument that I cannot change my posting. (In fact, the willingness to change the posted odds in the face of new evidence might be one of the things that distinguishes successful bookies!)

But we must be careful about the sort of changes that evidence can warrant. The case at hand is *not* one that can actually happen. The clue to this lies in the fact that the probabilities mentioned entail that $P(S \& -T) = 0.2$ exactly; the probability that I would assign to T after observing S can be construed as a constraint on the prior probabilities of S & T and S & $-T$. It is only where the strength rule is invoked that we can have a legitimate violation of confirmational conditionalization. It is easy to see this in the original example concerning the three measuring instruments A, B, and C. If we hang on to the probabilities that are determined by direct inference, there is NO adjustment of prior probabilities that will preserve coherence. To regain coherence, at least one direct inference must go!

The ideal Bayesian robot, to be sure, has no need of either direct inference or interval-valued probabilities. But his probabilities are at base a *priori* prejudices, compounded, to be sure, with observations. Alternatively, we may follow Fisher and Neyman in abandoning inverse inference. Especially once we have liberalized our notion of probability to accommodate intervals or sets of distributions, the loss of inverse inference is no loss.

The basic Bayesian Blunder does not lie in the use of Bayes' *theorem*. The use of Bayes' theorem is perfectly compatible with the principle that all probabilities, without exception, are obtained by direct inference. The blunder lies in the conviction that only by inverse inference proper can the knowledge needed for direct inference be obtained. But we can't get *acceptance* from inverse inference alone, so inverse probability doesn't solve that problem. And, worse, inverse inference is seriously incompatible with direct inference, which was where we started from. The whole idea of inverse inference was to complete and complement an acceptable theory of direct inference; what we find, when we develop inverse inference far enough is that we have little or nothing left of direct inference. We have undermined the foundation on which we tried to build.

REFERENCES

Fisher, R. A. (1930), "Inverse probability". *Proceedings of the Cambridge Philosophical Society* **26**, 528–535.

Fisher, R. A. (1936), "Uncertain inference". *Proceedings of the American Academy of Arts and Sciences* **71**, 245–258.

Kyburg, H. E. (1968), "Bets and Beliefs". *American Philosophical Quarterly* **5**, 54–63.

Kyburg, H. E. (1974), *The Logical Foundations of Statistical Inference*. Dordrecht: Reidel.

Levi, I. (1977), "Direct inference". *Journal of Philosophy* **74**, 5–29.

Levi, I. (1980). *The Enterprise of Knowledge*. Cambridge, MA: MIT Press.

Neyman, J. (1934), "On the two different aspects of the representative method". *Journal of the Royal Statistical Society* **97**, 558–625.

Neyman, J. (1941), "Fiducial argument and the theory of confidence intervals". *Biometrika* **32**, 128–150.

Brian Skyrms [1]

DYNAMIC COHERENCE

ABSTRACT

Ramsey (1931) and de Finetti (1937) founded the notion of personal probability on *static coherence*. Someone's betting quotients over a boolean algebra of propositions are coherent if and only if it is not possible for a bettor to make a *dutch book* against him by means of a finite number of bets; i.e. there are a finite number of bets which he judges fair or favorable such that his net payoff from them is negative in every possible outcome. The classical dutch book theorem shows that betting quotients are coherent in this sense if and only if they constitute a finitely additive probability measure (de Finetti, 1937; Kemeny, 1955; Lehman, 1955; Shimony, 1955). Adams (1962) showed that the result can be strengthened to require countable additivity if the concept of coherence is modified to allow countably many bets. These betting quotients can be thought of as degrees of belief, in the dispositional sense of belief, as Ramsey suggested. The degrees of belief are held fixed for the classical dutch book argument; we are dealing with degrees of belief at a time. The notion of coherence involved is static; all betting is done at the same time with reference to the same set of degrees of belief. What is established is that coherent degrees of belief, in the sense at issue, are probabilities.

What can be said about *changes* in degrees of belief? Rules for changing degrees of belief in response to new evidence, such as Bayes' rule of conditionalization, play a central role in Bayesian reasoning. What about coherence concepts and results for rules for changing degrees of belief? This is what I am calling the question of *dynamic* coherence. It appears that this question has not even been considered seriously until recently. Hacking (1967) called attention to the problem of providing a justification of Bayesian conditionalization by means of a coherence argument, but was skeptical of the possibility of a positive result. Since Hacking's discussion, questions of dy-

[1] Department of Philosophy, University of California at Irvine, Irvine, California 92717

I. B. MacNeill and G. J. Umphrey (eds.), Foundations of Statistical Inference, 233–243.

namic coherence have been discussed by both philosophers and statisticians, and some fundamental results have been established.

1. BAYES' RULE

How should one update a probability distribution, Pr_1, upon learning that an event, E, has occurred? Providing that the sole additional information is that E has occurred and that E is in the field of Pr_1, the answer that is almost universally given is to move to a new probability, Pr_2 by conditionalization on E:

$$\text{(BAYES' RULE)}: \; Pr_2(Q) = Pr_1(Q\&E)/Pr_1(E).$$

Let us consider a model situation which Bayes' rule fits exactly. Our subject, *the bookie,* has at time 1 a prior probability assignment, Pr_1, over a probability space and there is a countable partition, $\{e_i\}$, over that space such that each member of that partition has positive prior probability. At time 2, the true member of the partition is announced and the bookie moves to a posterior probability, Pr_2, according to a rule which treats the announced member of the partition as total input. Such a rule is a function, r, which maps members of $\{e_i\}$ onto posterior probability distributions. (Requiring it to be a function of members of the partition is a precise way of saying that to all intents and purposes the bookie learned nothing more than the true member of the partition.) The bookie's *strategy* consists of the pair, $\langle Pr_1, r\rangle$. We will allow a cunning bettor to place a finite number of bets with the bookie at both times. He must make bets that the bookie considers fair or favorable (non-negative expected utility) at the time. He is allowed to know the bookie's probabilities at the times of the bets and is allowed to know the bookie's strategy. The bettor's strategy consists of a pair of functions, $\langle f_1, f_2\rangle$, where f_1 maps the pair consisting of the bookie's $\langle Pr_1, s\rangle$ onto bets which for the bookie have non-negative expected utility at t_1, and f_2 maps $\langle Pr_1, s, e_i\rangle$ onto bets which for the bookie have non-negative expected utility at t_2 if e_i is the event announced. The bettor makes a *dynamic dutch book* against the bookie iff no matter what the true member, e_i, of the partition, he is left at t_2 with a finite number of bets whose net payoff is positive for each point in e_i. The bookie's strategy is *dynamically coherent* if there is no bettor's strategy that makes a dynamic dutch book against it.

I. The bookie's strategy, $\langle Pr_1, r\rangle$, is dynamically coherent if and only if r is Bayes' rule.

The "only if" is established by Teller (1973, 1976) by the presentation of a bettor's strategy (due to David Lewis) which will make a dutch book against any bookie's strategy for which r differs from Bayes' rule. The leading idea is straightforward. The bettor can achieve the effect of a bet conditional on a member of the partition, e_i, in two distinct ways: either (i) make a conditional bet at t_1 (which can, in effect, be achieved by a finite number of unconditional bets at t_1) or (ii) adopt a strategy such that a bet is made at t_2 iff e_i is announced as the true member of the partition. The bookie's fair betting rates for conditional bets made in these two ways will agree iff r is Bayes' rule. If they do not agree, the bettor can exploit the difference to achieve sure winnings if e_i is announced. An appropriate side bet against e_i assures the bookie of winnings no matter what member of the partition is announced. The "if" is just as straightforward. If r is Bayes' rule, then any payoff function that the bettor can achieve by betting at t_1 and t_2 can also be achieved by betting only at t_1, by replacing bets made at t_2 if e_i is announced by bets conditional on e_i made at t_1. The bookie can make a dynamic dutch book only if he can make a static one at t_1, which he cannot by the classic coherence theorem.

(Note: The essential insight is that a rule for changing degrees of belief of the type at issue is tantamount to the adoption of a set of conditional betting quotients. Conversely, the adoption of an appropriate set of betting quotients conditional on members of the partition to remain fixed for the duration of the game is tantamount to a rule such as r. Then de Finetti's (1937) discussion of conditional bets takes us the rest of the way.)

2. BAYES' METHOD

The foregoing addresses the problem in a form which focuses on belief change: $Pr_1(q \mid e_i)$ is perfectly well defined, and the question is whether $Pr_2(q)$ should be equal to it. Non-Bayesian statisticians, however, are most likely to object not to Bayes' rule itself, but rather to the assumption of the existence of a suitable input, $Pr_1(q \mid e_i)$. For instance, suppose that there is a parameter space, T, and an experimental outcome space, E, such that each t in T determines a sampling probability over E. An experimental result, e, is observed. Bayes' *method* is to calculate $Pr_1(t \mid e)$ by assuming a prior over T and using Bayes' theorem and then moving to Pr_2 by Bayes' rule. The objection will be to the existence of the prior over T.

To be responsive to this objection, we should consider a modified game where the prior is not known, the bookie has some rule of updating as a function of the observed e_i and the bettor is allowed only to bet at t_2. This sort of situation was already considered by Freedman and Purves (1969) for finite spaces S and T. They consider a game with three players: a master of ceremonies, a bookie and a bettor. The parameter space, X, indexes experimental probability distributions over the experimental space, X, in a way known to all participants in the game. The master of ceremonies selects "at his pleasure" a member of the parameter space, and then selects according to the experimental probability determined by his first selection, an experimental outcome. He announces the experimental outcome to both players. The bookie then posts odds on subsets of the parameter space and the bettor places a finite number of bets. The bookies odds are called *coherent* if it is not possible for the bettor to bet such that the bookie has a negative expected payoff at each point in the parameter space, the expectation being in terms of the experimental probability which is common knowledge. Freedman and Purves show that coherence requires the bookies odds coincide with those obtained by Bayes' method for some prior or other. Extensions of their results and related work are given by Heath and Sudderth (1972, 1978), Lane and Sudderth (1983), and Dawid and Stone (1972, 1973). These results have a static reading in terms of which conditional probabilities can come from a single joint distribution, but the game used to motivate the discussion clearly takes the dynamic point of view.

3. PROBABILITY KINEMATICS

Suppose, as before, that we have a probability, Pr_1, and a partition, $\{p_i\}$, such that each member of the partition has positive probability. We say, following Jeffrey (1965), that a probability, Pr_2, on the same space *comes from Pr_1 by probability kinematics on the partition $\{p_i\}$* iff:

$$\text{(Kinematics)}: Pr_2(q \mid p_i) = Pr_1(q \mid p_i)$$
$$\text{(for every measurable set } q \text{ and every } i\text{)}.$$

In statistical terminology, $\{p_i\}$ is a *sufficient partition* for the set of probabilities that can come from Pr_1 by probability kinematics on it. That set of probability measures is a convex set whose extreme points are the probabilities that come from Pr_1 by conditionalization on members of $\{p_i\}$.

Jeffrey's idea was that change by probability kinematics could handle more naturally some cases of probability change where application of Bayes' rule is strained. In particular, Jeffrey had in mind cases of uncertain observation where no proposition in the agent's language (measurable set in his probability space) adequately sums up what he learned. What he learned might nevertheless be about a certain partition (e.g., a partition of colors of an object) while falling short of certain knowledge of the true member of the partition.

One might suppose that he learned something else for certain, such as the final probabilities of the members of the partition, and that the agent should simply apply Bayes' rule in a larger probability space in which what he learned *does* correspond to a measurable set. This approach is pursued by Armendt (1980), Good (1981), and Skyrms (1980a,b). It is not, however, completely responsive to Jeffrey's concern, which was for the case in which there *is* no appropriate set on which to conditionalize. Is there a coherence argument for probability kinematics which models the situation that Jeffrey had in mind?

A coherence argument of this kind is put forward by Skyrms (forthcoming). Finding a formulation for the situation which generalizes the Lewis argument and remains true to Jeffrey is not completely straightforward, because according to Jeffrey the observational input is ineffable. The bookie's strategy cannot, then, be thought of as a *function*. How, then, are we to represent the status of a partition as the one such that the observation "is about it"? There is a way around these difficulties if we introduce not only the uncertain observation, but also a subsequent certain observation which reveals the true nature of the partition. Consider the following "Observation Game".

Suppose that the bookie is shown a jelly bean under dim light and on the basis of the observational interaction, must revise his prior probabilities on a finite space whose points are color-flavor pairs. Subsequently, he is told by the gamemaster the true color of the bean. He must post odds on subsets of the color-flavor space at three times: t_1, prior to any observation; t_2, after the uncertain observation; t_3, after the true color is announced. A bettor, who knows the bookie's strategy, can make a finite number of bets with the bookie at each time for whatever stakes he pleases. Finally, the true color is announced and the bets are settled.

What is the bookies strategy? It is not a function, but rather a relation: a set of quadruples $\langle Pr_1, Pr_2, \text{COLOR}, Pr_3 \rangle$, where COLOR is the color announced. All members of the set have the same Pr_1 which gives each atom of the space positive probability. The Pr_2's represent probabilities which are allowed as upshots of the ineffable observational interaction. Since it is an uncertain observation, we require that each Pr_2 give each atom positive probability. *How do we specify that that interaction only carries information about the color;* that, for instance, the bookie doesn't cheat and taste the beans. We require that his strategy be such that Pr_3 not depend on Pr_2, the thought being that the gamemaster's announcement of the true color swamps out any uncertain information about color gotten from the uncertain information. So we require that if two quadruples in the bookies' strategy agree on color, they agree on Pr_3. This is a kind of qualitative sufficiency condition which identifies color as the appropriate partition. The bookie must believe the gamemaster: his Pr_3 of a partition must give the announced color probability one. The bookie's strategy must allow for contingencies: for every Pr_2 allowed in any quadruple and any color, there must be a quadruple that contains both. There are no restrictions on what Pr_2's are in a strategy; that is up to the bookie. However, we call a strategy *catholic* if for every probability assignment on the colors which gives each color positive probability, there is a Pr_2 which gives that probability assignment to the partition of colors. *Catholicity* allows for all contingencies concerning uncertain observations of colors.

The bettor's strategy is then an ordered triple of functions, the first of which maps the bookies strategy onto a finite number of bets at time 1; the second of which maps the bookie's $\langle$Strat, $Pr_2\rangle$ onto a finite number of bets at t_2; the third of which maps the bookie's $\langle Pr_1, Pr_2,$ COLOR, $Pr_3\rangle$ onto a finite number of bets at t_3.

What is the appropriate notion of coherence for the bookie? We say that a bettor's strategy *scores against* that of a bookie iff both (i) there is a sequence of play in which the bettor ends up at time 3 with bets whose net result is positive for him for every flavor, (i.e., a dutch book against the bookie) and (ii) for every sequence of play, the bettors winnings are non-negative for every flavor. The bookie's strategy is *bulletproof* if no players strategy will score against it. Being bulletproof is the strong coherence concept adopted here. (We will have more to say later about its connection to the possibility of an absolute dutch book.)

Skyrms (forthcoming) shows that in the observation game:

(A) *A bookie's strategy that does not proceed by probability kinematics on the colors is not bulletproof and (B) a catholic bookie's strategy which proceeds by probability kinematics on the colors is bulletproof.*

The leading idea of the proof of (A) is that we apply the Lewis argument for conditionalization between time 1 and time 3, and between time 2 and time 3 to assure constancy of the relevant conditional probabilities between time 1 and time 2. The leading idea behind the proof of (B) is that such a strategy is related to a conditionalizing strategy in an expanded game with higher order probabilities such that a bettor can score against the kinematic strategy in the little game iff he can make a dutch book against the conditionalizing strategy in the big game.

4. HIGHER ORDER PROBABILITIES

Suppose that the bookie has at t_1 a prior probability over courses of play of the observation game, and will take bets on sets of courses of play accordingly. Then we can say something about a concept of coherence weaker than that of being bulletproof. We say that the bettor's strategy constitutes an *unconditional dynamic dutch book* against that of the bookie if for *every* sequence of play the bettor ends up at time 3 with a finite number of bets whose net winning are positive for every flavor.

Then, under modest measurability conditions, Skyrms (forthcoming) shows that:

> The bookie can have an unconditional dynamic dutch book made against him if his strategy does not proceed by probability kinematics almost everywhere (in Pr_1).

The leading idea is that if the set of courses of play which violate probability kinematics has positive Pr_1, suitable side bets allow the bettor to achieve net winnings whether or not those courses of play come about.

The introduction of probabilities of probabilities also imposes additional coherence constraints. Goldstein (1983) and van Fraassen (1984) both argued that coherence requires that Pr_1 be equal to the prior expectation of Pr_2 considered as a random variable. In the terminology of this paper, Goldstein (1983) showed:

> The bookies prior, Pr_1, must be such that for any subset, Q, of the color-flavor space, and any closed interval, I, such that $Pr_1[Pr_2(Q) \text{ in } I]$ is positive, $Pr_1[Q \mid Pr_2(Q) \text{ in } I]$ in I, or the bettor can make an unconditional dynamic dutch book against him.

The expectation principle follows immediately.

There is a difference between the notions of coherence used by van Fraassen and Goldstein that makes no real difference in the context of this theorem, but which bears comment. Like Lewis and this paper, van Fraassen worked with a notion of *dynamic coherence,* where the notion of coherence attaches to rules or strategies for changing degrees of belief. Goldstein worked with a notion of temporal coherence, wherein probabilities at one time are defined on a space which has probabilities at another time as a random variable, and the explicit temporal reference carries the connection. Actually, it is not the time but the reference that is crucial to his argument, so his notion is essentially one of *higher order coherence.* It does not, therefore, apply directly to the concerns of the last section wherein no probabilities of probabilities are assumed. On the other hand, if we are allowed a rich enough probability space with enough resources to state the proposition that the rule is followed we can (as van Fraassen suggests in personal correspondence) make a side bet against that proposition. Thus, in a rich enough context, any dynamic dutch book can be transformed into a higher order dutch book. The dynamic requirement that the rule be in force is replaced with the static higher order requirement that the bookie believes that the rule is in force with probability one. If the rule is in force, the dutch book goes as in the dynamic case. If not, the bettor wins the side bet which cost him nothing.

5. MAXENT, ETC.

Other models of learning situations, which raise new questions of dynamic coherence, are possible. Consider the following "Expectation Constraint Game". As in the observation game, there is betting at three times, but at t_2 the bookie does not make an uncertain observation but rather is presented by the gamemaster with an envelope containing the specification of a constraint, CON, on his Pr_2 in the form of the expected value that Pr_2 gives some random variable. (For definiteness, suppose that the random variable is just the assignment of different integers to the members of the partition of colors in the observation game.) At t_3 the bookie learns the true value of the random variable. The bookie's strategy must make his Pr_3 a function only of the announced true value of the random variable, as before. Betting proceeds as in the observation game.

The bookie might move from Pr_1 to Pr_2 by minimizing the discrimination information:

$$\text{SUM}_i\ Pr_2(s_i)\ \log\ [Pr_2(s_i) \mid Pr_1(s_i)]$$

(where the s_i's are the points of the finite probability space) subject to the announced constraint, and to Pr_3 by conditionalization on the true color. We call this the MAXENT strategy.

Minimizing the discrimination information was suggested and investigated by Kullback and Leibler (1951), and Kullback (1959). It is a special kind of belief change by probability kinematics. It results in change by probability kinematics on the partition consisting of the inverse images of the statistic of the constraint. It is a special kind of probability kinematics because in the general case there may be several changes by probability kinematics on that partition which satisfy the constraint with different associated discrimination information values. [This is discussed by Kullback (1959), Diaconis and Zabell (1982), Skyrms (1985), van Fraassen (1980), and Williams (1980).] The question arises as to whether there is a coherence argument for MAXENT of the kind that we have for conditionalization and probability kinematics.

Without attempting a full answer to that general question, we can remark that coherence in the "Expectation Constraint Game" sketched at the beginning of this section does not require MAXENT. Bookie's strategies in the expectation constraint game *are* just special sorts of strategies in the observation game. It is shown by Skyrms (forthcoming) that in the observation game if the bookie's strategy proceeds by probability kinematics on the partition of colors and his Pr_1 is representable as a mixture of his strategy's possible Pr_2's such that each has non-zero mixing coefficient, his strategy is bulletproof [Theorem VI]. It will come as no surprise to those who have studied the relation of MAXENT to conditionalization in a larger space that there are many strategies which conflict with MAXENT and yet satisfy these conditions for coherence (Dias and Shimony, 1981; Friedman and Shimony, 1971; Shimony, 1973; Skyrms, 1985; Seidenfeld, 1986). This is not to say that there may not be more special situations and interpretations of the constraint in which MAXENT is tied closely to conditionalization and thus to dynamic coherence (Tjur, 1974; van Campenhout and Cover, 1981; Zabell, 1974).

Questions of dynamic coherence relating to the combination of evidence (e.g., Dempster, 1967, 1968; Shafer, 1976) remain to be investigated. Dempster's rule is also related to probability kinematics (Shafer, 1981). Both MAXENT and Dempster's rule have been of special interest to the artificial intelligence community as ways of avoiding a Bayesian combinatorial explosion (e.g., see Cheesman, 1983). A fuller clarification of their status *vis a vis* dynamic coherence would be of general interest.

ACKNOWLEDGMENT

This research was partially supported by NSF Grant SES-84-09898.

REFERENCES

Adams, E. (1962), "On rational betting systems". *Archive fur Mathematische Logik und Grundlagenforshung* **6**, 7–29, 112–128.

Armendt, B. (1980), "Is there a dutch book argument for probability kinematics?" *Philosophy of Science* **47**, 583–588.

Cheesman, P. (1983), "A method of computing generalized Bayesian probability values for expert systems". *Proceedings of the Eighth International Joint Conference on Artificial Intelligence* Karlsruhe, West Germany, (International Joint Conferences in Artificial Intelligence, Inc., Los Altos, California) distributed by William Kaufmann, Inc., 198–202.

Dawid, A. P., and M. Stone (1972), "Expectation consistency of inverse probability distributions". *Biometrika* **59**, 486–489.

Dawid, A. P., and M. Stone (1973), "Expectation consistency and generalized Bayes inference". *Annals of Statistics* **1**, 478–485.

de Finetti, B. (1937), "La prevision: ses lois logiques, ses sources subjectives". *Annales de l'Institut Henri Poincare* **7**, 1–68. Translated as "Foresight: its logical laws, its subjective sources" in *Studies in Subjective Probability*, ed. H. E. Kyburg, Jr. and H. Smokler (Huntington, NY: Krieger, 1980).

Dempster, A. P. (1967), "Upper and lower probabilities induced by a multivalued mapping". *Annals of Mathematical Statistics* **38**, 325–339.

Dempster, A. P. (1968), "A generalization of Bayesian inference". *Journal of the Royal Statistical Society*, Series B **30**, 205–249.

Diaconis, P., and S. Zabell (1982), "Updating subjective probability". *Journal of the American Statistical Association* **77**, 822–830.

Dias, P. M. C., and A. Shimony (1981), "A critique of Jaynes' maximum entropy principle". *Advances in Applied Mathematics* **2**, 172–211.

Freedman, D. A., and R. A. Purves (1969), "Bayes' method for bookies". *Annals of Mathematical Statistics* **40**, 1177–1186.

Friedman, K., and A. Shimony (1971), "Jaynes's maximum entropy prescription and probability theory". *Journal of Statistical Physics* **3**, 381–384.

Goldstein, M. (1983), "The prevision of a prevision". *Journal of the American Statistical Association* **78**, 817–819.

Good, I. J. (1981), "The weight of evidence provided by uncertain testimony or from an uncertain event". *Journal of Statistical Computation and Simulation* **13**, 56–60.

Hacking, I. (1967), "Slightly more realistic personal probability". *Philosophy of Science* **34**, 311–325.

Heath, D., and W. Sudderth (1972), "On a theorem of de Finetti, oddsmaking and game theory". *Annals of Mathematical Statistics* **43**, 2072–2077.

Heath, D., and W. Sudderth (1978), "On finitely additive priors, coherence, and extended admissibility". *Annals of Statistics* **6**, 333–345.

Jeffrey, R. (1965), *The Logic of Decision*, 2nd edition. New York: McGraw-Hill. (University of Chicago Press: Chicago, 1983).

Kemeny, J. (1955), "Fair bets and inductive probabilities". *Journal of Symbolic Logic* **20**, 263–273.

Kullback, S. (1959), *Information Theory and Statistics*. New York: Wiley and Sons.

Kullback, S., and R. Leibler (1951), "On information and sufficiency". *Annals of Mathematical Statistics* **22**, 79–86.

Lane, D., and W. Sudderth (1983), "Coherent and continuous inference". *The Annals of Statistics* **11**, 114–120.

Lehman, R. (1955), "On confirmation and rational betting". *Journal of Symbolic Logic* **20**, 251–262.

Ramsey, R. P. (1931), "Truth and Probability". In *The Foundations of Mathematics and Other Essays*, ed. R. B. Braithwaite. N.Y.: Harcourt Brace. Also in *Studies in Subjective Probability*, ed. H. Kyburg and H. Smokler (Huntington, N.Y.: Krieger, 1980).

Seidenfeld, Teddy (1986), "Entropy and uncertainty". Theses Proceedings.

Shafer, G. (1976), *A Mathematical Theory of Evidence*. Princeton, N.J.: Princeton University Press.

Shafer, G. (1981), "Jeffrey's rule of conditioning". *Philosophy of Science* **48**, 337–362.

Shimony, A. (1955), "Coherence and the axioms of confirmation". *Journal of Symbolic Logic* **20**, 1–28.

Shimony, A. (1973), "Comment on the interpretation of inductive probabilities". *Journal of Statistical Physics* **9**, 187–191.

Skyrms, B. (1980a) *Causal Necessity.* New Haven: Yale University Press. Appendix 2.

Skyrms, B. (1980b), "Higher order degrees of belief". In *Prospects for Pragmatism*, ed. D. H. Mellor. Cambridge, MA: Cambridge University Press.

Skyrms, B. (1983), "Zeno's paradox of measure". In *Physics, Philosophy, and Psychoanalysis*, ed. R. S. Cohen and L. Lauden, pp. 223–254. Dordrecht, Holland: Reidel.

Skyrms, B. (1984), *Pragmatics and Empiricism.* New Haven, Conn.: Yale University Press.

Skyrms, B. (1985), "Maximum entropy inference as a special case of conditionalization". *Synthese* **63**, 55–74.

Skyrms, B. (forthcoming) "Dynamic coherence and probability kinematics". *Philosophy of Science*, forthcoming.

Teller, P. (1973), "Conditionalization and observation". *Synthese* **26**, 218–258.

Teller, P. (1976), "Conditionalization, observation, and change of preference". In *Foundations of Probability Theory, Statistical Inference, and Statistical Theories of Science*, ed. W. Harper and C. Hooker, pp. 205–253. Dordrecht, Holland: D. Reidel.

Tjur, T. (1974), *Conditional Probability Distributions*, Lecture Notes 2, Institute of Mathematical Statistics, University of Copenhagen, Sections 36, 37.

van Campenhout, J., and T. Cover (1981), "Maximum entropy and conditional probability". *IEEE Transactions on Information Theory* **IT-27**, 483–489.

van Fraassen, B. (1980), "Rational belief and probability kinematics". *Philosophy of Science* **47**, 165–187.

van Fraassen, B. (1981), "A problem for relative information minimizers in probability kinematics". *British Journal for the Philosophy of Science* **32**, 375–379.

van Fraassen, B. (1984), "Belief and the will". *Journal of Philosophy* **81**, 235–256.

Williams, P. M. (1980), "Bayesian conditionalization and the principle of minimum information". *British Journal for the Philosophy of Science* **31**, 131–144.

Zabell, S. (1974), "A limit theorem for conditional expectations with applications to probability theory and statistical mechanics". Ph.D. Thesis, Harvard University.

John L. Pollock [1]

SKETCH OF THE THEORY OF NOMIC PROBABILITY

1. INTRODUCTION

It has often been supposed that probability can be used to throw light on epistemology and the structure of epistemic justification. For reasons I will not pursue here, I do not think that can work. But I do think that the converse can work. That is, by appealing to a fairly sophisticated epistemology proceeding in terms of prima facie reasons and defeaters, one can make sense of objective probability concepts. The details of this are presented in a number of recent articles, and will be presented in full in a future book. My purpose here is to give a brief sketch of the theory of nomic probability. (See "Notes" at end of paper.)

Nomic probability is the kind of probability involved in statistical laws of nature. Nomic probabilities relate properties. I propose to think of the nomic probability of an F being a G ($\text{prob}(G|F)$) as the proportion of physically possible F's that would be G's. The theory of nomic probability builds upon this idea. The theory has two primitive parts and two derived parts. The primitive parts consist of an acceptance rule and a set of computational principles comprising a strengthened version of the probability calculus. The derived parts consist of a theory of direct inference, and a theory of induction.

2. AN ACCEPTANCE RULE

Rules telling us when it is rational to believe something on the basis of high probability are called *acceptance rules*. The acceptance rule upon which the theory of nomic probability is based is the following:

(A1) If F is projectible with respect to G and $r \geq .5$, then $\ulcorner G\underline{a}$ & $\text{prob}(F|G) > r\urcorner$ is a prima facie reason for believing $\ulcorner F\underline{a}\urcorner$, the strength of the reason depending upon the value of r.

[1] Department of Philosophy, University of Arizona, Tucson, Arizona 85721

I. B. MacNeill and G. J. Umphrey (eds.), Foundations of Statistical Inference, 245–257.

It is the projectibility constraint together with the fact that (A1) provides us with only a prima facie (or defeasible) reason that enables it to avoid the lottery paradox. Projectibility, as it is used in (A1), is the standard notion of projectibility occurring in discussions of induction. To illustrate how (A1) handles the lottery paradox, suppose we have a fair lottery consisting of one million tickets. The probability that a draw in such a lottery will be a draw of ticket n is .000001, so we have a prima facie reason for believing of each ticket that it will not be drawn. But the collection of all such conclusions is incompatible with what we already know, namely, that some ticket will be drawn. Thus, we do not want to be able to conclude of each ticket that it will not be drawn, and as there is nothing to favor one ticket over any of the others, we do not want to be able to conclude of any ticket that it will not be drawn. (A1) handles this properly by virtue of the fact that the reason it provides is only prima facie. We have a prima facie reason for believing that ticket n will not be drawn. But for each $m \neq n$ we have an equally good prima facie reason for believing that ticket m will not be drawn. This provides us with a counterargument supporting the conclusion that ticket n will be drawn:

Argument #1:

Ticket 1 will not be drawn.

Ticket 2 will not be drawn.

⋮

Ticket $n - 1$ will not be drawn.

Ticket $n + 1$ will not be drawn.

⋮

Some ticket will be drawn.

Therefore, ticket n will be drawn.

For each of the premises of this argument, our reason for believing it is as good as our reason for believing that ticket n will not be drawn, and an argument is as good as its weakest link, so this argument provides us with a reason for believing that ticket n will be drawn that is as good as our reason for believing that ticket n will not be drawn. Both reasons are prima facie reasons, so they simply defeat one another, leaving us with no justified conclusion about whether ticket n will be drawn.

It turns out that without the projectibility constraint in (A1), all cases of high probability could be given the structure of the lottery paradox, with the result that every use of (A1) would be defeated and the rule would be useless. Basically, the projectibility constraint is required to rule out arbitrary disjunctions. But it is a bit surprising that the required constraint should be one of projectibility. Projectibility has to do with induction. Why should it have anything to do with acceptance rules? As I have indicated, the theory of nomic probability contains a solution to the problem of induction. This solution results from the fact that principles of induction are derivable from the theory of nomic probability. (A1) plays a central role in that derivation, and the nature of the derivation is such that the constraints on the acceptance rule are inherited by the principles of induction. By definition, the proper constraint on induction is projectibility, so that must also be the constraint on the acceptance rule. In other words, the role of projectibility in induction is entirely derivative. Projectibility has first and foremost to do with acceptance rules, and it is only because of its role there that it becomes a constraint on induction.

The reason provided by (A1) is only a prima facie reason, and as such it is defeasible. As with any prima facie reason, it can be defeated by having a stronger reason for denying the conclusion. A reason for denying the conclusion constitutes a *rebutting defeater*. But prima facie reasons can also be defeated by defeaters attacking the connection between the reason and the conclusion rather than attacking the conclusion itself. These are *undercutting defeaters*. There is an important kind of undercutting defeater for (A1). In (A1), we infer the truth of $\ulcorner F\underline{a}\urcorner$ on the basis of probabilities conditional on a limited set of purported facts about $\underline{a}$. But if we know additional facts about $\underline{a}$ that lower the probability, that defeats the prima facie reason:

(D) If H is such that F is projectible with respect to $(G\&H)$ then $\ulcorner H\underline{a}$ & $\text{prob}(F|G\&H) < \text{prob}(F|G)\urcorner$ is an undercutting defeater for (A1).

This amounts to a kind of "total evidence requirement". It requires us to make our inference on the basis of the most comprehensive facts regarding which we know the requisite probabilities.

Given certain assumptions about the way in which possibly conflicting prima facie reasons interact to determine which beliefs are justified, it can be shown that (A1) is equivalent to the following acceptance rule, which looks initially stronger:

(A3) If F is projectible with respect to G then $\ulcorner \text{prob}(F|G) > r\urcorner$ is a prima facie reason for the conditional $\ulcorner G\underline{a} \supset F\underline{a}\urcorner$, the strength of the reason depending upon the value of r.

The defeater formulated in (D) is then a defeater for (A3) as well.

I take it that (A3) is actually quite an intuitive acceptance rule. It amounts to a rule saying that, when F is projectible with respect to G, if we know that most G's are F, that gives us a reason for thinking of any particular object that it is an F if it is a G. The only surprising feature of this rule is the projectibility constraint. (A3) is the basic epistemic principle from which all the rest of the theory of nomic probability is derived.

3. COMPUTATIONAL PRINCIPLES

Nomic probabilities are what I call *indefinite probabilities*. They relate properties. *Definite probabilities*, on the other hand, attach to propositions or states of affairs. Definite probabilities are sometimes called "single case probabilities". The fact that nomic probabilities attach to properties, including relational properties of arbitrarily many places, introduces logical relationships into the theory of nomic probability that are ignored in the classical probability calculus. One simple example is the principle:

(1) $\text{prob}(Axy|Rxy \;\&\; y = \underline{a}) = \text{prob}(Ax\underline{a}|Rx\underline{a})$.

This is an essentially relational principle and is not a theorem of the classical probability calculus, but it turns out to be true for nomic probabilities. The reason there can be general truths regarding nomic probability that are not theorems of the classical probability calculus is that, historically, the probability calculus was devised with definite probabilities in mind. The relationship between the calculus of indefinite probabilities and the calculus of definite probabilities is a bit like the relationship between the predicate calculus and the propositional calculus. Specifically, there are principles regarding relations and quantifiers that must be added to the classical probability calculus to obtain a reasonable calculus of nomic probabilities. To elicit these principles, I propose to begin by taking seriously (or at least semi-seriously) the explanation of nomic probability as a measure of the proportion of physically possible things of one sort that would be of another sort. This will not yield a logical analysis of nomic probability because in the end proportions turn out to be explained in terms of nomic probabilities. Nevertheless, treating probabilities in terms of proportions proves to be a useful approach for investigating the logical and mathematical structure of nomic probability.

Where A and B are sets, we can talk about "the proportion of members of B that are in A", $\text{p}(A|B)$. The proportion function is a kind of "relative measure function" in the mathematical sense of "measure function". It measures the size of one set relative to the size of another. Where F is a property, I propose to understand a physically possible F to be an ordered pair $\langle w, x\rangle$ such that w is a physically possible world (i.e., one having the same physical laws as the actual world) and x is an F at w. I suggest that (at least in the non-counterlegal case) we identify nomic probabilities with proportions as in:

(2) if $\underset{p}{\diamond}(\exists x)Gx$ then $\text{prob}(F|G) = \mathbf{p}(\mathbf{F}|\mathbf{G})$; where

$\mathbf{F} = \{\langle w, x\rangle \mid w$ is a physically possible world & x is F at $w\}$ and

$\mathbf{G} = \{\langle w, x\rangle \mid w$ is a physically possible world & x is G at $w\}$.

Some people will balk at the ontology of possible worlds and possible objects involved in (2), but this need not be taken entirely seriously. If you want, you can regard it as just a way of getting the mathematical structure right. That is the only role it plays in the theory.

Time precludes my going into the details of the theory of proportions here, so instead let me just describe the most important dividend of this approach. Naturally, we obtain versions of the standard axioms for the probability calculus. But in addition we obtain a very powerful theorem that plays a central role in founding both direct inference and induction. This is "the principle of agreement", which I will now explain. First take note of a combinatorial fact that tends to be quite surprising to the uninitiated in probability theory. Consider the proportion of members of a finite set B that are in some subset A of B. Subsets of B need not exhibit the same proportion of A's as does B itself, but it is a striking fact of set theory that subsets of B *tend* to exhibit *approximately* the same proportion of A's as does B, and both the strength of the tendency and the degree of approximation improve as the size of B increases. More precisely, symbolizing

(3) $x \underset{\delta}{\approx} y$ iff the difference between x and y is less than δ,

the following is a theorem of set theory:

(4) For every $\delta, \gamma > 0$, there is an n such that if B is a finite set containing at least n members then $\text{freq}[\text{freq}[A|X] \underset{\delta}{\approx} \text{freq}\,[A|B]|X \subseteq B] > 1-\gamma$. (The frequency with which the frequency of A's in subsets of B agrees to any given degree with the frequency of A in B itself goes to 1 as B gets larger.)

This suggests that when B becomes infinite, the proportion of subsets agreeing with B to any given degree of approximation should become 1. That is what I call "the principle of agreement for proportions". More accurately:

(5) If B is infinite and $\mathbf{p}(A|B) = p$ then for every $\delta > 0$:

$\mathbf{p}[\mathbf{p}(A|X) \underset{\delta}{\approx} p|X \subseteq B] = 1$ (the proportion of subsets of B for which the proportion of A's is within δ of p is 1).

This is a theorem of the theory of proportions upon which the theory of nomic probabilities is based. It implies an analogous principle regarding nomic probabilities. Given a property G, a *subproperty* of G is any stronger property that is not "counterlegal". Where "$\underset{p}{\diamond}$" and "$\underset{p}{\Box}$" symbolize physical possibility and physical necessity:

(6) $H \lesssim G$ iff $\underset{p}{\diamond}(\exists x)Hx$ & $\underset{p}{\Box}(\forall x)(Hx \subset Gx)$.

Then *The Principle of Agreement* is:

(AGREE) If F and G are properties and there are infinitely many physically possible G's and $\text{prob}(F|G) = p$ then for every $\delta > 0$:

$\text{prob}[\text{prob}(F|X) \underset{\delta}{\approx} p | X \lesssim G] = 1.$

(The probability is 1 that the probability of F on an arbitrary subproperty of G will be within δ of p).

In other words, the probability is 1 of a subproperty of G agreeing (in probability) arbitrarily closely with G itself. This principle plays a pivotal role in providing foundations for direct inference and in the justification for induction. It is the most important formal principle of nomic probability.

Although the calculus of nomic probabilities is stronger than the standard probability calculus in a number of important respects, it is still just a calculus of probabilities. From a philosophical point of view, these computational principles are not particularly weighty. It is noteworthy, then, that we need no assumptions specifically about probability beyond these computational axioms and the acceptance rule (A3) to found the entire theory of nomic probability, including the derivation of principles of induction.

4. DIRECT INFERENCE AND DEFINITE PROBABILITIES

The problem of direct inference is the problem of obtaining definite probabilities from indefinite probabilities. I will call this "classical direct inference". The basic idea behind standard theories of classical direct inference was first articulated by Hans Reichenbach: in determining the probability that an individual $\underline{a}$ has a property F, we find the narrowest reference class X for which we have reliable statistics and then infer that $\text{PROB}(F\underline{a}) = \text{prob}(Fx | x \in X)$. There is almost universal agreement that direct inference is based upon some such principle as this, but there is little agreement about the precise form the theory should take. It is interesting, then, that a theory of classical direct inference can be derived as a set of theorems in the theory of nomic probability. It is derived from the theory of what I call *non-classical direct inference*, which concerns inferences from indefinite probabilities to indefinite probabilities. The basic principle of non-classical direct inference is:

(DI) If F is projectible with respect to G then $\ulcorner H \lesssim G \;\&\; \text{prob}(F|G) = r \urcorner$ is a prima facie reason for $\ulcorner \text{prob}(F|H) = r \urcorner$.

(DI) amounts to a kind of principle of insufficient reason. (DI) tells us that if we have no reason to think otherwise, it is reasonable for us to anticipate that strengthening G will not affect the probability of F. (DI) is derived from the principle of agreement and the acceptance rule (A3). We have the following instance of (A3):

(7) If $\ulcorner \mathrm{prob}(F|X) \underset{\delta}{\approx} r \urcorner$ is projectible with respect to $\ulcorner X \lesssim G \urcorner$ then $\ulcorner \mathrm{prob}[\, \mathrm{prob}(F|X) \underset{\delta}{\approx} r | X$ scriptstyle$\lesssim G] = 1 \urcorner$ is a prima facie reason for $\ulcorner H \lesssim G \supset \mathrm{prob}(F|H) \underset{\delta}{\approx} r \urcorner$.

If we assume that the antecedent of (7) holds whenever F is projectible with respect to G, then it follows that:

(8) If F is projectible with respect to G then $\ulcorner \mathrm{prob}[\, \mathrm{prob}(F|X) \underset{\delta}{\approx} r | X \lesssim G] = 1 \urcorner$ is a prima facie reason for $\ulcorner H \lesssim G \supset \mathrm{prob}(F|H) \underset{\delta}{\approx} r \urcorner$.

By (AGREE),

(9) For each $\delta > 0$, $\ulcorner \exists_\infty G$ & $\mathrm{prob}(F|G) = r \urcorner$ entails $\ulcorner \mathrm{prob}[\, \mathrm{prob}(F| X) \underset{\delta}{\approx} r | X \lesssim G] = 1 \urcorner$ (where $\ulcorner \exists_\infty G \urcorner$ means $\ulcorner$There are infinitely many physically possible $G's \urcorner$)

so it follows that:

(10) If F is projectible with respect to G then for each $\delta > 0$, $\ulcorner \exists_\infty G$ & $\mathrm{prob}(F|G) = r$ & $H \lesssim G \urcorner$ is a prima facie reason for $\ulcorner \mathrm{prob}(F|H) \underset{\delta}{\approx} r \urcorner$.

Consider the requirement in (10) that $\exists_\infty G$. To require that there are infinitely many physically possible G's is to require very little more than that it is physically possible for there to be G's. It is extremely difficult to construct properties G that are not counterlegal but are such that there are only finitely many physically possible G's. In this connection, recall that a physically possible G is not just a physically possible object that is G in some world. Rather, a physically possible G is a pair $\langle w, x\rangle$ such that w is a physically possible world and x is G at w. Consequently, for there to be infinitely many physically possible G's it suffices for there to be infinitely many physically possible worlds at which there are G's. Any reasonable non-counterlegal property will satisfy this condition. Even if G is a property like that of *being Bertrand Russell*, which can only be possessed by a single object, there are infinitely many physically possible worlds at which Bertrand Russell has the property of being Bertrand Russell. It appears that the only way there can be physically possible G's but only finitely many of them is for G to be a very contrived property. For example, picking some particular possible world α, we might consider the property of *being Bertrand Russell and such that α is the actual world*. This peculiar property can only be possessed by Bertrand Russell, and it can only be possessed by him at the world

α. But any normal noncounterlegal property G will be such that there are infinitely many physically possible G's. In particular, any noncounterlegal "qualitative" property will satisfy this condition. Specifically, I will assume that any noncounterlegal projectible property satisfies this condition. It follows that if we are warranted in believing $\ulcorner \underset{p}{\diamond} \exists G \urcorner$ then we are warranted in believing $\ulcorner \exists_{\infty} G \urcorner$. But $\ulcorner H \lesssim G \urcorner$ entails $\ulcorner \underset{p}{\diamond} \exists G \urcorner$, so (10) can be simplified as follows:

(11) If F is projectible with respect to G then for each $\delta > 0$, $\ulcorner \text{prob}(F|G) = r \,\&H \lesssim G \urcorner$ is a prima facie reason for $\ulcorner \text{prob}(F|H) \underset{\delta}{\approx} r \urcorner$.

This is quite close to the principle (DI) of nonclassical direct inference, and (DI) can be derived from it by using some technical results that pertain to prima facie reasons. The idea is that we have a prima facie reason for believing that the probability lies within δ of p for every δ, and that gives us a prima facie reason for believing that it is exactly p. Note how the projectibility constraint in (DI) arises out of the projectibility constraint in (A3).

Intuitively, it seems clear that if $H \lesssim J \lesssim G$ and we know that $\text{prob}(F|J) \neq \text{prob}(F|G)$, that should preclude our using (D1) to infer that $\text{prob}(F|G) = \text{prob}(F|H)$. This is *the principle of subset defeat*:

(SD) If F is projectible with respect to J then $\ulcorner H \lesssim J \lesssim G \;\&\; \text{prob}(F|J) \neq \text{prob}(F|G) \urcorner$ is an undercutting defeater for (D1).

(SD) is derivable from the theory of nomic probability much as (DI) was. The essential ingredient in its derivation is the defeater (D) for the acceptance rule (A3).

I have been talking about nonclassical direct inference. Direct inference has traditionally been identified with classical direct inference, but I believe that it is most fundamentally nonclassical direct inference. If we let K be the conjunction of all the propositions we are warranted in believing, my proposal is that we can identify the definite probability $\text{PROB}(F|\underline{a})$ with the following indefinite probability:

(12) $\text{PROB}(F\underline{a}) = \text{prob}[Fx|x = \underline{a} \;\&\; K]$.

This is an indefinite probability, but a degenerate one. Philosophers have resisted defining definite probabilities in this simple manner because they have supposed probabilities like $\text{prob}[Fx|x = \underline{a} \;\&\; K]$ to be illegitimate on one of two grounds. It might be objected that such a probability must be either 1 or 0, because the reference class is a unit set. This objection is dispelled by taking direct inference to pertain to nomic probabilities rather than relative frequencies. The nomic probability need not be either 1 or 0, because it averages the behavior of $\underline{a}$ over many different possible worlds. A second objection is that there is no way we could know the value of $\text{prob}[Fx|x = \underline{a} \;\&\; K]$. Our basic way of knowing the values of nomic probabilities is by some kind of statistical induction. It is only possible to ascertain the value of $\text{prob}(F|G)$ inductively if the extension of G is large so that we can compile reliable statistics about the proportion of G's that are F. On the supposition that the only way to ascertain the value of a nomic probability is inductively, it follows that there is no way to ascertain the value of $\text{prob}[Fx|x = \underline{a}\&K]$. However, that supposition is mistaken. Nonclassical direct inference provides another way of ascertaining the values of nomic probabilities. Induction and direct inference jointly provide the logical or epistemological machinery for dealing with nomic probabilities. By induction, we learn the values of certain nomic probabilities, and then by direct inference we infer the values of others.

Given the reduction of definite probabilities to nomic probabilities, it becomes possible to derive principles of classical direct inference from the theory of nonclassical direct inference. In particular, we obtain a precise version of Reichenbach's theory, consisting most basically of the following two principles (where $\ulcorner W\phi \urcorner$ abbreviates $\ulcorner$We are warranted in believing $\phi\urcorner$):

(CDI) If F is projectible with respect to G then $\ulcorner \text{prob}(F|G) = r \;\&\; W(G\underline{a}) \;\&\; W(P \equiv F\underline{a})\urcorner$ is a prima facie reason for $\ulcorner \text{PROB}(P) = r\urcorner$.

(CSD) If F is projectible with respect to H then $\ulcorner \text{prob}(F|H) \neq \text{prob}(F|G) \;\&\; W(H\underline{a}) \;\&\; \Box\forall(H \supset G)\urcorner$ is an undercutting defeater for (CDI).

Principle (CSD) is called "the principle of subset defeat", because it says that probabilities based upon more specific information take precedence over those based upon less specific information. Note the projectibility constraint in these rules. I originally proposed that constraint in an *ad hoc* way in order to avoid various paradoxes of direct inference, but just as in the case of our acceptance rules, it is initially puzzling that there should be such a constraint. It is illuminating that it falls right out of the derivation of these principles.

To illustrate this account of classical direct inference, suppose we know that Herman is a 40 year old resident of the United States who smokes. Suppose we also know that the probability of a 40 year old resident of the United States having lung cancer is .1, but the probability of a 40 year old smoker who resides in the United States having lung cancer is .2. Intuitively, if we know nothing else that is relevant we will infer that the probability of Herman having lung cancer is .2. (CDI) provides us with one prima facie reason for inferring that the probability is .1 and a second prima facie reason for inferring that the probability is .2. However, the latter prima facie reason is based upon more specific information, and so by (CSD) it takes precedence, defeating the first prima facie reason and leaving us justified in inferring that the probability is .2.

(CDI) and (CSD) are the basic principles of classical direct inference, but they do not comprise a complete basis for the theory. I have argued elsewhere that there are a number of additional sources of defeat for (CDI), but these are also theorems of the theory of nomic probability.

5. INDUCTION

Our most fundamental source of knowledge regarding the values of nomic probabilities is *statistical induction*. In statistical induction, we observe the relative frequency of F's in a sample of G's, and then infer that $\text{prob}(F|G)$ is approximately equal to that relative frequency. One of the most interesting features of the theory of nomic probability is that precise principles of induction can be derived from (and hence justified on the basis of) the acceptance rules and computational principles that I have already endorsed.

The justification of statistical induction proceeds in terms of what I call *the statistical induction argument*. The details of this argument are complicated, but I will try to give you some feel for the way the argument goes. The simplest way to do that is to consider a special case of statistical induction. In a normal case, $\text{prob}(F|G)$ could have any value from 0 to 1. The fact that there are infinitely many possible values makes the argument complicated. For present purposes, let us suppose that we know somehow that $\text{prob}(F|G)$ has one of a finite set of values $p_1, \ldots, p_k$. Suppose we have observed a sample of n G's and noted that r of them are F's (where F and $\sim F$ are projectible with respect to G). Then the relative frequency of F's in the sample is r/n. From this we want to infer that $\text{prob}(F|G)$ is approximately r/n. Our reasoning proceeds in two stages, the first stage employing the theory of nonclassical direct inference, and the second stage employing the acceptance rule (A3).

Stage I

Let us abbreviate

(13) $x_1, \ldots, x_n$ are distinct & Gx_1 & ... & Gx_n & $\text{prob}(F|G) = p$

as $\ulcorner \theta_p \urcorner$. Nonclassical direct inference gives us a reason for believing that

(14) $\text{prob}[Fx_i | Fx_{i+1} \ \& \ \ldots \ \& \ Fx_r \ \& \ \sim Fx_{r+1} \ \& \ \ldots \ \& \sim Fx_n \ \& \ \theta_p] = \text{prob}[Fx_i | Gx_i \ \& \ \text{prob}(F|G) = p]$,

The following is a theorem of the calculus of nomic probabilities:

(PPROB) If $\diamond[(\exists x)Gx \ \& \ \text{prob}(F|G) = r]$ then $\text{prob}[F|G \ \& \ \text{prob}(F|G) = p] = p$.

Thus the probability in (14) equals p. Similarly, nonclassical direct inference gives us a reason for believing:

(15) $\text{prob}[\sim Fx_i | \sim Fx_{i+1} \ \& \ \ldots \ \& \ \sim Fx_n \ \& \ \theta_p] = 1 - p$.

Standard calculations in the probability calculus then allow us to compute:

$$(16) \qquad \text{prob}\left[\text{freq}[F|\{x_1, \ldots, x_n\}] \neq \frac{r}{n} | \theta_p\right] = 1 - \frac{n!p^r(1-p)^{n-r}}{r!(n-r)!}.$$

The reasoning involved in this calculation is parallel to some standard and quite familiar statistical reasoning, but that reasoning normally requires an assumption of statistical independence, which is a substantive probabilistic assumption. In effect, nonclassical direct inference gives us a prima facie defense of that assumption without requiring us to already have contingent knowledge about probabilities.

Stage II

I assume at this point that if F is projectible with respect to G then the probability expressed by (16) satisfies the projectibility constraint of our acceptance rule. For sizable samples, this probability is almost always close to 1. For instance, if $n = 50$, $r = 20$, and $p = .5$, the computed probability is .96. Thus by (A3) and (16), for each choice of p we have a prima facie reason for believing

(17) if our sample X consists of n G's and $\text{prob}(F|G) = p$ then $\text{freq}[F|X] \neq r/n$.

As we know that our sample does consist of n G's and $\text{freq}[F|X] = \frac{r}{n}$, this gives us a prima facie reason for believing that $\text{prob}(F|G) \neq p$. But we know that for some one of $p_1, \ldots, p_k$, $\text{prob}(F|G) = p_i$. This is much like the case of the lottery paradox. For each i we have a prima facie reason for believing that $\text{prob}(F|G) \neq p_i$, but we also have a counterargument for the conclusion that $\text{prob}(F|G) = p_i$, namely:

Argument #2:

$\text{prob}(F|G) \neq p_1$

$\text{prob}(F|G) \neq p_2$

$\vdots$

$\text{prob}(F|G) \neq p_{i-1}$

$\text{prob}(F|G) \neq p_{i+1}$

$\vdots$

$\text{prob}(F|G) \neq p_k$

For some j between 1 and k, $\text{prob}(F|G) = p_j$.

Therefore, $\text{prob}(F|G) = p_i$.

Notice that this is exactly parallel to Argument #1. There is, however, an important difference between the present case and the case of a fair lottery. For each i, we have a prima facie reason for believing that $\text{prob}(F|G) \neq p_i$, but these reasons are not all of the same strength because the probabilities assigned by (16) differ for the different p_i's. The counterargument is only as good as its weakest link, so the possibility arises that for some of the p_i's, the counterargument will not be strong enough to defeat the prima facie reason for believing that $\text{prob}(F|G) \neq p_i$. Thus, there may be a subset of $\{p_1, \ldots, p_i\}$—the *rejection class*—such that we will be justified in concluding that for each p in the rejection class, $\text{prob}(F|G) \neq p$, and hence that $\text{prob}(F|G)$ is not in the rejection class. Therefore, we are justified in believing that $\text{prob}(F|G)$ is in the complement of the rejection class, which I will call *the acceptance class*. The acceptance class will consist of those p_i's closest in value to r/n. Thus, we can think of the acceptance class as an interval around the observed frequency such that we are justified in believing that $\text{prob}(F|G)$ lies in that interval.

Only the discrete case in which we know that prob $(F|G)$ has one of a finite set of values has been considered here, but the argument can be generalized to apply to the continuous case as well. In the continuous case, the acceptance class is an interval, so the argument provides us with justification for believing that $\text{prob}(F|G)$ lies in a precisely defined interval around the observed relative frequency, the width of the interval being a function of the degree of justification. This can be made precise and the interval can be calculated exactly. As the size of the sample increases, the width of the interval shrinks dramatically. This is illustrated by Table 1.

Table 1. *The acceptance interval for relative frequency .5, as a function of size n of the sample and a parameter α measuring the degree of justification.*

		n: 10	10^2	10^3	10^4	10^5	10^6
	.1	[.196, .804]	[.393, .607]	[.466, .534]	[.489, .511]	[.496, .504]	[.498, .502]
α	.01	[.112, .888]	[.351, .649]	[.452, .548]	[.484, .516]	[.495, .505]	[.498, .502]
	.001	[.068, .932]	[.320, .680]	[.441, .559]	[.481, .519]	[.494, .506]	[.498, .502]

The statistical induction argument makes precise the way in which observation of the relative frequency of F's in our sample justifies us in thinking that $\text{prob}(F|G)$ is approximately the same as that relative frequency. This conclusion has been derived from the general theory of nomic probability, thus justifying statistical induction on the basis of that more general theory. The conclusions reached by this reasoning bear obvious similarities to the conclusions of orthodox statistical reasoning. What is novel about the present account is not the conclusions drawn but the arguments advanced on their behalf. This is, in effect, a logical defense of standard statistical intuitions.

NOTES

The articles are "Epistemology and probability," *Synthese* **55** (1983), 231–252; "Nomic probability," *Midwest Studies in Philosophy* **9** (1984), 177–204; "A solution to the problem of induction," *Nous* **18** (1984), 423–462; "Foundations for direct inference," *Theory and Decision* **17** (1984), 221–256; "The paradox of the preface," *Philosophy of Science*, forthcoming; "Probability and Proportions," in *Theory and Decision: Essays in Honor of Werner Leinfellner*, ed. H. Berghel and G. Eberlein, Reidel, Dordrecht, Holland.

Teddy Seidenfeld [1]

ENTROPY AND UNCERTAINTY

ABSTRACT

This essay is, primarily, a discussion of four results about the principle of maximizing entropy (MAXENT) and its connections with Bayesian theory. Result 1 provides a restricted equivalence between the two where the Bayesian model for MAXENT inference uses an *a priori* probability that is uniform, and where all MAXENT constraints are limited to 0–1 expectations for simple indicator-variables. The other three results report on an inability to extend the equivalence beyond these specialized constraints. Result 2 establishes a sensitivity of MAXENT inference to the choice of the algebra of possibilities even though all empirical constraints imposed on the MAXENT solution are satisfied in each measure space considered. The resulting MAXENT distribution is not invariant over the choice of measure space. Thus, old and familiar problems with the Laplacean principle of Insufficient Reason also plague MAXENT theory. Result 3 builds upon the findings of Friedman and Shimony (1971, 1973) and demonstrates the absence of an exchangeable, Bayesian model for predictive MAXENT distributions when the MAXENT constraints are interpreted according to Jaynes' (1978) prescription for his (1963) Brandeis Dice problem. Last, Result 4 generalizes the Friedman and Shimony objection to cross-entropy (Kullback-information) shifts subject to a constraint of a new odds-ratio for two disjoint events.

1. INTRODUCTION

Thirty-six years after Shannon (1948) and Wiener (1948) introduced their now familiar expression for the uncertainty captured in a probability distribution, entropy formalism is a thriving enterprise. Its advocates find applications in diverse settings, including problems of image restoration

[1] Department of Philosophy, Carnegie-Mellon University, Pittsburgh, Pennsylvania 15213

I. B. MacNeill and G. J. Umphrey (eds.), Foundations of Statistical Inference, 259–287.

(Frieden, 1972) and estimating missing proportions in contingency tables for socio-economic survey data (Denzau *et al.*, 1984). But I doubt there is a more staunch defender of the generality of entropy as a basis for quantifying (probabilistic) uncertainty than the physicist E. T. Jaynes.

Almost thirty years ago, Jaynes (1957) offered his celebrated papers on "Information Theory and Statistical Mechanics." There he argued that statistical mechanics is best understood as an instance of "inference," subject to inductive principles for maximizing uncertainty (measured by entropy), rather than as a "physical theory" in which, for example, the results of ergodic theory depend upon equations of motion and suspect assumptions about appropriateness of time-intervals (for use in identifying time frequencies and phase averages). In one fell-swoop Jaynes' approach reproduced a host of computational rules for determining statistical distributions, grounded on a simple rule for maximizing entropy. The conceptual innovation was to give this rule a wide scope, elevating it to a principle of inductive logic for assigning (subjective) probabilities in an observer invariant (objective) fashion. Investigators holding the same "evidence" agree in their determination of probabilities, provided they adhere to Jaynes' program for selecting a probability distribution which maximizes entropy subject to the constraints of the shared "evidence".

Consider a simple illustration, used by Jaynes (1963) in his Brandeis Lectures. Suppose we are faced with an ordinary six-sided die whose "bias" is stipulated to constrain our expectation for the next roll:

$$E[\text{ number of spots on next roll }] = 3.5. \tag{1}$$

The problem is to determine a (subjective) probability distribution for the set $X = \{1, \ldots, 6\}$ of possible outcomes. Shannon's formula for the uncertainty (entropy) in a discrete distribution (over n-states) is:

$$U_S = -\sum_{i=1}^{n} p_i \cdot \log(p_i). \tag{2}$$

Jaynes' principle of Maximizing Entropy (MAXENT) directs us to choose that distribution over $X(p_i \geq 0, \sum_i p_i = 1)$ which maximizes (2) subject to the constraint (1). That is, from among those distributions satisfying:

$$\sum_{i=1}^{6} i \cdot p(i) = 3.5,$$

maximize uncertainty. The solution is the uniform distribution, $p(i) =$

$1/6$ ($i = 1, \ldots, 6$).[2] If, instead the constraint specifies

$$E[\text{ number of spots on next roll }] = 4.5 \tag{3}$$

instead of the value 3.5 (for a fair die), the MAXENT solution (Jaynes, 1978) is (to five places):

$$\{p_1, \ldots, p_6\} = \{.05435, .07877, .11416, .16545, .23977, .34749\}. \tag{4}$$

Note that in (4) the probabilities are shifted away from the uniform distribution to lie on a smooth (convex) curve, increasing (decreasing) in p_i whenever the constraint fixes an expectation greater than (less than) 3.5—corresponding to the uniform distribution.

Why does Jaynes find the MAXENT principle compelling? Why should a rational person pick the uniform distribution from among the continuum of distributions satisfying (1), or choose the distribution (4) from among the continuum of distributions satisfying (3)? I can identify five reasons proposed by various authors:

(i) A pragmatic justification—in an impressive variety of empirical problems, researchers find MAXENT solutions useful (see Frieden, 1984).

(ii) An argument for the long-run—asymptotically, a MAXENT distribution is the focus of concentration among all distributions satisfying the given constraints. That is, if we use entropy to gauge "distance" between distributions, asymptotically, the class of distributions satisfying the given constraints concentrate sharply about the MAXENT solution (see Jaynes, 1979).

(iii) An *a priori* analysis—MAXENT is justified by axiomatic considerations of (necessary) conditions for representing uncertainty (see Shore and Johnson, 1980, 1981).

(iv) A defense of MAXENT through Insufficient Reason—MAXENT provides a consistent form of the Laplacean principle of Insufficient Reason; hence, it helps rehabilitate the classical interpretation of probability (see Jaynes, 1978).

(v) MAXENT justified as an extension of Bayesian theory—the Bayesian program for representing degrees of belief by probabilities and "updating" these through conditional probability (as regulated by Bayes' theorem) is a special case of MAXENT inference (see Jaynes, 1968, 1978, and 1981; Rosenkrantz, 1977; Williams, 1980).

---

[2] In this paper, footnotes from 2 on are found in a separate section before the references.

Not all who have examined these supporting arguments find them convincing. (See especially Dias and Shimony, 1981; Frieden, 1984; Friedman and Shimony, 1971; Rowlinson, 1970; Shimony, 1973. Jaynes offers selected rebuttal in (1978).) In what follows I present concerns I have primarily with the third, fourth, and fifth claims (above). I fear MAXENT is not as attractive as the advertising suggests. In particular, my doubts center on the assertion that MAXENT avoids the conceptual difficulties which plague simpler versions of Insufficient Reason. (This is discussed in Section 3. See, also my (1979).) A related argument (given in Section 4) undercuts the allegation that canonical applications of MAXENT have Bayesian models; in fact, it shows that all but the most trivial applications of MAXENT are un-Bayesian. Hence, there is solid ground for disputing the fifth claim (above). All of this is previewed in the discussion (Section 2.1) of the relation between Bayesian "conditionalization" and shifts which minimize changes in entropy—connected with an evaluation of claim (iii).

The scope of a single essay is insufficient also to address the first two arguments (justifications (i) and (ii)) in the detail they deserve. A pragmatic appeal to successful applications of MAXENT formalism cannot be dismissed lightly. The objections to MAXENT which I raise in this paper are general. Whether (and if so, how) the researchers who apply MAXENT avoid these difficulties remains an open question. Perhaps, by appeal to extra, discipline-specific assumptions, they find ways to resolve the conflicts within MAXENT theory. A case-by-case examination is called for.

Justification (ii) introduces a family of issues separate from those relevant to concerns (iii)–(v): when do asymptotic properties of an inductive principle warrant its use in the short run too? I offer some reflections on the "concentration" theorem in Section 5.

The reader will observe that throughout this essay I rely on Jaynes' prescriptions for the application and interpretation of the MAXENT formalism. Of course, my intent is to ask serious questions, not to hunt out minor inconsistencies in a scholar's writings spanning thirty years of active work. That is, I take Jaynes' papers on MAXENT to be the most thorough account available.

2. AXIOMATIC PROPERTIES CHARACTERIZING MAXENT AND ITS GENERALIZATION THROUGH KULLBACK-LEIBLER CROSS-ENTROPY.

2.1

Shannon (1948) proved an elegant uniqueness theorem establishing that U_S (2) is characterized by three simple properties:

(S_1) U_S is a continuous function of the p_i's.

(S_2) When $P = \{1/n, \ldots, 1/n\}$ is the uniform distribution on n-states, U_S is monotonically increasing in n, the number of states over which one is uncertain.

(S_3) U_S is additive over decomposition of the sample space of possible outcomes. That is, let $\Omega = \{s_1, \ldots, s_n\}$ be the set of (n) possible outcomes, and let Ω be partitioned into $m \leq n$ disjoint subsets $\Omega' = \{r_1, \ldots, r_m\}$, with r_i a subset of Ω. If P is a probability distribution over Ω, P' the corresponding distribution over Ω', and $P(\cdot \mid r_i)$ the conditional distribution (over Ω) given r_i , then:

$$U_S(P) = U_S(P') + \sum_{i=1}^{m} P'_i \cdot U_S(P(\cdot \mid r_i)). \tag{5}$$

A few remarks remind the reader why these three conditions are important for the MAXENT program. (S_1) is a structural assumption that guarantees MAXENT distributions shift smoothly with smooth changes in constraints. (S_2) is important since the uniform distribution $p_i = 1/n$ $(i = 1, \ldots, n)$ maximizes entropy over all distributions on n-states. Hence, (S_2) assures that, subject to MAXENT, uncertainty increases with the number of possibilities about which one is "ignorant." Last, (S_3) is reminiscent of the multiplication rule for probabilities:

$$P(A\&B) = P(A \mid B) \cdot P(B).$$

Condition (S_3) suggests a version of the Bayesian principle of conditionalization is satisfied by MAXENT (as I noted in (1979, p. 438, fn. 22)). Specifically, we have:

Result 1. Let P_0 be a MAXENT solution subject to the constraints $C_0 = \{c_1, \ldots, c_k\}$. If one adds the constraint that event e occurs (assumed consistent with C_0), then the new (updated) MAXENT distribution P_1 is the "old" conditional probability $P_0(\cdot \mid e)$ if and only if $P_0(\cdot \mid e)$ satisfies the constraints in C_0.

Proof ("if"). Use (S_3) by setting $\Omega' = \{e, \sim e\}$. Let $C_1 = \{c_1, \ldots, c_k, c_{k+1}\}$, where c_{k+1} is the constraint $E[I_e] = 1$, for the indicator variable

$$\begin{aligned} I_e &= 1 \text{ if } e \text{ occurs} \\ &= 0 \text{ otherwise} . \end{aligned}$$

Contrary to the conclusion, suppose P_1 (the MAXENT solution subject to C_1) is not equal to $P_0(\cdot \mid e)$. That is, suppose

$$U_S(P_1) > U_S(P_0(\cdot \mid e)). \tag{*}$$

Now, it is clear that $P_1(\cdot) = P_1(\cdot \mid e)$, since P_1 satisfies c_{k+1}. Define a probability $P_0'(\cdot)$ by $P_0'(\cdot) = P_0(e) \cdot P_1(\cdot \mid e) + P_0(\sim e) \cdot P_0(\cdot \mid \sim e)$. Then, by (S_3), $U_S(P_0') > U_S(P_0)$, in light of the inequality $(*)$. But P_0' satisfies C_0, contradicting the assumption that P_0 is the MAXENT solution for constraints C_0. To verify that P_0' satisfies C_0, note that the class of distributions satisfying a constraint set is convex (see Appendix A), and note that P_1 does (since it satisfies C_1) and that either $P_0(e) = 1$ whence $P_0' = P_1$, or else $P_0(\cdot \mid \sim e)$ satisfies C_0 since P_0 and $P_0(\cdot \mid e)$ do (and constraints are taken to be linear in probability—see Appendix A).

("only if"): This is trivial. Whenever $P_1 = P_0(\cdot \mid e)$, $P_0(\cdot \mid e)$ satisfies C_1 and hence satisfies C_0 also.

Result 1 provides, also, for the following:

Corollary. Where C_0 is vacuous and $\{C_i\}$ $(i = 1, \ldots,)$ is an increasing sequence of constraint sets, $C_i \subseteq C_{i+1}$, corresponding to a sequence $\{e_i\}$ of mutually consistent observations (measurable) in the initial sample space, then $P_i(\cdot) = P_0(\cdot \mid e_1, \ldots, e_i)$ is the MAXENT probability for constraints C_i.

Proof. C_i is summarized by the sole constraint: $I_{e_1 \cap \ldots \cap e_i} = 1$. Hence, $C_i = C_{i-1} \cup \{I_{e_i} = 1\}$. Then apply mathematical induction with Result 1.[3]

Whenever the constraints arise by observations of events (measurable) in the space X of P_0, the corollary establishes an equivalence of the MAXENT principle and Bayesian conditionalization with a uniform *a priori* probability over X. But before this equivalence is accepted as justification for the fourth or fifth claims (p. 4), two questions must be addressed:

(A) What is the relation between MAXENT and Bayesian solutions that use other than a uniform *a priori* probability over X?

(B) What is the relation between MAXENT and Bayesian solutions when other than indicator-variables appear among the constraints?

I discuss the first of these in Section 2.2, following. The significance of the second question is made evident by an example.

Recall that the unconstrained MAXENT solution for the six-sided die, $X = \{1, \ldots, 6\}$, is the uniform probability $p_i = 1/6$ $(i = 1, \ldots, 6)$. As this distribution satisfies the constraint $E[X] = 3.5$, we may take

$$C_0' = \{E[X] = 3.5\}$$

while preserving the uniform probability, $p_i = 1/6$, as the MAXENT solution $P_0'(\cdot) = P_0(\cdot)$. However, if we add the observation, e_1, that an odd-numbered side resulted on the roll, then the MAXENT solution for $C_1' = \{E[X] = 3.5, I_{e_1} = 1\}$ is *not* the uniform distribution over the three outcomes $\{1,3,5\}$—which is the conditional probability $P_0'(\cdot \mid e_1)$—but instead is the distribution (see Appendix).

$$P_1'(i) = \{.21624, .31752, .46624\} \qquad (i = 1,3,5) \tag{5a}$$

Likewise, had the observation been that the roll yielded an even numbered side, $I_{e_1} = 0$, the MAXENT solution for the constraint set $C_1'' = \{E[X] = 3.5, I_{e_1} = 0\}$ would be

$$P_1''(i) = \{.46624, .31752, .21624\} \qquad (i = 2,4,6) \tag{5b}$$

instead of the conditional probability $P_0'(\cdot \mid \bar{e}_1)$, uniform over $\{2,4,6\}$. Bayesian conditionalization requires that $P_{C_1'}(\cdot) = P_{C_0'}(\cdot \mid e_1)$ and that $P_{C_1''}(\cdot) = P_{C_0'}(\cdot \mid \bar{e}_1)$, both in conflict with (5a) and (5b). Expressed in still other words, the MAXENT solutions $P_1'(\cdot)$ and $P_1''(\cdot)$ are not the conditional probabilities $P_0'(\cdot \mid e_1)$, though the former correspond to an addition of new evidence e_1 or $\bar{e}_1$ to the constraints imposed on P_0'.

Of course, where $P_0(\cdot \mid e)$ fails to satisfy the old constraints, C_0, P_1 must differ from this conditional probability. Unfortunately, whenever the initial constraints C_0 include more than mere 0–1 expectations for indictors (measurable) in the space of P_0, there are events in the algebra of P_0 for which $P_0(\cdot \mid e)$ fail C_0. Hence, without the proviso that $P_0(\cdot \mid e)$ satisfies C_0, Bayesian conditionalization conflicts with shifts according to the MAXENT rule unless *all* constraints (in C_1) are mere 0–1 expectations for indicator variables.

Perhaps there is a way out of this difficulty by extending the algebra so that *all* constraints reduce to 0–1 expectations for indicator variables (measurable) in the extended algebra? This is discussed in Sections 3 and 4.

2.2

Aside on Kullback-information and its relation to (Shannon) uncertainty: There is an important generalization of $U_S(2)$, due to Kullback (1959), essential for a coherent account of "uncertainty" with continuous random variables and useful in widening the scope of the MAXENT principle even for discrete distributions. Let P^0 be an initial ("prior") distribution and P^1 some distribution to be compared with P^0. Define the Kullback-information in a shift from P^0 to P^1 by the formula

$$I_K(P^1, P^0) = \sum_{i=1}^{n} p_i^1 \cdot \log[p_i^1/p_i^0] \tag{6}$$

when P^0 is discrete, and by the analogous integral in densities

$$I_K(P^1, P^0) = \int_X p^1(x) \cdot \log[p^1(x)/p^0(x)]dx. \quad (7)$$

for continuous distributions.

In the case of discrete distributions, (6) is related to (2) in a straightforward fashion. Whereas U_S purports to measure the residual uncertainty in a distribution, i.e., U_S attempts to quantify how far a distribution is from certainty—how far a distribution is from 0–1 probability—I_K reports the *decrease* in uncertainty in shifting from P^0 to P^1. If we set P^U as the uniform distribution over the finite space X of P^0 (so that P^U is the MAXENT distribution (no constraints) over X), and if we set $P*$ as a 0–1, point distribution over X (so that $P*$ depicts a state of certainty with respect to X), then

$$U_S(P^1) = I_K(P*, P^U) - I_K(P^1, P^U). \quad (8)$$

(See Hobson and Cheng, 1973.) Moreover, Hobson (1971) showed that I_K is characterized by five properties (three of which parallel Shannon's conditions for U_S). To wit, (up to a constant) I_K uniquely satisfies

(K_1) I_K is a continuous function of P^0 and P^1.

(K_2) When $P^0 = \{1/n, \ldots, 1/n\}$ and $P^1 = \{1/m, \ldots, 1/m, 0, \ldots, 0\}$ ($m \leq n$) then I_K is increasing in n and decreasing in m.

(K_3) I_K is additive over decomposition of the sample space, analogous to (S_3).

(K_4) I_K is invariant over relabelling of the sample space.

(K_5) $I_K = 0$ just in case $P^0 = P^1$.

The remarks (pp. 263–264) about (S_1)–(S_3), and in particular the useful Result 1, apply to Kullback-information in parallel with the generalization of Shannon's three conditions by these five. Specifically, (K_3) (analogous to (S_3)) entails a restricted equivalence between Bayesian conditionalization and a minimum Kullback-information shift: where P^0 satisfies a constraint set C_0 and a minimum I_K-shift subject to the extra constraint of an event e_1 yields the revised probability P^1, then $P^1(\cdot) = P^0(\cdot \mid e_1)$ provided $P^0(\cdot \mid e_1)$ satisfies C_0.[4]

Just as in Result 1, this equivalence is relativized to cases where the conditional probability $P^0(\cdot \mid e_1)$ satisfies the initial constraints C_0. Where C_0 includes constraints other than the mere observation of events (measurable) in the space of P^0, the important proviso on $P^0(\cdot \mid e)$ fails for some events. Thus, unless the constraint set is restricted to 0–1 expectations for indicator variables, some (Bayesian) conditionalizations do not agree with the revision from P^0 to P^1 by minimizing the change in Kullback (or Shannon) information.

Besides generalizing U_S with discrete distributions, I_K affords a consistent extension of entropy to continuous distributions, unlike the (natural) continuous version of Shannon-uncertainty. That is, where we take a continuous version of (2) to be

$$U_S(P) = -\int_X p(x) \cdot \log[p(x)]dx \tag{9}$$

(with p the density for P), then it is well known (see Jaynes, 1963) this attempt fails to provide consistent results over smooth transformations of continuous random variables. For example, if $\mathbf{X}$ is confined to the unit interval [0,1], the use of (9) yields a MAXENT distribution uniform on [0,1]. However, if we consider the equivalent random variable $\mathbf{Z}$, defined by $z = x^3$, then $\mathbf{Z}$ (like $\mathbf{X}$) is a continuous variable on [0,1], and (9) generates a MAXENT distribution for $\mathbf{Z}$ uniform on [0,1]—in contradiction with the result for $\mathbf{X}$.

By contrast, if we use I_K to identify minimum information shifts, once P^0 is identified, I_K remains invariant over the class of random variables equivalent to the one chosen for identifying P^0. Of course, in the continuous case the MAXENT program then requires a supplementary principle to fix P^0, where P^0 depicts a "state of ignorance" prior to the introduction of "constraints." Jaynes (1968, 1978, 1980, for example), is favorably disposed towards Jeffreys' (1961) theory of Invariants for this component of his MAXENT program. Unfortunately, the policy of using Jeffreys' Invariants to fix such "prior" probabilities is inconsistent with basic Bayesian postulates. (See Seidenfeld, 1979.) Thus, it remains an open question how to determine an "ignorance" prior for continuous distributions in a fashion consistent with Bayesian theory. Since my discussion in this essay pertains to discrete distributions, we may bypass this problem and use I_K as a generalized account of minimum change in probability.[5]

3. ENTROPY AND INSUFFICIENT REASON: REPARTITIONING THE SAMPLE SPACE

A standard objection to the principle of Insufficient Reason is that it fails to provide consistent answers across simple reformulations of questions of interest. One cannot assign equal probability to disjoint events merely on the grounds that the question posed (together with tacit background assumptions—of fact) fails to include good deductive reason for selecting one answer over another. If you are "ignorant" about the outcome of a roll of a cubical die (with spots from 1 to 6 arranged in conventional order), then you may appeal to Insufficient Reason to assign each of the six outcomes: one-spot uppermost ,..., six-spot uppermost, equal probability (1/6). Or, you can cite Insufficient Reason to partition the outcomes in two: one-spot uppermost, more than one-spot uppermost, and assign these possibilities equal probability (1/2). On its face, Insufficient Reason does not dictate which of these contrary analyses is appropriate.

Nor will it do to give priority to the more refined partition of possibilities merely on the grounds that added possibilities indicate more information about the circumstances. The added refinement may be both irrelevant and nonsymmetric to the basic question. Consider the standard, cubical die arranged with six numbered spots so that opposite sides sum to seven.[6] A roll of a die typically provides an observer with either 2 or 3 visible surfaces. In addition to the single side showing uppermost, the die displays one to two vertical faces as well. Let us partition outcomes as follows: for each of the six sides showing uppermost, characterize the roll also according to whether the *sum* of the visible spots on the side (vertically showing) faces(s) (*a*) is greater than, (*b*) equals, or (*c*) is less than the number of spots showing on the top face. See Figure 1.

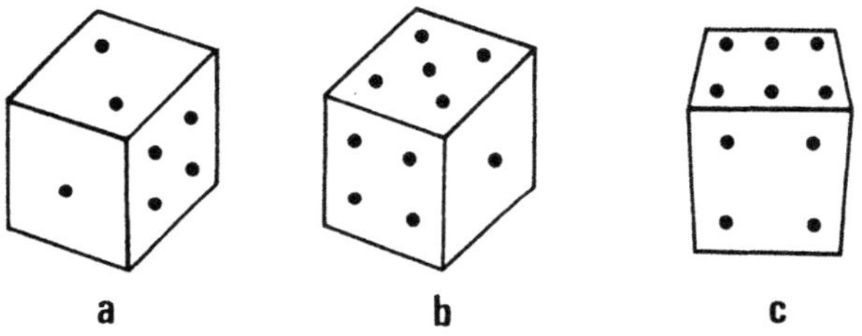

Figure 1. *Repartitioning the sample space for a roll of the die. Outcomes where the sum of visible side-faces: (a) exceeds; (b) equals, and (c) is less than the top-face.*

Instead of 6 outcomes, this partition rolls into 14 different possibilities (as displayed in Table 1).

These 14 possibilities constitute a partition of all rolls with a standardly numbered die. Are we to apply Insufficient Reason to this refined partition (of the six familiar events) leading to a probability distribution (.07142, .14285, .21428, .14285, .21428, .21428) over the basic six outcomes of the roll (how the die landed)? If we believe that added refinement of possibilities reflects more information, then the 14-fold partition of states has

Table 1. *('Yes'/'No' identifies which arrangements are possible.)*

# spots showing on upper face of die	Sum of spots visible on side face(s)		
i	$< i$	$= i$	$> i$
1	No	No	Yes
2	Yes	No	Yes
3	Yes	Yes	Yes
4	Yes	No	Yes
5	Yes	Yes	Yes
6	Yes	Yes	Yes

priority in the application of Insufficient Reason. Of course, what is lacking is a judgement of relevance of the refinement introduced by considering the (nuisance) factor: sum of spots showing on side face(s).

Does the MAXENT program offer new guidance in this old problem? We noted (in discussion of Shannon's condition (S_2), p. 263) the well known result that the uniform distribution $p_i = 1/n$ $(i = 1, \ldots, n)$ maximizes entropy over all discrete distributions with $\sum_{i=1}^{n} p_i = 1$. Thus, MAXENT faces the same sensitivity to repartitions of the sample space as does the simpler principle of Insufficient Reason. Perhaps, in the absence of any constraints other than the number of possibilities, advocates of MAXENT can argue that refinement of possibilities by an observable (as with the modified sample space for the die—Table 1) does constitute new, relevant information. Unfortunately, the problem is not restricted to *a priori* MAXENT probability assignments. That is, the question of which partition is appropriate for application of MAXENT arises even when "constraints" are imposed.

As in Jaynes' example from his Brandeis Lectures (1963), restated in greater detail 15 years later (1978), let us impose the "constraint"

$$E[\text{ number of spots showing }] = 55/14 \simeq 3.9285. \tag{10}$$

If we apply MAXENT to the partition of outcomes by number of spots showing (up), i.e., in the familiar six-fold partition, the distribution which maximizes entropy subject to (10) is (to five places—see the appendix)

$$(.11122, .12908, .14981, .17387, .20180, .23422), \tag{11}$$

where p_i ($i = 1, \ldots, 6$) is the probability of i-spots showing up.

However, since the alternative partition (Table 1) is a refinement of the six-fold partition used above, the constraint (10) applies there too. Specifically, define $f(\text{ state }_j)$ ($j = 1, \ldots, 14$—counting across possible states in Table 1) as follows:

$$\begin{aligned}
&f(\text{state}_1) = 1, f(\text{state}_2) = f(\text{state}_3) = 2,\\
&f(\text{state}_4) = f(\text{state}_5) = f(\text{state}_6) = 3,\\
&f(\text{state}_7) = f(\text{state}_8) = 4,\\
&f(\text{state}_9) = f(\text{state}_{10}) = f(\text{state}_{11}) = 5,\\
&\text{and}\\
&f(\text{state}_{12}) = f(\text{state}_{13}) = f(\text{state}_{14}) = 6.
\end{aligned}$$

Then (10) is equivalent to the constraint:

$$E[f] = 55/14. \tag{12}$$

But the distribution over the 14 states which maximizes entropy, subject to (12), is *not* one that yields a (marginal) distribution for the number of spots showing corresponding to (11). Instead MAXENT, applied to the refined partition, subject to the constraint (12) yields the (marginal) distribution for the number of spots showing:

$$(.07142, .14285, .21428, .14285, .21428, .21428). \qquad (13)^7$$

The difference between these solutions can be conceptualized in the following terms. When the empirical "constraints" all involve a quantity (parameter) of interest, the MAXENT distribution for the parameter of interest is sensitive to which (refined) algebra of possibilities the investigator uses to solve the problem. Even though the investigator professes "ignorance" about the nuisance factor, and bases the MAXENT solution on the empirical constraints (all of which involve the parameter of interest alone), the MAXENT solution (like the principle of Insufficient Reason) changes with the addition of a refined partition of possibilities.

A sufficient condition for ensuring the refinement does *not* affect the MAXENT solution is to make the refined algebra a product space in which the new factor (constituting the refinement) is probabilistically independent of the parameter of interest.[8] Where the nuisance factor is (for other reasons) required to be probabilistically dependent with the parameter of interest, this sufficient condition can be simulated by imposing a degenerate 0–1 marginal distribution on the nuisance factor. Then the nuisance factor is, in effect, a constant and constants are (vacuously) probabilistically independent of other variables. In Section 4, where MAXENT is contrasted with Bayesian inference, the device of using a degenerate 0–1 distribution with nuisance factors is key to understanding an important objection raised by Friedman and Shimony (1971).

Summary: The question addressed in this section is prompted by claim (iv), that the MAXENT program provides a satisfactory account of the Laplacean principle of Insufficient Reason. That principle, in its simplest version, succumbs to inconsistencies when the space of possibilities is repartitioned and Insufficient Reason is applied to both algebras of possibilities. The same inconsistencies can arise with the MAXENT principle (i) in the absence and (ii) in the presence of empirical constraints on the quantities of interest.

What is lacking is an account of how nuisance factors are judged for their relevance. Left to the MAXENT rule, the verdict is loaded in favor of relevance of the nuisance factor (since mere repartitioning is enough to affect MAXENT, as demonstrated above).[9] The problem is encapsulated in the following Result.

Result 2. Given constraints $C = \{c_i\}$ are a function of θ the parameter of interest alone, the MAXENT (marginal) distribution for θ may differ from the θ-marginalization of the (joint) MAXENT solution. That is, maximizing entropy in a marginal (average) distribution does not agree with marginalizing (averaging) the overall maximum entropy unless independence obtains between the parameter of interest and the nuisance parameter. The MAXENT solution is consistent with respect to marginalization only if the joint MAXENT distribution is a product of marginal MAXENT distributions.

Proof: by the construction above.

However, Result 2 does not apply to cross-entropy (Kullback-information) shifts, as shown in the lemma (Appendix B).

4. ENTROPY AND BAYESIAN THEORIES

Claim (v) (p. 261) asserts the thesis that MAXENT inference subsumes Bayesian theory as a special case. To assess the claim we need guidelines for what counts as Bayesian inference. The point is not moot. (See Good (1971) for some 4.7×10^4 varieties.) I rest content here with some core postulates for Bayesian theory:

(B_1) An agent's belief state is represented by a coherent, finitely additive conditional probability $P_{BK}(\cdot \mid \cdot)$—coherence.

(B_2) $P_{BK}(\cdot \mid \cdot)$ is relativized to background evidence BK (consistent and closed under entailment), where BK depicts the agent's *total background evidence.*

(B_3) As regulated by Bayes' theorem for conditional probability $P_{BK}(\cdot \mid \cdot \& e)$ is the agent's hypothetical belief state for the hypothesis that he accepts only the new (consistent) evidence e, i.e., under the hypothesis that BK is enlarged by addition of e (and its new consequences given BK)—conditionalization.[10]

We have Result 1 (p. 263 from Shannon's property (S_3)), establishing a restricted equivalence between revising probabilities through MAXENT and through conditionalization.[11] The restriction in Result 1 is that the "old" conditional MAXENT probability satisfy *all* the constraints and not merely the final constraint, newly imposed, which prompts the revision. Of course, if the restriction is satisfied then the constraint set is mutually consistent. Otherwise, not only is conditionalization at odds with a revised MAXENT solution, but where (in a sequence of shifts) constraints imposed at earlier stages are not retained in subsequent stages, the new Shannon or Kullback shift depends upon the order in which constraints are introduced and replaced. (See Tribus and Rossi, 1973.) Hence, if a net Shannon or Kullback change is to be path invariant over the order in which constraints are added, they must be mutually consistent—else, some constraints must be dropped before others are introduced. (See Shore and Johnson, 1981, property 13, and related discussion.)[12]

We may satisfy the restriction in Result 1 by limiting all constraints to 0–1 expectations for indictor variables (measurable) in the initial algebra of P_0. Then, by the corollary to Result 1, MAXENT reduces to Bayesian theory with a uniform *a priori* probability. If we use the Kullback-information generalization, the parallel result and corollary equates minimum-information shifts with Bayesian conditionalization from an arbitrary (*a priori*) probability.

Canonical illustrations of the MAXENT program, for instance Jaynes' Brandeis dice problem, use constraints which do *not* reduce to 0–1 expectations for indicator variables in the measure space of P_0. If, as a Bayesian, one hopes to understand a constraint as "evidence", then it is reasonable to ask whether, by extending the algebra, the constraint can be interpreted as an "event" in the larger algebra of possibilities. The question also holds out the hope that, in the extended algebra, there will be a Bayesian model for MAXENT formalism by application of Result 1 in the larger space of possibilities. Hence, if we are to consider the more interesting version of MAXENT theory (with an enriched language of constraints), the thesis that the MAXENT principle is coherent (from a Bayesian point of view) returns us to the question of the previous section. Under which conditions can we extend (refine) the field of events, while preserving MAXENT solutions for a given set of constraints?

It is from this perspective I propose we consider the interesting result of Friedman and Shimony (1971) (and Shimony's generalization of (1973)). Let me rehearse their analysis in some detail.[13] Suppose we require a MAXENT probability distribution for a discrete space $X = \{x_1, \ldots, x_n\}$ of n-states ($n \geq 3$) based on *a priori* considerations, i.e., in the absence of additional information about X. As noted (above), the uniform distribution $P(x_i) = p_i = 1/n$ ($i = 1, \ldots, n$) is the MAXENT solution. This result is not altered by adding the structure of distinct numerical magnitudes $f(x_i) = a_i$ to X, $a_i \neq a_j$ if $i \neq j$, so long as we profess ignorance about, e.g., an expected value $E[f]$.[14]

Next, suppose we imagine acquiring information about X, reported by a constraint $E[f] = r$ (where, of course, r lies between the minimum and maximum of the n-values $\{a_1, \ldots, a_n\}$). We may apply MAXENT to determine a new distribution $P_r(X)$, subject to the constraint $E[f] = r$. Friedman and Shimony ask, in effect, for necessary and sufficient conditions that $P_r(X)$ can be a conditional probability $P(X \mid$"$E[f] = r$"$)$ obtained by extending P to a field that includes the constraint as an event. Their findings are remarkable:

Theorem [FS] Subject to the conditions above, P can be so extended just in case $P($"$E[f] = (1/n) \cdot \sum_{i=1}^{n} a_i$"$) = 1$. In words, P can be extended if and only if the extension makes the constraint, $E(f) = r =$ average of the a_i's, practically certain.[15]

A simple example brings home the point. Following Frieden (1984), let us simplify the "dice" problem by collapsing the space of outcomes to the three-sided die with 1, 2, and 3 spots (respectively) on each face. (Just identify a roll of a usual six-sided die by the *minimum* of the horizontal faces.) Then the MAXENT distribution for a roll of the die, based on *a priori* information, is the uniform $(1/3, 1/3, 1/3)$ for each face. Suppose we quantify outcomes by identifying a state with the number of spots, $f(i\text{-spots}) = i$ ($i = 1, 2, 3$). As in the dice example of Section 1 , we can calculate a MAXENT solution for a constraint $E[f] = r$ ($1 \leq r \leq 3$), denoted by $P_r(i)$. Since the average $(1/3) \cdot \sum_{i=1}^{3} i = 2$, the FS Theorem dictates that, in extending P to make P_r a conditional probability, it is practically certain that $r = 2$.

If the constraint is interpretable as fixing the center of gravity of the die as belonging to a region that makes $E[f] = r$ a correct statement of "chance", then the FS result shows the MAXENT solution requires an *a priori* assignment of probability 1 to the empirical claim that the die is loaded so that $r = 2$. (See Shimony, 1973.) I doubt this is the intended interpretation Jaynes wants for the "constraint" in his Brandeis Dice problem.[16] So, instead, let us examine Jaynes' own interpretation of the constraint.

In (1978) he writes,

> When a die is tossed, the number of spots up can have any value i in $1 \leq i \leq 6$. Suppose a die has been tossed N times and we are told only that the average number of spots up was not 3.5 as we might expect from an "honest" die but 4.5. Given this information, *and nothing else*, what probability should we assign to i spots on the next toss? (p. 244 in (1983))

And in the discussion which follows, Jaynes uses the MAXENT distribution given the constraint to determine a predictive (subjective) distribution for the $(N+1)st$ roll. Thus, we can apply the FS result to the problem of the three-sided die, in accord with Jaynes' proposal for interpreting the constraint. Let r be the "sample average" of the first N rolls. Then, $P_r(i)$ is a conditional probability for the $(N+1)$st roll, in the extended product field X^{N+1} ($X = \{1, 2, 3\}$), just in case the *a priori* probability is 1 that $r = 2$.

A connection with the problem of Section 3 is obvious. The question posed by Friedman and Shimony addresses the coherence of the MAXENT program by providing necessary and sufficient conditions for interpreting the MAXENT solution as a conditional marginal distribution in a refined (product) algebra that includes the "constraint" as a conditioning event. Not surprisingly, since the constraint is a relevant bit of information for fixing the (marginal) distribution of the $(N+1)st$ roll, coherence is achieved by converting the nuisance parameter (r) into a constant, almost surely. (See the discussion on p. 271). That is, the problem of repartitioning the algebra to permit the same MAXENT distribution for the $(N+1)st$ roll—both in the minimal field of X *and* in the product X^{N+1}—admits only degenerate solutions for the nuisance parameter, r, defined on the sub-field X^N.

We can press the investigation further into the realm of Bayesian models. What if we allow the agent to hold an *exchangeable* probability for rolls of the die? (The probability P is exchangeable if, for any sub-sequence of n-trials, P is invariant under permutations of the order of outcomes.) Then even the FS solution is barred. That is:

Result 3. If $E(f) = r$ is a constraint imposed on the distribution for the $N+1st$ roll of an n-sided die ($n \geq 3$), based on the "sample average" from N (different) rolls, and P_r is this MAXENT solution, then there is no exchangeable Bayesian model which makes P_r a conditional probability with $P(i \mid \text{"sample average"} = r) = P_r(i)$ $(i = 1, \ldots, n)$.

Proof (outline): According to de Finetti's representation theorem, such an exchangeable P is a mixture if i.i.d. multinomial distributions (each on a sample space of n-outcomes) for some "mixing" prior π on the multinomial parameter. Recall, when $r = (n+1)/2$, that is when the "sample average" equals the average of the number of spots showing on the n faces of the die, then $P_r(i) = p_i = 1/n$, the uniform distribution. In other words, when the constraint satisfies $r = (n+1)/2$, the MAXENT distribution for the $(N+1)st$ roll is the uniform probability, independent of the sample size N. Let Π be the class of "mixing" priors that satisfy this restriction, i.e., where the conditional probability for the $(N+1)st$ outcome is uniform given that the "sample average" of the first N outcomes equals $(n+1)/2$, for each N. Then π^+, the (degenerate) "mixing" prior that assigns probability 1 to the multinomial parameter $(1/n, \ldots, 1/n)$, belongs to this class Π. Given N, verify that among $\pi \in \Pi, \pi^+$ maximizes the probability of the event $\{r = (n+1)/2\}$. But, for this 'mixing" prior (π^+), hence for all "mixing" priors in Π, the event $\{r = (n+1)/2\}$ has probability less than 1. This contradicts the Friedman-Shimony theorem, establishing Result 3.[17]

Summary (Section 4): In this section we investigate the coherence of MAXENT theory when the constraint set includes more than 0–1 expectations for indicator variables (measurable) in the initial space of possibilities. The question asked is whether, by extending the algebra, MAXENT solutions have Bayesian models. The Friedman-Shimony result (1971) shows that where we attend to MAXENT solutions with even a single constraint (not a 0–1 expectation for an indicator variable), only degenerate Bayesian models exist. The degenerate Bayesian model is one in which the "constraint" is a nuisance parameter having, *a priori*, a 0–1 distribution. This agrees with the findings from Section 3, dealing with repartitioning the sample space, where such degenerate solutions avoid the conflict reported in Result 2. Last, using Jaynes' recent (1978) presentation of his (1963) Brandeis Dice problem, we show there is no exchangeable (Bayesian) probability that preserves his recommended interpretation of the constraints—Result 3.

Whereas the MAXENT principle is sensitive to the choice of measure space (Result 2), that is not the case with cross-entropy (Kullback-information) shifts—see Appendix B. However, the phenomenon pointed out in the Friedman-Shimony theorem (to wit: there are only degenerate Bayesian models that make "constraints" into events and make the MAXENT distributions into conditional probabilities given the constraints), does generalize to cross-entropy. This is shown in Appendix B, Corollary 2 to Result 4. This finding uses a generalization of an observation due to van Fraasen (1981).

5. COMMENTS ON THE CONCENTRATION THEOREM (JAYNES, 1979 AND SEE (1963, PP. 51–52))

Theorem (Jaynes): Consider N repetitions of an experiment with n possible outcomes on a given trial. Let f_i $(1 \leq i \leq n)$ be the observed relative frequency of the ith outcome in these N trials. Then the class of sequences of possible outcomes (from the N trials), satisfying a set of m constraints linear in these frequencies, is asymptotically (with $N \to \infty$) concentrated as $\chi^2/(2N)$ (with $n - m - 1$ degrees of freedom) about the MAXENT distribution for the f_i's. Here the "metric" across possible sequences of outcomes is given by the difference in the entropy of the corresponding f_i's.

I have two remarks to make about this interesting result.

First, unless there is some connection drawn between the long-run and short-run properties of the principle under question, mere asymptotics are insufficient for justification. To cite two (well known) cases where asymptotic concerns prove inadequate because they lack relevance for the short-run: neither the limiting frequency definition of probability, nor the criterion of asymptotic consistency of point-estimates is well received. (See Fisher, 1973, pp. 34–35 and 148–149 for discussion of these two examples.) So, at least, the asymptotic argument needs to be supplemented with analysis of the rate at which concentration about the MAXENT distribution occurs. But then it is hard to understand how Jaynes can use the concentration theorem to defend application of MAXENT in statistical mechanics, since he will need to show how to resolve the very problem he used in 1957 to undercut the grounding of statistical mechanics on ergodic theory. That is, to apply the concentration theorem in statistical mechanics, Jaynes needs to show, for example, what are the appropriate time intervals to use to achieve "concentration" about the MAXENT distribution.

A second objection to the argument that seeks justification of the MAXENT rule by appeal to claim (ii) is based on consideration of how, in Jaynes' (1979) result, limiting frequencies from repeated trials are contrasted with a subjective, MAXENT probability for a single trial. The concentration theorem establishes that, relativized to the given constraints (interpreted with limiting frequencies as probabilities), the class of limiting frequencies concentrate (in the sense of having entropy) close to the MAXENT distribution for a single trial. Apart from the important question why "constraints" on a MAXENT probability for a single trial translate into parallel conditions on limiting frequencies from repeated trials (see also footnote 17), there is the following difficulty with the attempted justification.

After relativizing the class of possible limiting frequencies to those satisfying the given constraints, we are directed to count each *logically* possible sequence of repeated trials as a separate state. Then the concentration about the MAXENT probability is determined by the (asymptotic) *proportion* of these states with frequencies close to the MAXENT distribution. Why is this a problem? It is because, if the concentration about the MAXENT solution demonstrates how highly *probable* the MAXENT solution is then, as Jaynes points out (1979, p. 322), the argument equates possibility with probability. In other words, if the concentration theorem is to show how probabilistically atypical "low" entropy distributions are (in repeated trials), logically distinct sequences must be judged equally probable.

An assignment of equal probability to distinct states characterizes an extreme Carnapian method, $\lambda = \infty$, whose Bayesian description is of an i.i.d. process with uniform ($p = 1/n$) probability for each of the n-outcomes of a single trial. Recall that the *a priori* MAXENT probability over n-outcomes is the uniform distribution, $p_i = 1/n$ ($i = 1, \ldots, n$). By the strong law of large numbers, we know that in an i.i.d. process, with probability 1 the limiting relative frequencies concentrate about this *a priori* distribution. If we restrict the limiting relative frequencies, so they satisfy the constraints imposed on the MAXENT solution, then the concentration, given the constraints, is at the MAXENT distribution. But, even here the argument depends upon the uniform "prior" probability, corresponding to $\lambda = \infty$. (See Dias and Shimony, 1981, pp. 192–193, for related discussion.)

The point of the objection is that, were the argument modified by choosing a different "prior" in place of the uniform one, the law of large numbers would continue to hold—in an i.i.d. process there still would be a concentration of limiting frequencies about the *a priori* probability and a related, conditional concentration given frequency constraints. Of course, with the change in "prior", the concentration of frequencies would not be determined simply by the *proportion* of sequences close to the "prior", but by some weighted proportion in which sequences were assigned unequal probability as dictated by the "prior." In short, the concentration theorem singles out MAXENT whenever distinct sequences are counted equally; however, by tailoring the weights on sequences to the "prior" chosen, we can defend any "prior" by a concentration-of-frequencies result. How does claim (ii) distinguish MAXENT from rival (Bayesian) methods?

ACKNOWLEDGMENTS

I thank J. Kadane, I. Levi, and A. Shimony for their detailed, constructive comments on an earlier draft of this paper. Also, I have benefitted from discussions with: A. Denzau, C. Genest, P. Gibbons, E. Greenberg, E. Jaynes, M. Schervish, G. Tsebelis, B. Wise, the members of the Philosophy Department Colloquium at Carnegie-Mellon University, and other helpful critics at the 29th NBER-NSF Seminar on Bayesian Inference in Economics.

Support for this research came from the Department of Preventive Medicine, Washington University (St. Louis), and NSF grant SES-8607300.

A version of this paper will also appear in *Philosophy of Science,* Volume 53, 1986.

FOOTNOTES

[2]. The MAXENT formalism is discussed in the appendix.

[3]. Where the support for P_0 is a denumerable set, this argument depends upon σ-additivity to extend it to $C_\infty = \bigcup_{i<\omega} C_i$.

[4]. Williams (1980) establishes the special case of this result when C_0 is vacuous.

[5]. For example, I_K provides an account of a minimum shift from a prior probability P^0 which is *not* itself identified as a solution to a MAXENT problem. In his recent (1983) paper, "Highly Informative Prior Probabilities," Jaynes makes use of this generalization. If Objective Bayesian theory is modulated to admit arbitrary (coherent) "prior" information as part of the "well posed problem," then the basic dispute with subjectivist Bayesians (such as Savage) is resolved in favor of the latter point of view. That is, even Savage has no objection to a position that makes "objective" a posterior probability constrained by a prior probability and likelihood! Nonetheless, I remain dubious of the claim (v) that Bayesian theory is a special case of Kullback-information theory.

[6]. The arrangement of spots is further constrained so that a pair of dice may sum to 7 on each of the six pairs of parallel faces, i.e., dice are uniformly oriented.

[7]. That is, the MAXENT solution to this problem corresponds to the uniform distribution over the 14 states. That (13) is the MAXENT solution follows directly from the fact that $(1/n)\sum_{j=1}^{14} f(\text{state}_j) = 55/14$. The uniform distribution over the 14 states in Table 1 satisfies condition (13). Recall that the uniform distribution maximizes entropy over all discrete distributions with sample space confined to (a subset of) 14 elements.

[8]. In their application of MAXENT to estimating frequencies in contingency tables, subject to constraints of lower dimensional contingency tables, Denzau, *et al* (1984) note this independence is necessary for a coherent solution.

[9]. This policy, to presume that changes which result from refinement of the algebra of possibilities reflect added *relevant* information in the refinement, seems to underlie Jaynes' (1980) analysis of the "marginalization

paradoxes" (due to Dawid *et al.*, 1973). As Dawid *et al.* use their anomalies to question this policy (Does it work consistently for Bayesians using "improper" distributions?), it comes as no surprise to me that the involved parties accuse each other of missing the point (see the discussion and rebuttal to Jaynes, 1980).

For an alternative account of these "paradoxes," based on an interpretation of improper distributions as finite but not countably additive probabilities, see Sudderth (1980) and Kadane *et al.* (1981).

[10]. This brief statement of the core postulates rides roughshod over several important subtleties in a proper formulation. In particular, I have not attended to temporal versus atemporal interpretations of (B_3)-conditionalization. See Levi (1981), and references cited there, for a careful discussion of such matters.

[11]. Recall, too, this restricted equivalence extends to the Kullback-information approach—see Section 2.2.

[12]. This problem is exacerbated by the unpleasant fact that I_K induces a semi-metric only—it does not satisfy the triangle inequality in general. (See Burbea and Rao, 1982, for additional results.) In his (1968, §III) example of the distribution of impurities in a crystal lattice, Jaynes constructs a "prior" MAXENT solution from a constraint that is *not* satisfied by the "posterior" he obtains through data from a subsequent (neutron reflection) experiment. Thus, the question raised in this note has a basis in the current application of the MAXENT program. (I thank Prof. E. Greenberg (Economics, Washington University) for the last reference.)

[13]. I bother with the particulars since Jaynes [1978, pp. 249–251 in (1983)] finds the F-S argument unacceptable. His complaint is that they use ill-defined constraints. In an otherwise patient review of several objections to the MAXENT program, he writes (following brief but general remarks about the difference between testable constraints and conditioning events),

> Of course, it is as true in probability theory as in carpentry that introduction of more powerful tools brings with it the obligation to exercise a higher level of understanding and judgment in using them. If you give a carpenter a fancy new power tool, he *may* use it to turn out more precise work in greater quantity; or he may just cut off his thumb with it. It depends upon the carpenter.
>
> The FS article led to considerable more discussion...in which severed thumbs proliferated like hydras; but the level of confusion about the points already noted is such that it would be futile to attempt any analysis of the FS arguments.
>
> [pp. 250–251 in (1983)]

[14]. In the (1971) version of this argument, there is the added premise

that for one state, say the mth, $f(x_m) = (1/n) \cdot \sum_{i=1}^{n} f(x_i)$. That is, in the (1971) formulation, it is supposed there is one state whose magnitude a_m equals the average of the n magnitudes $a_m = (1/n) \cdot \sum_{i=1}^{n} a_i$. This condition is relaxed in Shimony's (1973) generalization. For the example which follows, involving the trinomial "die," the (1971) applies. At the expense of complicating the calculations, Shimony's (1973) version is applicable to Jaynes' Brandeis Dice example as presented in Jaynes' [1978, pp. 243–245 in (1983)].

[15]. The Friedman-Shimony proof uses disintegrability of P in the partition by r, the constraint. This assumption is not guaranteed for a general, finitely additive probability. But in the application to the Dice problem (and its generalizations), where r is the "sample average" in the first N rolls, this problem does not arise since, given N, r has a finite sample space.

[16]. Though his response to Rowlinson's (1970) question concerning Wolf dice data suggests the geometric interpretation above. (See Jaynes, 1978, pp. 258–268, in (1983).)

[17]. Frieden (1984) considers the case of a trinomial die with a uniform "prior" distribution for the multinomial parameter. He shows the interesting result that $P(i \mid r = m = 2) = 1/4$ for $i = 1, 3$ and $P(i \mid r = m = 2) = 1/2$ for $i = 2$, for all $N \geq 3$. The uniform prior corresponds to a Carnapian confirmation function c^* ($\lambda = 3$ in his continuum of inductive methods). Thus, it is enlightening to compare Frieden's analysis with the Dias-Shimony result [1981, Appendix B (B.5a) and (B.5b)], for this case. Their results are in agreement, of course.

For contrast, I note that with an "improper" prior (whose density is $1/w_1 \cdot w_2 \cdot w_3$ for the multinomial parameter (w_1, w_2, w_3)), the predictive $P(i \mid r = m = 2)$ is likewise "improper," with all its mass concentrated at the extreme point $(0, 1, 0)$, $i = 1, 2, 3$.

Note also, Result 3 (like the FS-theorem) depends upon the assumption $n \geq 3$. For $n = 2$, the "sample average" is a sufficient statistic with an exchangeable P (unlike the case with $n > 2$). Then the predictive probability $P_r(i) = P(i \mid r)$ has a Bayesian model with the "improper" prior (whose density is $1/w_1 \cdot w_2$), corresponding to the "straight-rule" in Carnap's [1951] continuum of inductive methods ($\lambda = 0$). But for $n = 2$, $P_r(i)$ then is determined *without* appeal to entropy considerations, since the class of distributions satisfying the constraint $E[i] = r$ is a unit set!

Last, Dias and Shimony (1981) proved a restricted agreement between MAXENT and Bayesian methods for the case of the trinomial die. Their theorem, §IV (4.10) shows that the extreme Carnapian method ($\lambda = \infty$), $c^\dagger$, is in asymptotic agreement (for increasing population sizes) with MAXENT solutions to select problems of direct inference. Result 3 demonstrates this agreement cannot be extended to simple problems of predictive inference.

(Recall, $\lambda = \infty$ corresponds to the point-probability 1 for the multinomial parameter $(1/n, \ldots, 1/n)$ in de Finetti's representation of Carnapian methods.)

I thank Prof. E. Greenberg for alerting me to Frieden's recent work.

APPENDIX A: ON THE MAXENT FORMALISM

Here we review some of the mathematics for calculating MAXENT solutions. Following Shore and Johnson (1980), a constraint is an expectation (linear in probability) for a bounded function of the state variables. (We use only linear, equality constraints, $E[f] = c$, instead of the more general class including inequalities too.) Hence, the class of distributions satisfying a (finite) set of constraints is convex. Thus $c_j = \sum_{i=1}^{n} p_i f_j(x_i)$ is the jth constraint.

With k constraints, $c_1, \ldots, c_k$, the matter of choosing a distribution which satisfies these constraints and maximizes entropy is a variational problem (familiar in physics), solved by the device of Lagrange multipliers. (See Courant and Hilbert, 1963, pp. 164–174.) The formal solution obeys:

$$p_i = P(x_i) = [Z(\lambda_1, \ldots, \lambda_k)]^{-1} \cdot \exp[-\lambda_1 f_1(x_i) - \cdots - \lambda_k f_k(x_i)] \quad \text{(A1)}$$

where

$$Z(\lambda_1, \ldots, \lambda_k) = \sum_{i=1}^{n} \exp[-\lambda_1 f_1(x_i) - \cdots - \lambda_k f_k(x_i)], \quad \text{(A2)}$$

and the λ's are the Lagrange multipliers, chosen to satisfy the k constraints, i.e.,

$$c_j = -\frac{\partial}{\partial \lambda_j} \log Z. \quad \text{(A3)}$$

In the case of Jaynes' Brandeis Dice problem, there is one constraint arising from the expectation for the function $f(i\text{-spots}) = i$ $(i = 1, \ldots, 6)$, so that

$$c_1 = \sum_{i=1}^{6} i \cdot p(i\text{-spots}). \quad \text{(A4)}$$

As Jaynes' shows (see (1978), p. 244 in (1980)),

$$Z(\lambda_1) = \sum_{i=1}^{6} e^{-\lambda_1 i} = x(1 - x^6)/(1 - x), \quad \text{(A5)}$$

where $x = e^{-\lambda_1}$. (The r.h.s. of (A5) is by the usual rule for geometric series.) Then, by (A3) and (A4)

$$-\frac{\partial}{\partial \lambda_1} \log Z = \left(1 - 7x^6 + 6x^7\right) / \left[(1-x)(1-x^6)\right] = c_1. \qquad \text{(A6)}$$

In the problem discussed on p. 269, (10) sets the constraint: $c_1 = 55/14$. Solving (A6) for this value yields:

$$x \simeq 1.160601, \qquad Z \simeq 10.43509 \qquad \text{(A7)}$$

(as obtained on my TI 58C). This results in the MAXENT distribution (11) in accord with (A1). The MAXENT distributions (5a) and (5b) are calculated in the identical manner.

It is interesting to note, as reported by Denzau *et al.* (1984), the MAXENT solution (A1) is associated with a LOGIT model by a simple reidentification of parameters. (See the interesting papers in Manski and McFadden (1981) for a very helpful discussion of the role played by LOGIT models in econometric models of composite data from individual decision problems.)

APPENDIX B: ON MINIMUM INFORMATION SHIFTS ARISING FROM THE SPECIFICATION OF NEW CONDITIONAL PROBABILITIES

Recall, the entropy in a distribution P is given by

$$-\Sigma_i p(x_i) \cdot \log p(x_i),$$

and the cross-entropy (or Kullback-information) in a shift from P^0 to P^1 is given by

$$\Sigma_i p^1(x_i) \cdot \log \left[p^1(x_i)/p^0(x_i)\right].$$

Result 4. Let $X = \{x_1, \ldots, x_n\}$ with $x_i \cap x_j = \phi$ for $i \neq j$ and $n \geq 3$. Let $E_1, E_2 \subset X$ with $E_1 \cap E_2 = \phi$ and $X - (E_1 \cup E_2) = E_3 \neq 0$. Let $N = \{1, \ldots, n\}$, and choose $I_1, I_2 \subset N$, with $I_1 \cap I_2 = \phi$ so that $E_i = \cup_{i \in I_i} x_i (i = 1, 2)$. Assume $\mid E_1 \mid = k$ and $\mid E_2 \mid = m$ so $k + m < n$. Specify a constraint $c : p(E_1)/p(E_2) = (1-\alpha)/\alpha$. If P is the MAXENT solution subject to constraint c, then

either $P(E_3) > [n - (k+m)]/n$,

or else $\alpha = m/(k+m)$ when P is the uniform distribution and $P(E_3) = [n - (k+m)]/n$.

In other words, Result 4 establishes that the MAXENT probability P subject to a conditional probability, $P(E_2 \mid [E_1 \cup E_2]) = \alpha$, requires an *increase* in the probability of the complementary event E_3 over the value it carries under a uniform $P^U(E_3) = [n - (k + m)]/n$, unless the constraint is irrelevant to the uniform distribution and $P = P^U$.

Corollary 1. Let P^0 be a probability on X. Let E_1 and E_2 be as above. Let P^1 be a minimum Kullback-information (cross-entropy) shift from P^0 subject to the constraint c, as above. Then $P^1(E_3) > P^0(E_3)$, unless $P^1 = P^0$. In other words, the corollary says that the same phenomenon occurs with minimum cross-entropy shifts regardless whether P^0 is the uniform probability P^U or some other distribution on X. (Note: van Fraassen (1981) gives a direct argument for this corollary in the special case $P^0(E_1) = P^0(E_2) = .25$. His analysis makes tacit use of the lemma (below).)

Proof of Corollary 1. The corollary follows from Result 4 and a simple lemma about cross-entropy.

Lemma. Let P^0 be a distribution on X, and let P^1 be a minimum cross-entropy shift from P^0 subject to the set of constraints $\mathcal{C}$. Let Y be a refinement of X, i.e., $\forall x \in X$ $(x \subset Y)$. Let P^0_Y be a distribution on Y that agrees with P^0 on X, and let $\mathcal{C}_Y$ be the reformulation of $\mathcal{C}$ in the measure space generated by Y. If P^1_Y is the minimum cross-entropy shift from P^0_Y subject to $\mathcal{C}_Y$, then P^1_Y agrees with P^1_Y agrees with P^1 on X. In other words, minimum cross-entropy shifts are invariant over refinements of the original algebra. (I thank Ben Wise, of C.-M.U., for raising the question of this lemma.)

Proof of the Lemma. This is immediate from the additive decomposition of a cross-entropy shift from P^0 to P^1 into a sum of a marginal cross-entropy shift and an expected P^1 shift in conditional probability. In particular, decompose the shift from P^0_Y to P^1_Y into a sum of a marginal shift with respect to X, and an expected (P^1_Y) shift in conditional probability given $x_i \in X$. Then, the shift from P^0_Y to P^1_Y is minimized by having these two agree on all conditional probabilities given $x_i \in X$. (Since a mere change in *these* conditional probabilities does not affect the satisfaction of the constraints and adds to the cross-entropy in the overall shift). Thus, the second term in the decomposition is 0 and the overall shift from P^0_Y to P^1_Y, subject to $\mathcal{C}$, is minimized by minimizing the shift from P^0_X to P^1_X.

Without loss of generality, assume P^0_X is rational-valued. (Otherwise, consider a sequence $\langle P^0_{ix} \rangle$ of rational-valued probabilities converging to P^0_X

and use the argument which follows to establish the desired property for each P^1_{iX}. By continuity of cross-entropy shifts, the desired property obtains for their limit, P^1_X.) Refine X to Y so that P^0_Y is uniform on Y. (This is possible by the assumption that P^0_X is rational-valued.) Reformulate the constraint c in the measure space $(Y, \mathcal{Y})$. By the lemma (above), the minimum cross-entropy shift from P^0_Y to P^1_Y agrees with the minimum cross-entropy shift from P^0_X to P^1_X on X. But, with P^0_Y uniform on Y, the minimum cross-entropy shift is just the MAXENT distribution P, in the measure space $(Y, \mathcal{Y})$, subject to the constraint c. Then apply Result 4 to show that P^1_Y has the desired property on E_3. To wit: $P^1_Y(E_3) > P^0_Y(E_3)$, unless $P^1_Y = P^0_Y$.

Proof of Result 4. Let X, E_1, E_2, and E_3 be as stated. Introduce the constraint of new conditional odds via "called-off" bets. That is, define

$$\begin{aligned} f(x_i) &= -\alpha && \text{if } x_i \in E_1 \\ &= (1-\alpha) && \text{if } x_i \in E_2 \\ &= 0 && \text{if } x_i \in E_3. \end{aligned}$$

The constraint c, then, is formulated by: $E[f] = 0$. Distributions satisfying this constraint also satisfy $P(E_1)/P(E_2) = (1-\alpha)/\alpha$. The MAXENT distribution subject to c, denoted by P, is determined through the equation

$$P(x_i) = e^{-\lambda f(x_i)} \cdot Z^{-1}, \tag{B1}$$

where

$$Z(\lambda) = ky^{-\alpha} + my^{(1-\alpha)} + (n-k-m) \tag{B2}$$

for

$$y = e^{-\lambda}$$

and

$$c = 0 = -\frac{d \log Z(\lambda)}{d\lambda}. \tag{B3}$$

Then

$$y = (\alpha/1-\alpha) \cdot (k/m), \tag{B4}$$

Substituting (B4) into (B1), we arrive at

$$\begin{aligned} P(E_1 \cup E_2) \cdot Z &= k[m(1-\alpha)/(k\alpha)]^{\alpha} + m[k\alpha/(m(1-\alpha))]^{1-\alpha} && \text{(B5)} \\ &\leq k + m. && \text{(B6)} \end{aligned}$$

The inequality in (B6) is strict unless $\alpha = m/(k+m)$, when P is the uniform distribution, P^U, on X.

The inequality (B6) is demonstrated as follows. Let

$$k = rm, \tag{B7}$$

so $P^U(E_1)/P^U(E_2) = r$. Substituting (B7) into (B5), we obtain

$$P(E_1 \cup E_2) \cdot Z = mr^{1-\alpha}(1/[\alpha^\alpha + (1-\alpha)^{1-\alpha}]). \tag{B8}$$

The inequality (B6) obtains just in case

$$1/\left[\alpha^\alpha + (1-\alpha)^{1-\alpha}\right] \leq (1+r)/r^{1-\alpha}\left\{= r^{\alpha-1} + r^\alpha\right\}. \tag{B9}$$

Taking the derivative (with respect to r) of the r.h.s. of (B9) and setting it equal to 0 yields the value

$$r = (1-\alpha)/\alpha\,\{= P(E_1)/P(E_2)\} \tag{B10}$$

as the minimum for the r.h.s. of (B9), which makes (B9) into an equality. But for this value of α, $P(E_1)/P(E_2) = r$ and P is P^U. Thus, P is the uniform probability P^U unless the inequalities (B6) and (B9) are strict.

Corollary 2. With X, E_1, E_2, P^0_X and c as above, the only coherent probability that makes P^1_X into a conditional probability P_W $(\cdot \mid$ "c"$)$ in some measure space $(W, \mathcal{W})$ which extends $(X, \mathcal{X})$, is where

$$P_W\left((1-\alpha)/\alpha = P^0_X(E_1)/P^0_X(E_2)\right) = 1.$$

That is, for coherence, with probability 1 the constraint c is irrelevant to P^0. This corollary augments the Friedman-Shimony (1971) and Shimony (1973) theorems by generalizing their criticism to cross-entropy shifts from arbitrary probability distributions.

Proof. Note that for any value of α other than the one irrelevant to P^0 the P^1-probability of the event E_3 increases. Thus, no probability mixture of the conditional probabilities $P_W(E_3 \mid$ "c"$)$ can equal the unconditional probability $P_W(E_3)(= P^0_X(E_3))$ unless "c" is irrelevant to $P_W(E_3)$ almost surely.

REFERENCES

Burbea, J., and C. R. Rao (1982), "On the convexity of some divergence measures based on entropy functions." *IEEE Transactions on Information Theory* **28**, 489–495.

Carnap, R. (1952), *The Continuum of Inductive Methods.* Chicago: Chicago University Press.

Courant, R., and D. Hilbert (1963), *Methods of Mathematical Physics*, Volume 1, 4th printing. New York: Interscience Publishers.

Dawid, A. P., M. Stone, and J. V. Zidek (1973), "Marginalization paradoxes in Bayesian and structural inference." *Journal of the Royal Statistical Society*, Series B **35**, 189–233 (with discussion).

Denzau, A. T., P. C. Gibbons, and E. Greenberg (1984), "Bayesian estimation of proportions with an entropy prior." Department of Economics, Washington University, St. Louis 63130.

Dias, P. M., and A. Shimony (1981), "A critique of Jaynes' maximum entropy principle." *Advances in Applied Mathematics* **2**, 172–211.

Fisher, R. A. (1973), *Statistical Methods and Scientific Inference*, 3rd edition. New York: Hafner.

Frieden, B. R. (1972), "Restoring with maximum likelihood and maximum entropy." *Journal of the Optical Society of America* **62**, 511–518.

Frieden, B. R. (1984), "Dice, entropy and likelihood." Optical Sciences Center, University of Arizona, Tucson, Arizona 85721.

Friedman, K., and A. Shimony (1971), "Jaynes's maximum entropy prescription and probability theory." *Journal of Statistical Physics* **3**, 381–384.

Good, I. J. (1983), "46656 varieties of Bayesians." *Journal of the American Statistical Association* **25**, 62–63. Reprinted in *Good Thinking*, Minneapolis: University of Minnesota.

Hobson, A. (1971), *Concepts in Statistical Mechanics.* New York: Gordon and Breach.

Hobson, A., and Bin-Kang Cheng (1973), "A comparison of the Shannon and Kullback information measures." *Journal of Statistical Physics* **7**, 301–310.

Jaynes, E. T. (1957), "Information theory and statistical mechanics. I, II". *Physical Review* **106**, 620–630; **108**, (1957), 171–190. Reprinted in Jaynes (1983).

Jaynes, E. T. (1963), "Information theory and statistical mechanics." In *1962 Brandeis Summer Institute in Theoretical Physics*, ed. K. Ford. New York: Benjamin. Reprinted in Jaynes (1983).

Jaynes, E. T. (1968), "Prior probabilities." *IEEE Transactions on Systems Science and Cybernetics* **SSC-4**, 227–241. Reprinted in Jaynes (1983).

Jaynes, E. T. (1978), "Where do we stand on maximum entropy?". In *The Maximum Entropy Formalism*, ed. R. D. Levine and M. Tribus, pp. 15–118. Cambridge, MA: MIT Press. Reprinted in Jaynes (1983).

Jaynes, E. T. (1979), "Concentration of distributions at entropy maxima." In Jaynes (1983).

Jaynes, E. T. (1980), "Marginalization and prior probabilities." In *Bayesian Analysis in Econometrics and Statistics*, ed. A. Zellner. Amsterdam: North-Holland. Reprinted in Jaynes (1983).

Jaynes, E. T. (1981), "What is the question?" In *Bayesian Statistics*, ed. J.M. Bernardo, M. H. DeGroot, D. V. Lindley, and A. F. M. Smith, pp. 618–629. Valencia, Spain: Valencia University Press. Reprinted in Jaynes (1983).

Jaynes, E. T. (1983), *Papers on Probability, Statistics and Statistical Physics*, ed. R. Rosenkrantz. Dordrecht: D. Reidel.

Jaynes, E. T. (1983), "Highly informative priors." In *Bayesian Statistics 2*, ed. J.M. Bernardo, M. H. DeGroot, D. V. Lindley, and A. F. M. Smith, pp. 329–352. Valencia, Spain: Valencia University Press.

Jeffreys, H. (1961), *Theory of Probability*, 3rd edition. Oxford: Oxford University Press.

Kadane, J., M. Schervish, and T. Seidenfeld (1981), "Statistical implications of finitely additive probability." Forthcoming *de Finetti Festschrift*, Goel, ed.

Kullback, S. (1959), *Information Theory and Statistics*. New York: Wiley and Sons.

Levi, I. (1981), "Confirmational conditionalization." *Philosophy of Science* **48**, 532–552.

Manski, C. F., and D. McFadden, eds. (1981), *Structural Analysis of Discrete Data with Econometric Applications*. Cambridge, MA: MIT Press.

Rosenkrantz, R. (1977), *Inference, Method and Decision*. Dordrecht, Holland: Reidel.

Rowlinson, J. (1970), "Probability, information and entropy." *Nature* **225**, 1196–1198.

Seidenfeld, T. (1979), "Why I am not an objective Bayesian; some reflections prompted by Rosenkrantz." *Theory and Decision* **11**, 413–440.

Shannon, C. (1948), "A mathematical theory of communication." *Bell System Technical Journal* **27**, 379–423; 623–656.

Shimony, A. (1973), "Comment on the interpretation of inductive probabilities." *Journal of Statistical Physics* **9**, 187–191.

Shore, J., and R. Johnson (1980), "Axiomatic derivation of the principle of maximum entropy and the principle of minimum cross-entropy." *IEEE Transactions on Information Theory* **IT-26**, 26–37.

Shore, J., and R. Johnson (1981), "Properties of cross-entropy minimization." *IEEE Transactions on Information Theory* **IT-27**, 472–482.

Sudderth, W. (1980), "Finitely additive priors, coherence and the marginalization paradox." *Journal of the Royal Statistical Society*, Series B **42**, 339–341.

Tribus, M., and R. Rossi (1973), "On the Kullback information measure as a basis for information theory: comments on a proposal by Hobson and Chang." *Journal of Statistical Physics* **9**, 331–338.

van Fraasen, B. C. (1981), "A problem for relative information minimizers in probability kinematics." *British Journal for the Philosophy of Science* **32**, 375–379.

Wiener, N. (1948), *Cybernetics*. New York: Wiley.

Williams, P. M. (1980), "Bayesian conditionalization and the principle of minimum information." *British Journal for the Philosophy of Science* **31**, 131–144.

THE UNIVERSITY OF WESTERN ONTARIO SERIES IN PHILOSOPHY OF SCIENCE

A Series of Books in Philosophy of Science, Methodology, Epistemology, Logic, History of Science, and Related Fields

1. J. Leach, R. Butts, and G. Pearche (eds.), *Science, Decision and Value*. 1973, vii + 219 pp.
2. C. A. Hooker (ed.), *Contemporary Research in the Foundations and Philosophy of Quantum Theory*. 1973, xx + 385 pp.
3. J. Bub, *The Interpretation of Quantum Mechanics*. 1974, ix + 155 pp.
4. D. Hockney, W. Harper, and B. Freed (eds.), *Contemporary Research in Philosophical Logic and Linguistic Semantics*. 1975, vii + 332 pp.
5. C. A. Hooker (ed.), *The Logico-Algebraic Approach to Quantum Mechanics*. 1975, xv + 607 pp.
6. W. L. Harper and C. A. Hooker (eds.), *Foundations of Probability Theory, Statistical Inference, and Statistical Theories of Science*. 3 Volumes. Vol. I: *Foundations and Philosophy of Epistemic Applications of Probability Theory*. 1976, xi + 308 pp. Vol. II: *Foundations and Philosophy of Statistical Inference*. 1976, xi + 455 pp. Vol. III: *Foundations and Philosophy of Statistical Theories in the Physical Sciences*. 1976, xii + 241 pp.
7. C. A. Hooker (ed.), *Physical Theory as Logico-Operational Structure*. 1979, xvii + 334 pp.
8. J. M. Nicholas (ed.), *Images, Perception, and Knowledge*. 1977, ix + 309 pp.
9. R. E. Butts and J. Hintikka (eds.), *Logic, Foundations of Mathematics, and Computability Theory*. 1977, x + 406 pp.
10. R. E. Butts and J. Hintikka (eds.), *Foundational Problems in the Special Sciences*. 1977, x + 427 pp.
11. R. E. Butts and J. Hintikka (eds.), *Basic Problems in Methodology and Linguistics*. 1977, x + 321 pp.
12. R. E. Butts and J. Hintikka (eds.), *Historical and Philosophical Dimensions of Logic, Methodology and Philosophy of Science*. 1977, x + 336 pp.
13. C. A. Hooker (ed.), *Foundations and Applications of Decision Theory*. 2 volumes. Vol. I: *Theoretical Foundations*. 1978, xxiii + 442 pp. Vol. II: *Epistemic and Social Applications*. 1978, xxiii + 206 pp.
14. R. E. Butts and J. C. Pitt (eds.), *New Perspectives on Galileo*. 1978, xvi + 262 pp.
15. W. L. Harper, R. Stalnaker, and G. Pearce (eds.), *Ifs. Conditionals, Belief, Decision, Chance, and Time*. 1980, ix + 345 pp.
16. J. C. Pitt (ed.), *Philosophy in Economics*. 1981, vii + 210 pp.
17. Michael Ruse, *Is Science Sexist?* 1981, xix + 299 pp.

18. Nicholas Rescher, *Leibniz's Metaphysics of Nature*. 1981, xiv + 126 pp.
19. Larry Laudan, *Science and Hypothesis*. 1981, x + 258 pp.
20. William R. Shea, *Nature Mathematized*. Vol. I, 1983, xiii + 325 pp.
21. Michael Ruse, *Nature Animated*. Vol. II, 1983, xiii + 274 pp.
22. William R. Shea (ed.), *Otto Hahn and the Rise of Nuclear Physics*. 1983, x + 252 pp.
23. H. F. Cohen, *Quantifying Music*. 1984, xvii + 308 pp.
24. Robert E. Butts, *Kant and the Double Government Methodology*. 1984, xvi + 339 pp.
25. James Robert Brown (ed.), *Scientific Rationality: The Sociological Turn*. 1984, viii + 330 pp.
26. Fred Wilson, *Explanation, Causation and Deduction*. 1985, xviii + 385 pp.
27. Joseph C. Pitt (ed.), *Change and Progress in Modern Science*. 1985, viii + 398 pp.
28. Henry B. Hollinger and Michael John Zenzen, *The Nature of Irreversibility*. 1985, xi + 340 pp.
29. Kathleen Okruhlik and James Robert Brown (eds.), *The Natural Philosophy of Leibniz*. 1985, viii + 342 pp.
30. Graham Oddie, *Likeness to Truth*. 1986, xv + 218 pp.
31. Fred Wilson, *Laws and Other Worlds*. 1986, xv + 328 pp.
32. John Earman, *A Primer on Determinism*. 1986, xiv + 273 pp.
33. Robert E. Butts (ed.), *Kant's Philosophy of Physical Science*. 1986, xii + 363 pp.
34. Ian B. MacNeill and Gary J. Umphrey (eds.), Vol. I, *Applied Probability, Stochastic Processes, and Sampling Theory*. 1987, xxv + 329 pp.
35. Ian B. MacNeill and Gary J. Umphrey (eds.), Vol. II, *Foundations of Statistical Inference*. 1987, xvii + 287 pp.
36. Ian B. MacNeill and Gary J. Umphrey (eds.), Vol. III, *Time Series and Econometric Modelling*. 1987, xix + 394 pp.
37. Ian B. MacNeill and Gary J. Umphrey (eds.), Vol. IV, *Stochastic Hydrology*. 1987, xv + 225 pp.
38. Ian B. MacNeill and Gary J. Umphrey (eds.), Vol. V, *Biostatistics*. 1987, xvi + 283 pp.
39. Ian B. MacNeill and Gary J. Umphrey (eds.), Vol. VI, *Actuarial Science*. 1987, xvi + 250 pp.

GPSR Compliance

The European Union's (EU) General Product Safety Regulation (GPSR) is a set of rules that requires consumer products to be safe and our obligations to ensure this.

If you have any concerns about our products, you can contact us on ProductSafety@springernature.com

In case Publisher is established outside the EU, the EU authorized representative is:

Springer Nature Customer Service Center GmbH
Europaplatz 3
69115 Heidelberg, Germany

Batch number: 10830943

Printed by Printforce, the Netherlands